Boston Studies in the Philosophy and History of Science

Volume 344

The series *Boston Studies in the Philosophy and History of Science* was conceived in the broadest framework of interdisciplinary and international concerns. Natural scientists, mathematicians, social scientists and philosophers have contributed to the series, as have historians and sociologists of science, linguists, psychologists, physicians, and literary critics.

The series has been able to include works by authors from many other countries around the world.

The editors believe that the history and philosophy of science should itself be scientific, self-consciously critical, humane as well as rational, sceptical and undogmatic while also receptive to discussion of first principles. One of the aims of Boston Studies, therefore, is to develop collaboration among scientists, historians and philosophers.

Boston Studies in the Philosophy and History of Science looks into and reflects on interactions between epistemological and historical dimensions in an effort to understand the scientific enterprise from every viewpoint.

Paulo Castro • John W. M. Bush • José Croca
Editors

Advances in Pilot Wave Theory

From Experiments to Foundations

Springer

Editors
Paulo Castro
Centre for Philosophy of Sciences
of the University of Lisbon
University of Lisbon
Lisbon, Portugal

John W. M. Bush
Department of Mathematics
Massachusetts Institute of Technology
Cambridge, MA, USA

José Croca
Centre for Philosophy of Sciences
of the University of Lisbon and Faculty
of Sciences, Department of Physics
University of Lisbon
Lisbon, Portugal

ISSN 0068-0346 ISSN 2214-7942 (electronic)
Boston Studies in the Philosophy and History of Science
ISBN 978-3-031-49860-2 ISBN 978-3-031-49861-9 (eBook)
https://doi.org/10.1007/978-3-031-49861-9

This work was supported by Fundação para a Ciência e tecnologia (FCT / UIDB/00678/2020).

This Springer imprint is published by the registered company Springer Nature Switzerland AG
The registered company address is: Gewerbestrasse 11, 6330 Cham, Switzerland

Contents

Chapter 1
Pilot Wave Theory in the Twenty-First Century

Paulo Castro, John W. M. Bush, and José R. Croca

This book is the result of the International Conference on Advances in Pilot Wave Theory, hosted by the Centre for Philosophy of Sciences of the University of Lisbon (CFCUL), Portugal, held from 26 to 30 July 2021 (International Conference on Advances in Pilot Wave Theory—Concurrently Hosting Hydrodynamic Quantum Analogs HQA-2021 n.d.; Advances in Pilot Wave Theory & HQA2021—YouTube n.d.). The conference concurrently hosted the Hydrodynamic Quantum Analogs meeting (HQA-2021) owing to the common interests of these seemingly disparate fields. The meeting took place online due to the COVID Pandemic lockdown and attracted participants from distant points of the globe, including the United States, Canada, Brazil, Mexico, Portugal, the United Kingdom, Italy, France, the Netherlands, Germany, Russia, Australia, and Israel.

It was a friendly and highly constructive meeting between physicists and philosophers of science that generally remain dissatisfied with the unintelligibility of the explanations provided by the Copenhagen interpretation of quantum mechanics. The meeting invoked the merits of the Pilot wave theoretical approach to quantum

---

P. Castro (✉)
Centre for Philosophy of Sciences of the University of Lisbon (CFCUL), University of Lisbon, Lisbon, Portugal
e-mail: jpcastro@ciencias.ulisboa.pt

J. W. M. Bush
Department of Mathematics, Massachusetts Institute of Technology (MIT), Cambridge, MA, USA
e-mail: bush@math.mit.edu

J. R. Croca
Department of Physics of the Faculty of Sciences of the University of Lisbon and Centre for Philosophy of Sciences of the University of Lisbon (CFCUL), University of Lisbon, Lisbon, Portugal
e-mail: jncroca@ciencias.ulisboa.pt

P. Castro et al. (eds.), *Advances in Pilot Wave Theory*, Boston Studies in the Philosophy and History of Science 344,
https://doi.org/10.1007/978-3-031-49861-9_1

dynamics that Louis de Broglie put forward almost a century ago in his double-solution theory.

Pilot wave theory vindicates an ontological belief about the nature of quantum phenomena, wherein corpuscles and waves are two very real physical entities, coexisting and interacting at all times. The wave guides or pilots the quantum corpuscle, and in turn, the corpuscle reenforces the wave, acting as a moving oscillator. This physical picture of the quantum world is very different from that put forward by the Bohrian complementarity view, according to which the undulatory and corpuscular aspects always present themselves in a mutually exclusive way. Louis de Broglie ontology is also distinct from that put forward by David Bohm; specifically, Bohmian mechanics does not posit the physical existence of a particle-generated quantum wave: the pilot-wave is the wave function of standard quantum theory and only in de Broglie's ontology would it make sense to discuss the possibility of a quantum wave detection. Something that is addressed in this volume (Croca, ch. 17) and that has also been recently treated in another publication (Croca et al. 2023).

Although this is not a book strictly about Hydrodynamic Quantum Analogs (HQA), it seems epistemologically pertinent to say a few words about the field. Until 2005, when Yves Couder, S. Protière, Emmanuel Fort and A. Boudaoud (Paris, France) published their findings (Couder et al. 2005), the pilot wave phenomena did not seem to be a feasible possibility, despite the physical diversity of the world. However, it just happened that Couder *et al*, in an astonishing act of experimental insight, found that under proper conditions, an oil droplet could generate a quasi-monochromatic wave field on an oil bath surface. This wave field propagates with the droplet, making it follow a certain trajectory, by way of a piloting or guiding effect. In certain circumstances, the droplet follows a path quite different from the usual Newtonian trajectory expected in the absence of the piloting wave effect. The discovery of the pilot wave hydrodynamic system (Bush 2015) can be said to represent a considerable historical milestone in modern physics. Its importance lies in the fact that it testifies, without a shred of doubt, that de Broglie's insight about a pilot-wave effect provides valuable insight into quantum phenomena. While it is still a macroscopic scale system described using Newtonian physics, it engenders new physics that can serve as an analogical guide to counter the opacity of the Bohrian ontological black box approach to quantum processes. Over the last 15 years, researchers in the emergent field of Hydrodynamic Quantum Analogs (HQA) have been studying, through experimental and computational means, systems that can be seen as being analogous to quantum phenomena (Bush and Oza 2021). Examples include the tunneling effect (Eddi et al. 2009; Tadrist et al. 2020), the quantum corral effect (Harris et al. 2013; Cristea-Platon et al. 2018), Landau levels (Fort et al. 2010; Oza et al. 2014), Friedel oscillations (Sáenz et al. 2020), diffraction from slits (Couder and Fort 2006; Pucci et al. 2018; Ellegaard and Levinsen 2020), spin states (Labousse et al. 2016; Oza et al. 2018), Zeeman splitting (Eddi et al. 2012), the quantum harmonic oscillator (Kurianski et al. 2017; Perrard and Labousse 2018), and the Hong-Ou-Mandel effect (Valani et al. 2018).

The most important and striking feature in Hydrodynamic Pilot-wave phenomena is the fact that the information about the past trajectory of the droplet is embedded in the wave field. Thus, as the droplet bounces and creates new waves, these overlap with earlier waves, producing an overall field that in turn alters the droplet's future path. This feature of the pilot-wave hydrodynamic system implies the existence of a path wave-memory for the droplet. It also represents a new example of physical structures capable of storing information about the behavior of other physical structures, something typically associated with biology or computation. In HQA experiments the oil bath must be acted on from below by a vibrating force, producing an amplitude acceleration below the Faraday threshold acceleration. This is the acceleration value above which the pilot wave phenomena will no longer exist, giving way to standing Faraday waves on the oil surface (Couder et al. 2005). Related with the path wave-memory property, another revealing fact about pilot wave hydrodynamics is that the decay time of the past waves defines the field's memory storage capacity, which is experimentally controlled by the vibrational acceleration of the bath. If, on one hand, the decay time is very rapid (low acceleration), the pilot wave effect will be less stringent. On the other hand, if the decay time is extended (using vibrational accelerations either near or above the Faraday threshold), the pilot wave phenomena will exhibit other behaviors. These will include dynamic regimes analogous to the semi-classical quantum chaos situation (Stöckmann 1999), where a droplet can bounce unpredictably over the bath. All this suggests that the wave memory encoding process has a storage limit, strongly influencing the droplet's behavior. HQA research seems to point to a new way of thinking about the ontological status of the quantum wave, in so much that it can be addressed as a kind of "nomological" memory (nomological, from the Greek: *nomos*, law, and *logos*, reason, order), containing information about the possible behaviors of corpuscles. An idea that it is also developed in the present volume (Castro, ch. 12).

The book begins with contributions focusing on the Physics of Pilot wave theory, assuming a more philosophical and historical tone as it progresses to the end. The second chapter covers the state-of-the-art in the Hydrodynamic Quantum Analogs (HQA) research field, and how it can help solve conceptual difficulties concerning non-locality and quantum entanglement (Bush, Frumkin and Papatryfonos, ch. 2). The next two chapters, also coming from HQA, provide a computational analysis about long-range pilot-wave interactions (Nachbin, ch. 3) and a deterministic hydrodynamically-inspired ensemble interpretation for free relativistic particles (Dagan, ch. 4). From the Stochastic mechanics approach we get, in chapter five, an answer to a long-standing criticism that multi-time correlations in stochastic mechanics differ from those predicted in quantum theory (Derakhshani and Bacciagaluppi, ch. 5). Within the de Bohm-Broglie framework we find in chapter six a version of de Broglie's double solution theory, reproducing Landau's quantization in a uniform magnetic field (Jamet and Drezet, ch. 6) and, in chapter seven, a comparative study between the deterministic de Broglie-Bohm formalism and the stochastic formalism of Bohm, Hiley and Nelson, concerning quantum equilibrium (Hatifi, Willox and Durt, ch. 7). Chapter eight features a work about configuration-space density frameworks, relating pilot-wave theory trajectories and configuration space representations (Roser and Scoggins, ch. 8). Chapter nine presents us with a description about quantized behaviors at the macroscopic level, observable in what is called the Doubochinski pendulum (Tennenbaum, ch. 9).

The more philosophically focused contributions start in chapter ten with a proposal to complete the quantum ontology with the electromagnetic zero-point field (de la Peña, Cetto, ch. 10). The next chapter provides an analysis of the possibility of hidden variables theories (Vervoort, ch. 11). In chapter twelve we find the Wave-memory interpretation of quantum mechanics, according to which the quantum wave is a four-dimensional memory structure, carrying probabilistic information about the corpuscle behavior. An alternative to the Copenhagen interpretation is suggested, trying to provide a conceptual basis to develop a four-dimensional realistic theory of quantum mechanics (Castro, ch. 12). Chapter thirteen critically addresses the measurement problem in quantum mechanics, arguing that it is a false problem for scientific realists. An historical account due to the positivistic climate of the time, is given for why a three-dimensional, microscopic ontology of quantum mechanics was not adopted. A possible alternative for such an ontology is speculated as way to avoid most paradoxes and puzzles of quantum mechanics (Allori, ch. 13). The next chapter offers several historical parallels between modern quantum theory and thermodynamics in the 1800s, which are relevant to the hidden variable debate (Kay, ch. 14). Chapter fifteen suggests the use of a general methodological approach, engendered in the pilot-wave framework, to study other domains of Nature, namely, biology (Magalhães, ch. 15). Chapter sixteen presents us with John Bell's unpublished notes on de Broglie's Pilot Wave hypothesis (Garuccio and Laurora, ch. 16). The book ends with an account of the experimental feasibility of quantum waves detection, including a history of the subject and a set of state-of-the-art experiments designed to accomplish such a feat (Croca, ch. 17).

Being a nonsystematic gathering of new ideas from physicists and philosophers of science about Nature itself, this volume should be considered a work in Natural Philosophy about quantum phenomena. It should also be read as a contemporary sign that the pilot wave approach to quantum phenomena is very much alive and that it can be developed and advanced, to the point that it can also address entanglement. This much, either through the existence of a wave-memory encoded correlation, transmitted by the wave field from one corpuscle to the other (Bush et al, ch. 2; Nachbin, ch. 3; Vervoort, ch. 11), or otherwise accepting a type of non-local necessitarianism from the nomological information encoded in the wave, as suggested in a more philosophical approach to the topic (Castro, ch. 12).

The reader is free to start from any chapter in the book, as each contribution is self-contained, although multiple relations can be established between them. With the present volume we hope to provide inspirational motivation to think beyond the restrictions imposed by the Copenhagen School interpretation. We also hope that in the near future a more realistic approach to quantum waves may emerge, from both an experimental and theoretical point of view, favoring new ways to look at Nature and allowing for new technological advances prompted by an improved understanding of the quantum realm.

Acknowledgements This work was supported by national funds through the FCT - Fundação para a Ciência e a Tecnologia, I.P., within the scope of the R&D Unit Centro de Filosofia das Ciências da Universidade de Lisboa (CFCUL), the strategic project with the reference FCT I.P. and DOI: UIDB/00678/2020, DOI 10.54499/UIDB/00678/2020 [https://doi.org/10.54499/UIDB/00678/2020].

References

Advances in Pilot Wave Theory & HQA2021—YouTube. n.d. Retrieved February 14, 2023, from https://www.youtube.com/playlist?list=PLXKRQSlglfTWfdIeG2OmnLmWBM3kx6jfH

Bush, J.W.M. 2015. Pilot-wave hydrodynamics. *Annual Review of Fluid Mechanics* 47: 269–292.

Bush, J.W.M., and A.U. Oza. 2021. *Hydrodynamic quantum analogs. Reports on Progress in physics*. IOP Publishing.

Couder, Y., and E. Fort. 2006. Single particle diffraction and interference at a macroscopic scale. *Physical Review Letters* 97: 154101.

Couder, Y., S. Protière, E. Fort, and A. Boudaoud. 2005. Walking and orbiting droplets. *Nature* 437: 208.

Cristea-Platon, T., P.J. Sáenz, and J.W.M. Bush. 2018. Walking droplets in a circular corral: Quantisation and chaos. *Chaos* 28: 096116.

Croca, J.R., P. Castro, M. Gatta, and R. Moreira. 2023. Proposed experiments to clarify the real nature of the quantum waves. *Foundations of Physics* 53: 14.

Eddi, A., E. Fort, F. Moisy, and Y. Couder. 2009. Unpredictable tunneling of a classical wave–particle association. *Physical Review Letters* 102: 240401.

Eddi, A., J. Moukhtar, S. Perrard, E. Fort, and Y. Couder. 2012. Level splitting at a macroscopic scale. *Physical Review Letters* 108: 264503.

Ellegaard, C., and M.T. Levinsen. 2020. Interaction of wave-driven particles with slit structures. *Physical Review E* 102: 023115.

Fort, E., A. Eddi, A. Boudaoud, J. Moukhtar, and Y. Couder. 2010. Path-memory induced quantization of classical orbits. *Proceedings of the National Academy of Sciences of the United States of America* 107: 17515–17520.

Harris, D.M., J. Moukhtar, E. Fort, Y. Couder, and J.W.M. Bush. 2013. Wavelike statistics from pilot-wave dynamics in a circular corral. *Physical Review E* 88: 011001.

International Conference on Advances in Pilot Wave Theory—Concurrently Hosting Hydrodynamic Quantum Analogs (HQA-2021). n.d.. Retrieved February 14, 2023, from https://pilotwave.campus.ciencias.ulisboa.pt/

Kurianski, K.M., A.U. Oza, and J.W.M. Bush. 2017. Simulations of pilot-wave dynamics in a simple harmonic potential. *Physical Review Fluids* 2: 113602.

Labousse, M., S. Perrard, Y. Couder, and E. Fort. 2016. Self attraction into spinning eigenstates of a mobile wave source by its emission back-reaction. *Physical Review E* 94: 042224.

Oza, A.U., D.M. Harris, R.R. Rosales, and J.W.M. Bush. 2014. Pilot-wave dynamics in a rotating frame: On the emergence of orbital quantization. *Journal of Fluid Mechanics* 744: 404–429.

Oza, A.U., R.R. Rosales, and J.W.M. Bush. 2018. Hydrodynamic spin states chaos. *Chaos: An Interdisciplinary Journal of Nonlinear Science* 28: 096106.

Perrard, S., and M. Labousse. 2018. Transition to chaos in wave memory dynamics in a harmonic well: Deterministic and noise-driven behavior. *Chaos* 28: 096109.

Pucci, G., D.M. Harris, L.M. Faria, and J.W.M. Bush. 2018. Walking droplets interacting with single and double slits. *Journal of Fluid Mechanics* 835: 1136–1156.

Sáenz, P.J., T. Cristea-Platon, and J.W.M. Bush. 2020. A hydrodynamic analog of Friedel oscillations. *Science Advances* 6: 20.

Stöckmann, H.-J. 1999. *Quantum chaos: An introduction*. Cambridge: Cambridge University Press.
Tadrist, L., T. Gilet, P. Schlagheck, and J.W.M. Bush. 2020. Predictability in a hydrodynamicpilot-wave system: Resolution of walker tunneling. *Physical Review E* 102: 013104.
Valani, R.N., A.C. Slim, and T. Simula. 2018. Hong–Ou–Mandellike two-droplet correlations. *Chaos* 28: 096104.

Chapter 2
The State of Play in Hydrodynamic Quantum Analogs

John W. M. Bush, Konstantinos Papatryfonos, and Valeri Frumkin

Abstract We describe the manner in which the physical picture emerging from hydrodynamic quantum analogs (HQAs) may serve to resolve some of the long-standing difficulties of quantum mechanics. We enumerate some of the most significant intellectual cul-de-sacs of quantum mechanics, and the manner in which HQA suggests a route past them. Particular attention is given to enumerating the many guises of quantum nonlocality as it appears in the standard quantum interpretations. We illustrate how one might misinfer such nonlocality from the walking-droplet system if one had an incomplete description of the system dynamics, if the variables required for its complete description were hidden rather than in plain sight. We highlight recent work that illustrates how phenomena typically attributed to nonlocality in quantum systems may be rationalized in terms of classical, pilot-wave dynamics. Finally, we define the frontiers of the field of hydrodynamic quantum analogs, including attempts to achieve classical entanglement by demonstrating Bell violations in pilot-wave hydrodynamics, and attempts to develop a model of quantum dynamics informed by the walking-droplet system.

Keywords Pilot-wave hydrodynamics · Walking droplets · Quantum analogs · Local realism

J. W. M. Bush (✉) · V. Frumkin
Department of Mathematics, Massachusetts Institute of Technology, Cambridge, MA, USA
e-mail: bush@math.mit.edu

K. Papatryfonos
Department of Mathematics, Massachusetts Institute of Technology, Cambridge, MA, USA

Gulliver UMR CNRS 7083, ESPCI Paris, Université PSL, Paris, France

P. Castro et al. (eds.), *Advances in Pilot Wave Theory*, Boston Studies in the Philosophy and History of Science 344,
https://doi.org/10.1007/978-3-031-49861-9_2

2.1 Introduction

The walking-droplet system discovered in 2005 by Yves Couder and Emmanuel Fort (Couder et al., 2005; Couder and Fort, 2006) has since captured many features previously thought to be exclusive the the microscopic, quantum realm (Bush, 2015; Bush and Oza, 2020). The growing grocery list of hydrodynamic quantum analogs includes single-particle diffraction and interference (Couder and Fort, 2006), quantized orbits (Fort et al., 2010; Perrard et al., 2014), unpredictable tunneling (Eddi et al., 2009a), Friedel oscillations (Sáenz et al., 2020), spin lattices (Sáenz et al., 2021), statistical projection effects (Sáenz et al., 2018) in corrals (Harris et al., 2013), surreal trajectories (Frumkin et al., 2022) and superradiance (Papatryfonos et al., 2022a; Frumkin et al., 2023). This pilot-wave hydrodynamic system has extended the range of classical systems and so provided a platform for delineating between what can and cannot be understood about quantum systems from a classical perspective.

The hydrodynamic pilot-wave system consists of a millimetric liquid droplet suspended on the surface of a vibrating liquid bath (Couder et al., 2005) (Fig. 2.1a–c). In certain parameter regimes (specifically, for certain drop sizes, liquid properties, and vibrational accelerations), the bouncing droplet achieves resonance with the bath's most unstable Faraday wave mode and destabilizes into a walking state in which it is guided or 'piloted' by its own wave field. The resulting 'walker' then consists of a droplet self-propelling along the bath surface, dressed in a quasi-monochromatic pilot-wave field (Fig. 2.1c). The key features of the system are two-fold (Bush and Oza, 2020). First, the resonance between the droplet and its wave ensures that

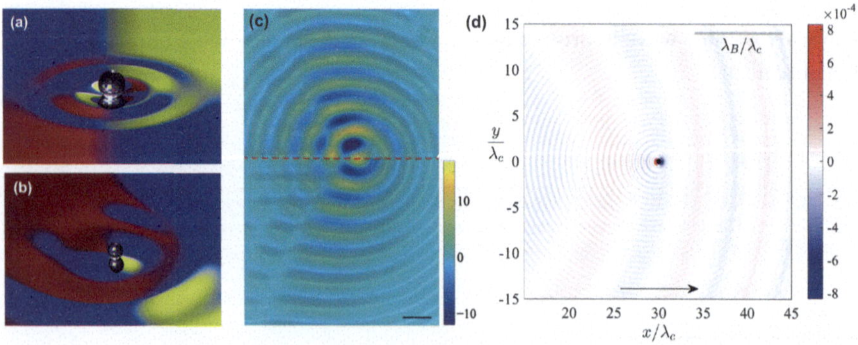

Fig. 2.1 (**a**) A millimetric droplet bounces in place on a vibrating bath. (**b**) When the droplet's bouncing frequency matches that of the vibrating bath's most unstable Faraday mode, the drop destabilizes into a 'walker', and is self-propelled across the bath by its pilot wave. (**c**) Comparison between experimental measurements (top) and simulation (bottom) of the wave field accompanying a walker moving from left to right (Damiano et al., 2016). Color bar indicates the wave height in microns. (**d**) The pilot-wave field computed for a quantum particle propelled by a wave field that evolves according to the Klein-Gordon equation and is excited locally by the particle vibrating at the Compton frequency, $\omega_c = mc^2/\hbar$ (Durey and Bush, 2020)

the pilot-wave field is quasi-monochromatc, a feature that is responsible for the emergence of quantized states in many settings. Second, the persistence of this pilot-wave field renders the droplet dynamics non-Markovian: computing its trajectory requires that one consider its history. The droplet is thus endowed with 'path-memory' (Eddi et al., 2011), and navigates a quasi-monochromatic potential of its own making, as is ultimately responsible for all of the system's emergent quantum features (Bush and Oza, 2020).

The most valuable aspect of the walking-droplet system is that it furnishes a macroscopic example of wave-particle interaction, and so a physical picture for how quantum dynamics might conceivably look. Importantly, this physical picture, of a vibrating particle moving in resonance with its own wave field, is not new and has features of several extant realist theories of quantum dynamics, including de Broglies double-solution pilot-wave theory (de Broglie, 1926, 1970, 1987), stochastic electrodynamics (de la Peña et al., 2015) and the Zitterbewegung theory of quantum mechanics (Hestenes, 1990). Each of these theories invokes the particles internal vibration at the Compton frequency as the source of its guiding wave. The commonality of this physical picture has allowed the HQA community to connect to others attempting to make sense of quantum mechanics. Moreover, it has motivated the development of a generalized classical pilot-wave theory that allows one to explore parameter regimes inaccessible in the laboratory with the walking-droplet system (Bush, 2015; Oza et al., 2018; Durey and Bush, 2021), as well as forge links with and extend extant quantum pilot-wave theories (Dagan and Bush, 2020; Durey and Bush, 2020).

The formalisms of pilot-wave hydrodynamics and quantum mechanics are markedly different. In the former, the HQA community has developed a hierarchy of progressively more sophisticated theories to describe the droplet and wave dynamics, both of which are required for an adequate desription of the droplet's trajectory (Turton et al., 2018). The resulting particle statistics are viewed as an emergent feature, the rationalization of which is not always straightforward. Conversely, quantum mechanics provides an explicit theory for the evolution of the system's statistics, as prescribed by the wave function. According to the standard Copenhagen Interpretation, there is no notion of an underlying dynamics, so no need for a trajectory equation. Bohmian mechanics furnishes a trajectory equation by positing that the particle moves in response to the quantum wavefunction (Holland, 1995). Throughout this review, we describe recent attempts to reconcile the very different theoretical formalisms developed to describe pilot-wave hydrodynamics and quantum mechanics. Specifically, we describe recent attempts to develop a statstical theory for pilot-wave hydrodynamics and a dynamical theory for quantum mechanics informed by the walker system.

We proceed by enumerating a number of concepts that may seem beguiling from the point of view of quantum mechanics, but become less problematic when considered from the new perspective offered by HQA. We advance in increasing order of difficulty, commencing with notions that transform from inscrutable to trivial, such as wave-particle duality and wave-function collapse. We move on to show how quantized states and coherent wave-like statistics naturally emerge from classical pilot-wave dynamics. We highlight the manner in which quantum non-locality in its various guises might be misinfered from the walking-droplet system. Finally, we discuss current attempts to demonstrate violation of Bell's inequality with the pilot-wave hydrodynamic system, which remains the central challenge to any local realist theory of quantum mechanics. We conclude with a discussion of the benefits of the perspective and enhanced physical intuition offered by HQAs on quantum mechanics and quantum foundations.

2.2 Wave-Particle Duality and Complementarity

> Both matter and radiation possess a remarkable duality of character, as they sometimes exhibit the properties of waves, at other times those of particles. Now it is obvious that a thing cannot be a form of wave motion and composed of particles at the same time—the two concepts are too different.
>
> – Heisenberg, 'The Physical Principles of the Quantum Theory' (Heisenberg, 1930).

Wave-particle duality is a notion common in both optics and quantum mechanics. It first arose in the debate over the nature of light, where it became apparent that light sometimes behaves like a particle, other times like a wave. Huygens and Fresnel were two of the most prominent proponents of the wave nature of light. In 1678, Huygens proposed that every point on a wavefront of a light beam may be seen as a new source, emitting spherical waves in the forward direction (Miller, 1991). These secondary wave sources interfere with each other to produce the advancing wavefront. Huygens thus explained linear and spherical wave propagation and derived the laws of reflection and refraction, but failed to rationalize other optical phenomena, such as the diffraction from an edge or an aperture (Miller, 1991). More than a century later, Fresnel combined his own theory of interference with Huygens's principle, which enabled him to rationalise these diffraction effects (Santos et al., 2018). Newton championed the corpuscular view, that light consisted of a series of discrete, localized corpuscles, now known as photons, 'skipping on the ether like stones on a pond' (Newton, 1704). Subsequently, Thomas Young's ripple tank experiments demonstrated that the diffraction of light was consistent with its having a wave nature. The wave view became dominant when Maxwell (1873) demonstrated that all forms of light (infrared, visible and ultraviolet) could be described as electromagnetic waves oscillating at different frequencies. When the wave-particle debate seemed all but settled, it was again revived in 1905 by Einstein's explanation of the photoelectric effect in terms of 'light quanta', a critical step in the development of quantum mechanics (McKagan et al., 2009).

The concept of complementarity was introduced by Niels Bohr as an essential feature of quantum theory (Folse, 1985). It asserts that a complete knowledge of quantum phenomena requires a simultaneous description of particle and wave properties, hence a quantum version of wave-particle duality. Specifically, prior to being measured, a single quantum particle is described in terms of its wave function that evolves according to Schrodinger's equation. Measurement forces the wave to collapse to a particle, for example when particles passing through slits arrive at the detection screen. Bohr asserted that it is impossible to observe both wave and particle aspects simultaneously, but both notions need be retained. Finally, quantum systems have certain pairs of 'complementary' properties that cannot be observed simultaneously. These are also known as non-commuting observables and include, for example, position and momentum, and different components of a particle's spin.

De Broglie's theory of matter waves was built upon his premise that the universe is composed of two elements, light and matter. Since light has both corpuscular and wave aspects, so too must matter. His was an attempt to reconcile quantum mechanics with Einstein's theory of relativity. He proposed that a particle of mass m has an internal frequency that may be deduced by equating its rest mass energy mc^2 to its wave energy $\hbar\omega$. The resulting de Broglie-Einstein equation defines the frequency of particle vibration to be the Compton frequency, $\omega_c = mc^2/\hbar$. He thus envisioned microscopic particles generating, then moving in response to, their own wave fields. He proposed that a free particle moves in response to gradients in the phase of its monochromatic guiding or 'pilot' wave, with a wavelength prescribed by the de Broglie relation, $p = \hbar k$. He imagined, but never proved, that this pilot-wave dynamics might give rise to statistical behavior consistent with that described by the standard formulation of quantum mechanics. His theory thus involved two waves, the real pilot-wave responsible for guiding the particle, and the emergent statistical wave, and his unfinished theoretical program was known as the double-solution pilot-wave theory (Hatifi et al., 2018; Colin et al., 2017). De Broglie's answer to the question of 'particle or waves?' was thus simply 'particle and wave' (Bell, 1987). On the basis of this physical picture, he predicted electron diffraction and was awarded the Nobel prize in 1929. Nevertheless, this physical picture has been largely ignored by the physics community (Bell, 1987) until its recent revival by the HQA community (Bush, 2015; Bush and Oza, 2020).

The walking droplet is inarguably a classical realization of wave-particle duality (Figs. 2.1b and 2.2). The 'walker' has both particle and wave aspects. Without the droplet, there would be no source of waves, and without the accompanying wave field, the droplet wouldn't self propel. The walker is, moreover, an embodiment of the physical picture proposed by de Broglie in his double-solution pilot-wave theory (Bush, 2015). In the walker system, the Faraday frequency plays the role of the Compton frequency, the Faraday wavelength that of the de Broglie wavelength. The walker moves in response to gradients in the wave amplitude rather than phase, but the resonance condition respected by the free walker effectively renders this distinction a moot point. The pilot-wave of the walker is quasi-monochromatic, of a distinctive, horseshoe-like form (Fig. 2.1c), while de Broglie envisaged a monochromatic plane wave of the form illustrated in Fig. 2.1d (Durey and Bush, 2020).

2.3 Quantized States

> I wish first to show in the simplest case of the hydrogen atom that the usual rates for quantization can be replaced by another requirement, in which mention of "whole numbers" no longer occurs. Instead the integers occur in the same natural way as the integers specifying the number of nodes in a vibrating string.
>
> – Schrödinger, 'Quantisierung als Eigenwertproblem' (Schrödinger, 1926).

In the pilot-wave hydrodynamic system, quantized dynamical states are a central feature, apparent in both the structure of droplet aggregates (Fig. 2.2) and orbital dynamics (Fig. 2.3). Bouncing droplets may form either static bound states, pairs, trios, rings or lattices (Eddi et al., 2009b), or dynamic bound states comprised of ratcheting (Eddi et al., 2008; Galeano-Rios et al., 2018) orbiting (Couder et al., 2005; Protière et al., 2008; Oza et al., 2017) or promenading pairs (Borghesi et al., 2014; Arbelaiz et al., 2018). In all such states, the interdrop distance is quantized, prescribed by the Faraday wave length. In both rings and lattices, the drop aggregates are most stable when the droplets bounce in local minima of their collective wave field (Couchman and Bush, 2020). A critical requirement for the emergence of quantization is the synchronization of the droplets, which ensures a coherent quasi-monochromatic wave form that serves as the trapping self-potential of the aggregate.

Quantization is also a key feature of orbital pilot-wave dynamics, as arises when walkers move subject to constraints. Three such systems have been explored

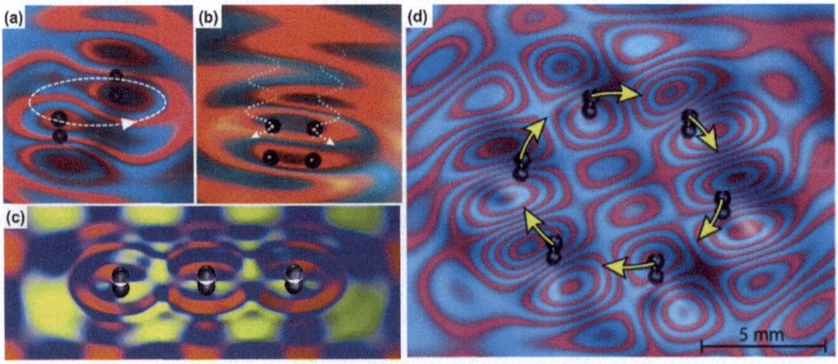

Fig. 2.2 Quantized bound states, both static and dynamic, arise owing to the quasi-monochromatic wave field generated by the droplets bouncing in resonance with the bath's most unstable Faraday mode. (**a**) Two walkers locked into a circular orbit. (**b**) A promenading pair: drops move together in the same direction, with the lateral distance between them varying periodically with time (Arbelaiz et al., 2018). (**c**) A colinear trio of stationary bouncers. (**d**) Rings of bouncing droplets may form with quantized radii. We show here the circular motion arising at the onset of instability of a stable ring induced as the driving acceleration exceeds a critical value (Couchman and Bush, 2020)

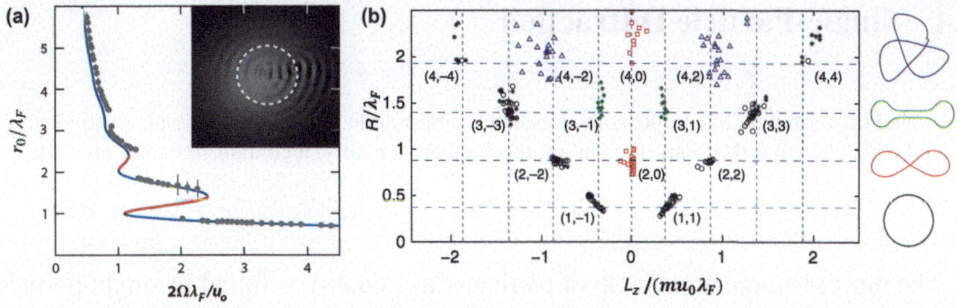

Fig. 2.3 Quantized orbital states arise for (**a**) walkers in a rotating frame or (**b**) walkers confined by a central spring force. (**a**) The solution curve for the orbital radius, r_0, as a function of rotation rate, Ω, has both stable (blue) and unstable (red) branches. The absence of stable solutions at certain radii leads to orbital quantization (Fort et al., 2010; Harris and Bush, 2014; Oza et al., 2014a). Inset: the walker moves along a circularly corrugated wave field, whose form imposes the quantization (Fort et al., 2010). (**b**) When a walker is confined by a radial spring force, a variety of periodic orbits are accessible, all of which are quantized in both mean radius $\bar{R} = R/\lambda_F$ and mean orbital angular momentum $\bar{L}_z = L_z/(mu_0\lambda_F)$ (Perrard et al., 2014; Labousse et al., 2014a)

both experimentally and theoretically. When walkers move in a rotating frame, the Coriolis force plays a role analogous to the Lorentz force acting on a charge moving in a uniform magnetic field (Fort et al., 2010). The expected continuum of circular inertial orbits arising at low memory are replaced by quantized circular orbits as the memory increases (Fort et al., 2010; Harris and Bush, 2014). The requirement for orbital quantization is that the orbital period of the drop is less than the memory time; thus, the drop feels its own capillary wake, which serves to quantize its orbital radius. The drop thus navigates its own potential, a circularly symmetric wave field with the Faraday wavelength centered on the orbital center (Fig. 2.3a, inset). In these quantized circular orbits, the Faraday wavelength plays a role analogous to the de Broglie wavelength in Landau levels (Fort et al., 2010).

When walkers move in a central force, in addition to quantized circular orbits, a family of more elaborate orbits arise, including lemniscates and trefoils (Perrard et al., 2014; Labousse et al., 2014a). As in the 2D quantum harmonic oscillator, these orbits are quantized in both energy and mean angular momentum. A similar progression of quantized orbits arise when a walker is confined to a small corral at relative low memory, when the bounding geometry plays the role of the confining potential (Cristea-Platon et al., 2018). Once again, the key requirement for the emergence of quantized states is resonance between drop and bath, as assures a quasi-monochromatic self-potential. Quantized orbital states emerge when the memory time exceeds the orbital period, so that the drop continuoulsy navigates a highly structured potential of its own making.

2.4 Single-Particle Diffraction

> A phenomenon which is impossible, absolutely impossible to explain in any classical way, and which has in it the heart of quantum mechanics. In reality, it contains the only mystery.
>
> – Richard Feynman (Feynman, 1964).

The most prominent example of particle-wave duality is found in single-particle diffraction. The first experimental investigation of single-photon diffraction was undertaken by a graduate student, G.I. Taylor, who subsequently went on to become a prominent and influential fluid and solid mechanician (Batchelor, 2008). The question posed him by his supervisor, J.J. Thomson, was whether a diffraction pattern would emerge when light passed through a slit, even when the photons passed through one at a time. The results of his experiments were conclusive: a diffraction pattern ultimately emerged. So, despite the photons passing through the slits one at a time, their superposition corresponded to a continuous wave pattern (Taylor, 1909), indicating single-particle diffraction.

Feynman's (Feynman, 1964) insistence on the inscrutability of the electron double-slit experiments is somewhat puzzling in light of de Broglie's work on electron diffraction (de Broglie, 1926; Bell, 1987). The relevant experiments were first performed with electrons diffracting around a pair of obstacles, the complement of the double slit (Tonomura et al., 1989). The build-up of the resulting diffraction pattern, into alternating bright and dark bands, has been numbered among the most beautiful experiments in the history of physics, the beauty presumably being rooted in the common conception that the phenomenon is impossible to understand from a classical perspective. The mystery is not the particular form of the diffraction pattern: the mystery is that the pattern emerges at all.

The original experiments of walkers passing through slits (Fig. 2.4) and accompanying simulations of Couder and Fort (2006) clearly showed a diffraction pattern. As the drop approaches the slit, the distortion of its pilot wave by the barriers causes the drop to be deflected, with certain deflection angles being preferred. A more exhaustive series experiments performed in different laboratories have made clear that single-particle diffraction is indeed a robust feature of pilot-wave hydrodynamics (Andersen et al., 2015; Pucci et al., 2018; Ellegaard and Levinsen, 2020). Moreover, the presence of a second slit alters the emergent diffraction pattern, another keystone of its quantum counterpart. The fact that the emerging diffraction pattern does not generally conform to the Fraunhofer pattern, is neither troubling nor surprising when one considers that the pilot-wave form is markedly different in the walker systems than what one would expect to arise in de Broglie's mechanics (Fig. 2.1c, d). The diffraction of walking droplets reminds us that the physical picture of de Broglie's is sufficient to understand the basic mysteries of electron diffraction.

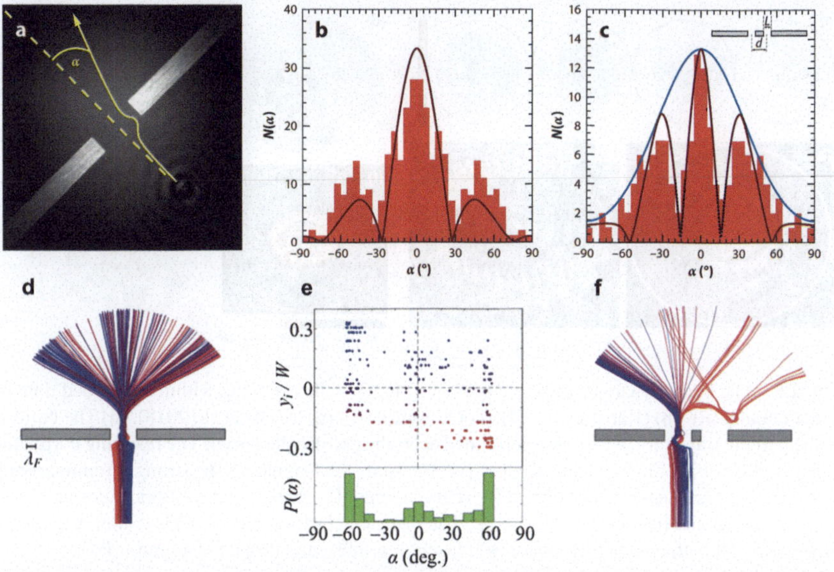

Fig. 2.4 Single-particle diffraction and interference of a walking drop. (**a**) A walker passing through a single slit is deviated from its initial path owing to the influence of the submerged topography on its pilot wave. Histogram for the final deflection angle α in (**b**) the single-slit experiments and (**c**) the double-slit arrangement. The droplet trajectories emerging from (**d**) the single-slit and (**f**) the double slit experiments of Pucci et al. (2018). (**e**) The distribution of deflection angles in the single-slit arrangement. In the double-slit arrangement, the presence of the second slit alters the diffraction pattern of walkers through the first slit because the pilot wave is influenced by both slits (Pucci et al., 2018). Panels (**a**)–(**c**) adapted from Couder and Fort (2006); panels (**d**)–(**f**) from Pucci et al. (2018)

2.5 Wave-Like Statistics

> The wave character of light is not vibrating stuff like a wave of water but rather a wavelike function encoding information about where you'll find the photon of light once it is detected.
>
> – Marcus du Sautoy, 'The Great Unknown: Seven Journeys to the Frontiers of Science' (Sautoy, 2017).

One of the most compelling features of pilot-wave hydrodynamics is that it naturally leads to the emergence of wave-like particle statistics reminiscent of those arising in many quantum systems. We have already seen the wave patterns arising in walker diffraction through slits (Fig. 2.4). Superpositions of dynamical states have been demonstrated in a number of settings involving chaotic pilot-wave hydrodynamics. In the orbital pilot-wave systems first considered by Couder and Fort, specifically walker motion in a rotating frame (Fort et al., 2010) and a central spring force (Perrard et al., 2014), analogs of a charge moving in a uniform magnetic field and the 2D quantum harmonic oscillator, respectively, quantized orbital states

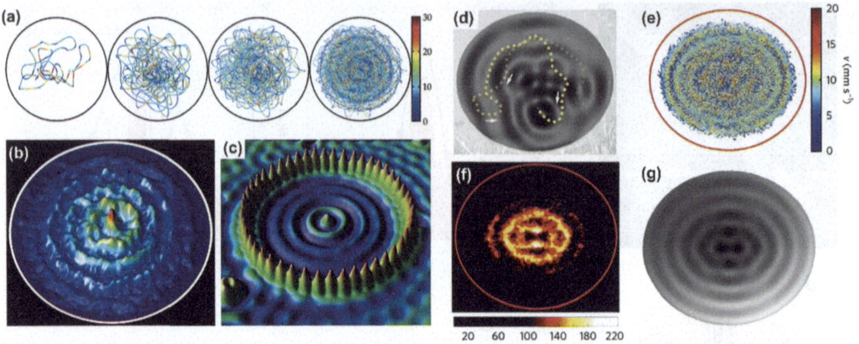

Fig. 2.5 The hydrodynamic corral consists of a single walker moving within a bounded domain, either circular (**a**)–(**b**) (Harris et al., 2013) or elliptical (**d**)–(**g**) (Sáenz et al., 2018). (**a**) The build-up of the particle trajectory, which is color-coded according to droplet speed. The resulting correlation between droplet position and speed gives rise to a statistical signature (**b**) strongly reminiscent of that arising in its quantum counterpart (**c**) (Crommie et al., 1993b). In the elliptical corral, similar speed maps (**e**) and histograms (**f**) emerge. The instantaneous wave field (**d**) is complex and time-dependent, while the mean pilot-wave (**g**) closely resembles the droplet histogram (**f**)

emerge as the memory is increased (Fig. 2.3). Eventually, at sufficiently high memory, these orbital states destabilize, giving rise to chaotic states marked by the drop switching intermittently between a number of finite accessible quantized orbital states (Harris and Bush, 2014; Oza et al., 2014b; Labousse et al., 2014a). The emergent wave-like statistics thus reflect a superposition of dynamical states, and the precise form of the emergent statistics reflects the relative instability of the accessible unstable orbits (Harris and Bush, 2014; Oza et al., 2014a; Labousse et al., 2014a).

Robust wave-like statistics also arose in Harris et al. (2013)'s experimental investigation of a walker in a circular corral (Fig. 2.5a–c). In the high-memory limit arising just below the Faraday threshold, the drop executes an erratic, chaotic trajectory. Ultimately, the correlation between drop position and speed give rises to a statistical signature of comparable form to the most unstable Faraday mode of the cavity. The emergent statistics in this ergodic system is virtually identical to that arising when electrons are confined to the quantum corral (Crommie et al., 1993b; Fiete and Heller, 2003), with the Faraday wavelength again playing a role analogous to the de Broglie wavelength. The walker corral is marked by three distinct timescales, those of droplet bouncing ($\sim$0.01 s), droplet translation ($\sim$2 s) and statistical convergence ($\sim$1 h). Given the vast difference in scales between this experiment and its quantum counterpart (e.g. the corral diameter is 3 cm rather than 75 Angstrom), the ability to resolve all three timescales in the laboratory is quite remarkable.

While theoretical models have yet to capture satisfactorily the wave-like statistical behavior in corrals (Durey et al., 2020), its robustness was demonstrated in a subsequent study of an elliptical corral (Sáenz et al., 2018), where additional

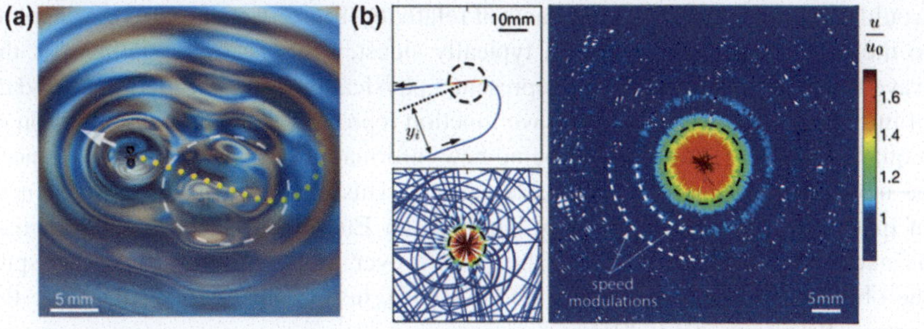

Fig. 2.6 The hydrodynamic analog of Friedel oscillations (Sáenz et al., 2020). (**a**) When a walker approaches a deep well, it is drawn inward along a spiral path, then exits the well radially. (**b**) In-line speed oscillations along its outgoing path are evident in the droplet's speed map. The associated correlation between droplet speed and radial position leads to a coherent statistical signature similar in form to Friedel oscillations (Crommie et al., 1993)

quantum-like features were revealed, including superposition of statistical states and statistical projection effects (Fig. 2.5d–g). Moreover, it was noted that the mean pilot wave was very similar in form to the emergent particle histogram, a result later rationalized by Durey et al. (2018, 2020), who demonstrated that the mean pilot-wave field for either periodic or ergodic walker mation may be deduced from a convolution of the particle histogram and the stationary bouncer wave field. This result provides the means to deduce the particle statistics from the mean pilot-wave form, so plays a role analogous to that of Born's Rule in quantum mechanics if one identifies the mean pilot wave with the wave function (Kutz et al., 2023). Bush and Oza (2020) discuss the relation between the mean-pilot-wave potential in the walker system and the quantum potential in Bohmian mechanics.

Robust wave-like statistics were also revealed in the hydrodynamic analog of Friedel oscillations (Sáenz et al., 2020, Fig. 2.6), waves in the density of states surrounding impurities in the electron sea on a metal surface (Crommie et al., 1993). The analog system consisted of a walker interacting with a deep well. The drop was drawn inwards along a spiral path until crossing the center of the well then exiting radially. As the drop exited the well, in-line speed oscillations were excited, and the resulting correlation between radial position and speed along the outgoing trajectory gave rise to a statistical signature identical to that arising in Friedel oscillations, with the Faraday wavelength again playing the role of the de Broglie wavelength. With this concrete physical mechanism for the emergent statistics in the analog quantum corral and Friedel oscillations, one can imagine that a similar mechanism might also be at play in their quantum counterparts. At the very least, one can conclude that the emergent statistics in both systems are not inconsistent with the notion of particle trajectories.

According to the Copenhagen Interpretation, the act of measurement causes the wave function to collapse to a particular eigenstate of the system, the associated probability cloud describing the particle position to collapse instantaneously to a point. Given the spatially extended nature of the associated wave form, this collapse would seem to violate the basic tenets of relativity (Einstein et al., 1935). Adherents to the Copenhagen Interpretation typically sidestep this issue by stating that the wave function collapse does not represent a physical process, but merely an update of information. However, if the wave function represents a complete description of a quantum system, there should be no new information according to which it need be updated (Hance and Hossenfelder, 2022). This criticism of the completeness of quantum theory was originally launched by Einstein et al. (1935), prompting an exchange now referred to as the Debate over the Nature of Reality. Despite the obscurity of Bohr's response (Bohr, 1935), history has judged him to be the winner (Bricmont, 2017, 2016).

2.6 The Measurement Problem

> The electron, as it leaves the atom, crystallises out of Schrödinger's mist like a genie emerging from his bottle.
>
> – Sir Arthur Stanley Eddington, Gifford Lectures (Eddington, 1927).

The state of a quantum system is described by the wave function, a vector in a Hilbert space that evolves deterministically in time according to the linear Schrodinger equation:

$$i\hbar\frac{\partial\psi}{\partial t} = \hat{H}\psi \tag{2.1}$$

where $\hat{H}$ is the system's Hamiltonian. Due to the linearity of the Schrodinger equation, any superposition (linear combination) of solutions will also be a solution and evolve in time according to Eq. (1). During measurement, the wave function "collapses" onto an eigenstate of the Hamiltonian that corresponds to a particular measurement outcome, the value measured in the laboratory (Griffiths, 2014). Unlike the time evolution described by Eq. (2.1), the collapse process is non-linear, non-deterministic, and typically non-local (Bassi et al., 2013). Importantly, the theory does not specify what precisely constitutes a measurement, which raises the following fundamental questions. What exactly happens during measurement? Is the "collapse" a physical process or merely an update of information? These puzzles in the foundations of quantum mechanics are collectively referred to as 'the measurement problem', and embodied in widely known paradoxes such as Schrodinger's cat (Schrödinger, 1935) and Wigner's friend (Wigner, 1995).

Attempts to solve the measurement problem have led to the various interpretations of quantum mechanics. In some, such as the de Broglie–Bohm theory (Bohm, 1952a, 1925b; Holland, 1995) (a.k.a. Bohmian mechanics, not to be confused with de Broglie's original double solution pilot-wave theory) and the Many Worlds Interpretation (Everett, 1957), there is no measurement problem since the wave function never undergoes a collapse process. In others, like the objective-collapse models, a stochastic term is added to the Schrodinger equation in order to induce a spontaneous collapse of the wavefunction at some characteristic time scale. The result is an emergent "classical" behavior for many particle systems, and an approximate quantum behavior for microscopic isolated systems (Bassi and Ghirardi, 2003). Thus, while collapse is continuously occurring, it is not the result of interaction with a measurement device.

An attempt to reconcile the Copenhagen interpretation with the measurement problem has given rise to the notion of decoherence. Roughly speaking, decoherence posits that, through interaction with the environment, a quantum system loses its coherent properties, resulting in a classical superposition of probabilities. In terms of the density matrix formalism, the off-diagonal terms that represent quantum interference disappear due to interaction with the environment, yielding a diagonal matrix with the Born probabilities as its entries (Zurek, 2003). Decoherence thus attempts to explain how quantum states evolve into classical probabilities, and so rationalize why we do not observe quantum superpositions in the laboratory. However, it does not explain why we do not observe classical superpositions instead. Schrodinger's cat evolves from being simultaneously dead and alive, to being half dead and half alive. However, when we observe the cat, it is either 100% dead or 100% alive, corresponding to a density matrix consisting of a single non-zero entry on its diagonal. Thus, while decoherence describes how quantum probabilities become classical, it does not provide an entirely satisfactory resolution of the measurement problem (Adler, 2003).

If we adopt the physical picture suggested by de Broglie and the walking droplets, the measurement problem vanishes from consideration. In particular, wave function collapse becomes a nonproblem (Bush and Oza, 2020). Consider the robust wave-like statistics emerging in the hydrodynamic corrals (Harris et al., 2013; Sáenz et al., 2018) (Fig. 2.5) or the analog Friedel oscillations (Fig. 2.6). If one were to assert that this statistical waveform were a complete description of the system, then its collapse into a discrete droplet in response to the act of observation might be troubling. With the knowledge of the underlying pilot-wave dynamics, it is obvious that wave function collapse is a feature of any statistical theory. Bush and Oza (2020) refer to this as 'statistical nonlocality', the misinference of nonlocality owing to one's insistence on the completeness of a statistical theory. Wave-function collapse may thus be seen as a dilemma only for those insistent on the completeness of a statistical theory, be it quantum or classical. Ditto for the measurement problem.

2.7 Quantum Superposition

> One cannot in the classical sense picture a system being partly in each of two states and see the equivalence of this to the system being completely in some other state. There is an entirely new idea involved, to which one must get accustomed and in terms of which one must proceed to build up an exact mathematical theory, without having any detailed classical picture.
>
> – P.A.M. Dirac, Principles of Quantum Mechanics (Dirac, 1958).

Due to the linearity of the Schrodinger equation (Eq. (2.1)), any linear combination of its solutions also constitutes a solution of the equation, and thus represent a valid quantum state. These linearly combined states are called superpositions, and they play a key role in the formalism and phenomenology of quantum mechanics. Formally, if ψ_1 and ψ_2 represent two different eigenstates of the Hamiltonian, then the superposed state, $\psi = a\psi_1 + b\psi_2$, is also a solution of the Schrodinger equation, with a, b being complex numbers such that $|a|^2 + |b|^2 = 1$. According to the Copenhagen interpretation, during measurement the wave function collapses onto one of the eigenstates ψ_1, ψ_2 with probabilities $|a|^2$, $|b|^2$ respectively. Notably, the coefficients a, b are generally complex numbers, and can assume negative values. As a result, the two eigenstates constituting the superposed state can destructively interfere with one another, yielding phenomena such as single particle interference as seen in the double-slit experiment (Sect. 2.4).

In their study of walker motion in an elliptical corral (Fig. 2.5d–g), Sáenz et al. (2018) demonstrated that in the high-memory, chaotic regime, the emergent statistics may be simply expressed in terms of the superposition of two cavity modes, one being the corrals most unstable Faraday mode at the systems driving frequency, the other being the most unstable mode at a nearby frequency. This then represents a superposition of statistical rather than dynamical states. By using bottom topography with high symmetry, they demonstrated that the relative weights of the two modes could be tuned. Specifically, by placing a submerged well at the focus of the ellipse, it favored one mode over the other. The analogous procedure in the quantum corral (Fiete and Heller, 2003), undertaken by manipulating magnetic impurities on the metal surface, leads to so-called 'statistical projection effects' (Moon et al., 2008; Manoharan et al., 2000). When an impurity is located at the focus of an ellipse, the preferred mode is that with extrema at the foci, leading to a projection effect referred to as the 'quantum mirage' (Manoharan et al., 2000). The walker system thus provides a rational means of interpreting both statistical projection and mirage effects.

We have seen that pilot-wave hydrodynamics can account for wave-particle duality and related phenomena such as single-particle interference. An important open question is thus whether one can derive a description of the emergent statistics equivalent to the Schrodinger formalism in quantum mechanics, from the dynamical description of the walker in HQA. The fact that the walker system exhibits both single-particle interference phenomena and the superposition of states suggests that such may be the case. In the HQA community, efforts are currently being made to develop a theory of walker statistics through consideration only of the pilot-wave field (Kutz et al., 2023). Specifically, for a walker confined to a one-dimensional well, the evolution of the pilot-wave field is characterized as the system memory (vibration forcing) is increased progressively. A discrete set of wave modes emerge sequentially, analogous to the new eigenstates of the wavefunciton arising as the particle energy is increased. Finally, the mean wavefield is inverted to yield the particle statistics, using Durey's convolution result (Durey et al., 2018, 2020). This recent study takes one step closer towards one current goal of the HQA community, developing a wave theory for the statistics of walking droplets (Kutz et al., 2023).

2.8 Nonlocality

That particular interpretation (Bohmian mechanics) has indeed a grossly non-local structure. This is characteristic, according to the result to be proved here, of any such theory which reproduces exactly the quantum mechanical predictions.

– John S. Bell, On the Einstein Podolsky Rosen paradox (Bell, 1964).

The inference made from the experimental violation of Bell's Theorem (see Sect. 2.10) has generally been that any hidden variable theory of quantum mechanics must be non-local; thus, non-locality is seen by some as being a necessary feature of a theory of quantum dynamics (Maudlin, 2014). For example, having seen the experimental violations of Bell's Inequality (Aspect et al., 1982a), Bell was an advocate of Bohmian mechanics on the grounds that it was non-local (Bricmont, 2016). Nonlocality has different guises in the various interpretations of quantum mechanics. For example, we have already seen in §2.5 that proponents of the Copenhagen Interpretation must contend with the instantaneous collapse of the wave function, a form of nonlocality simply rationalized from the new perspective offered from HQAs (Bush and Oza, 2020).

Another manifestation of quantum nonlocality takes the form of an apparent action at a distance, as arises in Bohmian mechanics. Specifically, if the position of one particle in an entangled pair is changed, the position of its entangled counterpart is instantaneously affected. In HQA, a number of systems have been considered in which one would infer action-at-a-distance if the pilot-wave dynamics were not adequately resolved. For example, in the double-slit walker diffraction experiments, the change prompted by the influence of the second slit on a droplet passing through the first would be considered nonlocal (Fig. 2.4d). When walkers approach submerged pillars (Harris et al., 2018) and wells (Sáenz et al., 2020, Fig. 2.6), they experience long-range lift forces that depend in a particular fashion on the distance from the obstacle. Were it not known that such forces are wave-mediated, one might misinfer that they imply action at a distance.

In single-particle Bohmian mechanics, nonlocality enters through the quantum potential. In classical mechanics, Newtons law of gravitation and Coulombs law both express non-local force laws, specifically, action at a distance. While the motion of a massive or charged particle in response to either force field is local (in the sense that it responds only to the local field), the field itself is nonlocal, in the sense that its origins cannot be rationalized without appealing to deeper theoretical developments, specifically, quantum electrodynamics or general relativity. The same could be said of Bohmian mechanics, according to which a particle moves in response to the quantum potential, whose form is prescribed by the quantum wave function. While the particle's motion may be considered local, it is moving in response to a nonlocal field imposed by fiat.

A recent example illustrates how the HQA perspective allows one to achieve quantum effects and rationalize them without appealing to nonlocality. 'Surreal trajectories' is a term coined by Englert et al. (1992) (ESSW) to describe Bohmian

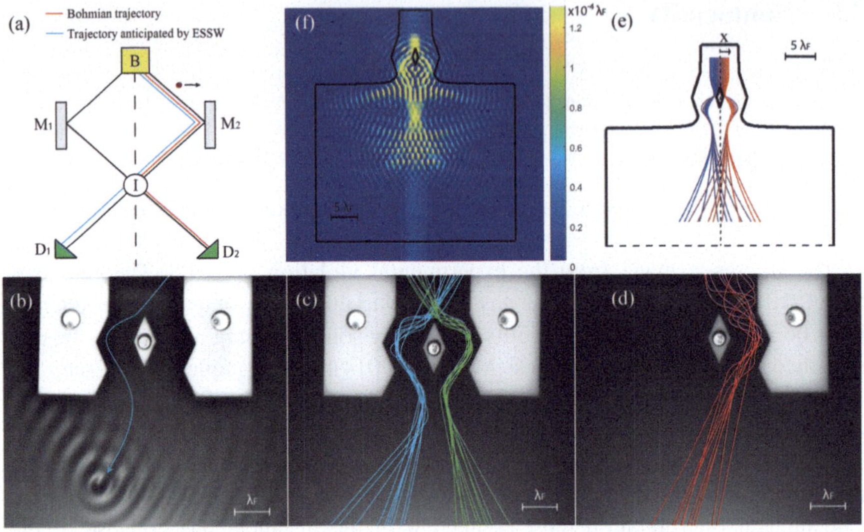

Fig. 2.7 Surreal trajectories in the hydrodynamic pilot-wave system (Frumkin et al., 2022). (**a**) A variant of the interferometer setup considered by ESSW (Englert et al., 1992). (**b**) A single particle trajectory, along with the instantaneous pilot wave field. (**c**) In a symmetric arrangement, the droplet enters the right or left channel with equal probability, after which it is deflected away from the system centerline, resulting in a 'surreal' trajectory. Twenty such trajectories are shown. (**d**) When one of the barriers is removed, the symmetry of the system is broken. The walking droplet is then reflected away from the remaining barrier, resulting in the trajectory that one might expect. (**e**) Numerical simulations of an ensemble of initially vertical trajectories with different values of the impact parameter x. (**f**) The mean pilot-wave field generated by averaging simulated droplet trajectories with a Gaussian distribution of initial impact parameters

trajectories predicted to arise in the interferometer arrangement illustrated in Fig. 2.7a. The term was intended to point out that the trajectories predicted by Bohmian mechanics are counterintuitive, and so cannot be real. Mahler et al. (2016) measured mean trajectories in the geometry proposed by ESSW via weak measurement, and found that they were consistent with those predicted by Bohmian mechanics. The authors concluded that 'the trajectories seem surreal only if one ignores their manifest nonlocality'. In Bohmian mechanics, surreal trajectories arise as a result of the particles being guided by the quantum potential, a non-local field. In the standard formulation, there is no notion of trajectories, but the experimental observations of mean trajectories consistent with surreal trajectories led (Mahler et al., 2016) to seek a rationale in terms of quantum nonlocality, specifically entanglement with the measurement device. In the walker system, 'real surreal trajectories' arise naturally from non-Markovian, classical dynamics in which the droplet navigates its pilot-wave field, a local potential of its own making. Nonlocality need not be invoked. Our study showed that the designation of Bohmian trajectories as surreal is based on misconceptions concerning the limitations of classical dynamics and a lack of familiarity with pilot-wave hydrodynamics (Frumkin et al., 2022). Moreover, it made clear that the physical picture furnished by the walker system allows one to see how to side-step the invocation of the non-local quantum potential required in Bohmian mechanics.

HQA suggests that one can avoid the invocation of a nonlocal field, as is done in Bohmian mechanics, provided one follows the suggestion of Holland (1995): "We can envisage a more active role for the particle, something which is not even admitted as conceivable in the conventional view. This may, for instance, enter as a source of the pilot-wave field through an inhomogeneous term in the wave equation". This conceptual leap would constitute a conformance to de Broglie's double-solution theory rather than the provisional theory now known as de Broglie-Bohm theory or Bohmian mechanics, and would seem to be a critical step in restoring locality to quantum pilot-wave theories. Considering the particle as the source of its own wave renders the resulting dynamics local: the particle generates its pilot wave and moves in response to it. This approach was recently followed by Dagan and Bush (2020) and Durey and Bush (2020), who considered particles with an internal oscillation at the Compton frequency generating, then moving in response to, waves satisfying a forced Klein-Gordon equation. While exploratory, the results yielded a striking result, a physical picture for the origins of the de Broglie relation. If a particle vibrating at the Compton frequency is dressed in a quasi-monochromatic wavefield that is a solution of the Klein-Gordon equation, then the group velocity of that wavefield must match the particle speed; thus, $p = \hbar k$. The emergent pilot-wave form for a free, uniformly translating particle, illustrated in Fig. 2.1d, roughly conforms to that envisaged by de Broglie (1987). The potential and limitations of this approach, of developing a model of quantum dynamics informed by the walker system, are currently being explored more widely.

2.9 Non-separable States

> Often a pair of quantum systems may be represented mathematically (by a vector) in a way each system alone cannot: the mathematical representation of the pair is said to be non-separable: Schrödinger called this feature of quantum theory entanglement.
>
> – Richard Healey, The Quantum Revolution in Philosophy (Healey, 2017).

Non-separable states arise in multi-partite systems when the state of the whole cannot be simply defined in terms of the state of its subsystems (Horodecki et al., 2009). For example, if a quantum system S is comprised of two subsystems, A and B, then the system will be considered non-separable if $\psi_S \neq \psi_A \otimes \psi_B$, where ψ_S, ψ_A and ψ_B are the wave functions of the respective systems, and $\otimes$ represents a tensor product. A canonical example is that of a singlet state, namely, two particles with opposite spins: $\psi = \frac{1}{\sqrt{2}}\left(|\downarrow\uparrow\rangle - |\uparrow\downarrow\rangle\right)$. Such states cannot be factored into a product of two independent states, specifically, there are no complex numbers a, b, c, d such that: $\psi = (a\,|\downarrow\rangle + b\,|\uparrow\rangle) \otimes (c\,|\downarrow\rangle + d\,|\uparrow\rangle)$. The singlet state is one of the maximally entangled Bell states (Bohm, 1951), and has been used historically to illustrate the difference between classical and quantum correlations through violations of Bell's inequality (see Sect. 2.10). The outcome of a spin measurement on one of these particles will necessarily be correlated with the outcome of a measurement on the other. When non-separability persists over arbitrarily large distances (Weihs et al., 1998; Yin et al., 2017), it is considered to be a signature of entanglement. According to the Copenhagen interpretation (Home and Nair, 1994), particles in a singlet state do not have a well-defined spin prior to measurement. Once one of the two particles is measured, the wave function collapses instantaneously for both. One might then ask how the second particle, that was not measured, knows that a measurement took place at the location of the distant first particle. The Copenhagen Interpretation suggests that there is an instantaneous transfer of information between the two particles, spcifically the information that one of the particles was measured. This peculiar "non-local" feature of entangled states in quantum mechanics, is what Einstein referred to as "spooky action at a distance" (Einstein et al., 1935).

In recent years, it has been shown that non-separability is not limited solely to the quantum realm. Indeed, it is possible to achieve non-separable states with different degrees of freedom of a classical electromagnetic field, and to use these non-separable states to violate Bell's inequality (Qian et al., 2015; Spreeuw, 1998). Non-separable acoustical states were also recently demonstrated in coupled elastic waveguides (Hasan et al., 2019), extending the notion of non-separability to the field of phononics. Due to the commonalities with the standard formalism of quantum entanglement, the above demonstrations have been described in terms of 'classical entanglement' (Spreeuw, 1998; Qian et al., 2015; Hasan et al., 2019). However, it has been convincingly argued that a different term, specifically 'non-separable states', would be preferable to distinguish these classical states from those exhibiting standard bipartite quantum entanglement (Karimi and Boyd, 2015).

Prior work on classical non-separable systems has made clear the distinction between classical and quantum superpositions of states: while non-separability may arise in both, entanglement (specifically, the persistence of non-separable states to arbitrary distances Weihs et al., 1998; Yin et al., 2017) is peculiar to quantum mechanics. While the classical non-separable states discovered to date cannot be used for testing local realism or quantum computing, they may be useful for a number of other quantum information related applications (Schmid et al., 2010; Simon et al., 2010; Töppel et al., 2014; Pinheiro et al., 2013). Unlike the aforementioned examples of classical non-separability, the bouncing droplet system allows for the possibility of non-separable states in a spatially separated bipartite system (Nachbin, 2022), and so for the application of these states in quantum-inspired classical computing.

In quantum mechanics, tunneling is unpredictable, but the probability of its occurrence is well-defined by the system geometry (Papatryfonos et al., 2015). Eddi et al. (2009a) and later Tadrist et al. (2020) demonstrated that such is also the case for bouncing droplets, and Nachbin et al. (2017) captured this behavior using a theoretical model for one-dimensional walker motion over a vibrating liquid bath with complex topography. Subsequently, Nachbin (2018) extended this model to study correlations between two droplets separated by an intervening cavity. The coupling strength was prescribed by the geometry of the central cavity, which served as a nearly resonant transmission line. Using this system, the author showed that the two-particle system is non-separable, by demonstrating that each particle's phase space dynamics is given by the system as a whole and cannot be described separately. He further showed that after the correlations were established, removing one of the particles from the system resulted in a drastic alteration of the phase space picture. The correlations arising in this model persisted even when the droplets were separated by large distances. Most recently, this model was extended to combine the aspects of unpredictable tunneling and bipartite non-separability in order to demonstrate a hydrodynamic analog of superradiance (Papatryfonos et al., 2022a).

Superradiant photon emission is the anomalous emission arising when a pair or assemblage of ions is in close proximity. Superradiance was originally attributed to quantum interference arising from entangled atoms (DeVoe and Brewer, 1996; Makarov and Letokhov, 2004; Karnieli et al., 2021), but has since been rationalised in terms of electromagnetic wave interference (Tanji-Suzuki et al., 2011). Papatryfonos et al. (2022a) have recently used the walker system to develop a hydrodynamic analog of superradiance in a numerical model of bipartite walker tunneling. Their system consists of a pair of walkers, each in a subsystem consisting of a pair of cavities across which the walker may tunnel between ground (outer) and excited (inner) states (see Fig. 2.8). While the bottom topography precludes the drops from escaping their respective subsystems, the two subsystems communicate through the wave field spanning the coupling cavity, allowing for strongly correlated states induced by the wave-mediated forces. By identifying the tunneling transitions between excited and ground states with photon emission events, they demonstrated that the wave-induced droplet tunneling may lead to an analog of superradiant photon emission. Moreover, they demonstrated that the emission rate varies sinusoidally with distance between the two subsystems, as is also the case in superradiant photon emission from ion pairs (DeVoe and Brewer, 1996).

2.10 Entanglement

> What Bell's theorem, together with the experimental results, proves to be impossible ... is not determinism or hidden variables or realism but locality, in a perfectly clear sense. What Bell proved, and what theoretical physics has not yet properly absorbed, is that the physical world itself is non-local.
>
> – Tim Maudlin, "What Bell Did", J. Phys. A: Mathematical and Theoretical (Maudlin, 2014).

Those that assert that quantum mechanics in its standard form is incomplete appeal to 'hidden variable theories' for its completion. The hidden variables are those variables needed to provide a complete description of the quantum dynamics underlying quantum statistics. For example, in pilot-wave theories of the form

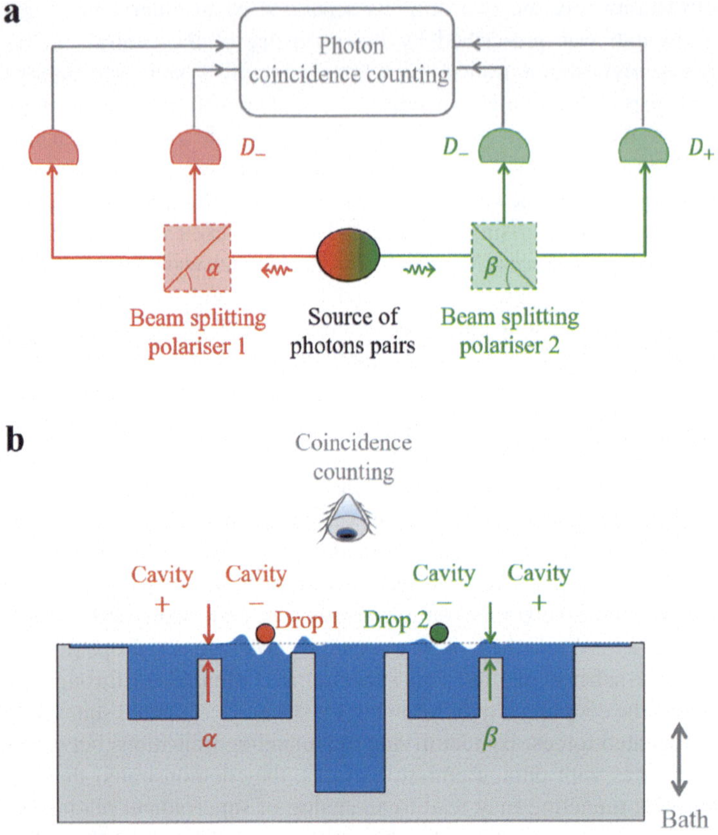

Fig. 2.8 Photonic and hydrodynamic Bell test arrangements. (**a**) For optical Bell tests, pairs of entangled photons are produced at the source and sent in opposite directions to two measurement stations. At each station one of two measurement settings is selected randomly and a measurement is performed. The measurement outcomes, which can take on two possible values, +1 or −1, are noted. (**b**) The hydrodynamic Bell test consists of a pair of drops (red and green) walking on the surface of a vibrating fluid bath (blue) that spans the solid substrate (grey). Each drop is confined to its subsystem, a pair of wells separated by barriers across which they may tunnel unpredictably at a rate prescribed by the barrier depths α and β, as may assume values of a, a' or b, b', respectively. The two subsystems are coupled through the intervening wave field. For a limited range of measurement settings (barrier depths), Bell violations are obtained, with a maximum violation of 2.49. (Figure adapted from Ref. Papatryfonos et al., 2022b)

suggested by pilot-wave hydrodynamics, the hidden variables would be the particle position and momentum and the form of the real pilot-wave field, as distinct from the wave function. In 1964, John Bell derived an inequality, the experimental violation of which is widely taken to indicate that no local hidden variable theory can account for the correlations apparent in quantum spin measurements (Bell, 1964). We refer here to a variant of Bell's theorem that was introduced by Clauser, Horne, Shimony, and Holt (CHSH) (Clauser et al., 1969; Clauser and Horne, 1974). Consider two entangled particles in a singlet state that are measured at two spatially separated detectors, A and B. Each detector can have two possible measurement settings, $\alpha \in \{a, a'\}$ for A, and $\beta \in \{b, b'\}$ for B. If $E(a, b)$ denotes the expectation value of the product of outcomes of measurements (+1 for spin-up and -1 for spin-down) with settings a and b, then the CHSH-Bell inequality states that any local hidden variable theory must satisfy the condition:

$$|S| = \left|E(a, b) + E(a, b') + E(a', b) - E(a', b')\right| \leq 2 \tag{2.2}$$

where S quantifies the correlations between measurement outcomes at the two detectors. One can demonstrate theoretically that for certain choices of measurement settings, the singlet state can violate the CHSH-Bell inequality, with a maximal correlation value of $S = 2\sqrt{2}$.

The predicted violation of the CHSH-Bell inequality by quantum mechanics has been verified experimentally in a wide range of physical systems (Brunner et al., 2014), thus leading to the conclusion that quantum mechanics does not satisfy at least one of the assumptions made in the derivation of Bell's theorem. Given that these assumptions include the basic tenets of reality (a physical world exists independent of human observation) and locality (nothing travels faster than the speed of light), some have even proclaimed Bell's Theorem to be 'the most profound in science' (Stapp, 1975). The experimental violation thereof by Aspect et al. (1982a,b) was rewarded with the most recent Nobel prize in physics, and was interpreted by the Nobel Prize Committee as proving that 'quantum mechanics cannot be replaced by a theory that uses hidden variables'.

While the experimental violation of Bell's inequality is widely taken to imply that quantum mechanics is inescapably non-local, less drastic conclusions are currently being explored through careful scrutiny of the assumptions made, either explicitly or implicitly, in the derivation of Bell's inequality. In particular, Vervoort (2018) has questioned whether the assumption of 'measurement-independence', according to which the hidden variables that prescribe the measurement outcomes are independent of α and β, is valid in systems with a background field. Vervoort argues that such may not be the case in pilot-wave systems, wherein the hidden variables characterizing the pilot-wave field might in principle be influenced by the analyzer settings. A similar line of questioning was originally put forth by workers in stochastic electrodynamics (de la Peña et al., 1972).

To inform this debate, we have recently adopted the model geometry used in the hydrodynamic analog of superradiance (Papatryfonos et al., 2022a) to administer the first static Bell test in the hydrodynamic pilot-wave system (see Fig. 2.8). To that end, we identified the location of the drop (inner or outer cavity) with the dichotomic property, corresponding to spin (+1 or -1) in traditional Bell tests. We identified the barrier depths in the two subsystems with the measurement settings (α, β). Judicious choice of combinations of measurement settings, including one for which strongly synchronized tunneling arises, allowed for violation of Bell's Inequality, with a maximum violation of 2.49 ± 0.04. This violation is made possible by the fact that in the walker system, the assumption of measurement-independence is violated. Specifically, owing to the wave-mediated coupling between the two subsystems, the probability of a droplet occupying a given cavity in either of the subsystems depends on both measurement settings (Papatryfonos et al., 2022b). Pilot-wave hydrodynamics thus illustrates how static Bell violations may arise in a classical setting. This particular system also shows how non-separable states with non-classical correlations can be established in classical bipartite systems.

The question remains open as to whether the violations deduced in the static Bell test will survive isolation of the two subsystems, as would effectively eliminate communication between them. A promising result in this direction has recently been demonstrated by Nachbin (2022), who considered correlations of two droplets confined to individual cavities linked through a coupling cavity. He demonstrated that the wave-mediated forces between the droplets allowed them to achieve strongly correlated energetic states inaccessible to either droplet in the other's absence. By decreasing the depth of the coupling cavity during the course of the simulations, he demonstrated that these correlated excited states could survive even when the droplets are no longer communicating. We are currently taking a similar approach to the two-cavity geometry considered in our static Bell test that will determine whether violations of Bell's inequality can persist even when the two subsystems are isolated. Such tests should settle the matter as to whether memory might provide a viable mechanism for establishing classical entanglement. Such violations might in principle be rationalized through the measurement-independence loophole (Vervoort, 2018): the hidden variables (specifically, the droplet trajectories and pilot-wave field) depend on both measurement settings, whose influence is imposed prior to, but persists after, isolation of the subsystems. Concurrently, numerous efforts are being made in the HQA community towards achieving Bell violations experimentally with the walker system.

2.11 Conclusion

> I think it very probable that the solution to our problem will come through the back door: some person who is not addressing himself to these difficulties with which I am concerned will probably see the light. An analogy that I like is that of a fly buzzing against a window when the door is open.
>
> – John. S. Bell (Davies and Brown, 1993).

Several pronouncements concerning the inscrutability of quantum mechanics, made by some of its most distinguished practitioners, have been revisited in light of insights gained from pilot-wave hydrodynamics. Many now fall flat. If one is unwilling to make one's peace with quantum nonlocality, there would appear to be two options. First, one may ascribe to one of the more extravagant quantum interpretations that side-step the nonlocality problem by offering up an even less appealing prospect, for example, the Multiverse (Everett, 1957). Second, one may follow the progressive approach offered by HQA, and explore the possibility of a hidden variable theory that violates Bell's inequality without being explicitly nonlocal, specifically without having to appeal to superluminal signaling or a nonlocal quantum field. The authors are currently following this approach, and so persist in the stubborn view that while nonlocality is undoubtedly a feature of the current quantum theory, it need not be a feature of an adequately resolved, complete theory of quantum physics (Kupczynski, 2020).

The proliferation of quantum interpretations has arisen because there is presently no experimental means by which to distinguish between them. Conversely, theories of quantum mechanics can be falsified or verified via experimental test. The hope is that HQAs will provide additional insight into the quantum pilot-wave theories of de Broglie (1987), de la Peña et al. (2015), and Hestenes (1990), that will ultimately be tested experimentally. Given that all such theories appeal to an unresolved dynamics on the Compton scale, it is conceivable that future experimental investigations, achieved for example in electron channeling (Catillon et al., 2008), might validate or refute the physical picture suggested by walking droplets.

Finally, it is noteworthy that the field of hydrodynamic analogs of general relativity was initiated by Unruh (1981) and Schützhold and Unruh (2002) and has subsequently been well-established (Euve et al., 2016; Rousseaux and Kellay, 2020). The field of HQAs is relatively recent, but burgeoning. A primary difficulty in modern physics is reconciling general relativity and quantum mechanics. Should the subject of fluid mechanics provide the church at which to marry the two, one imagines that the subject might become a more prominent component of the modern undergraduate physics curriculum.

References

Adler, S.L. 2003. Why decoherence has not solved the measurement problem: A response to P.W. Anderson. *Studies in History and Philosophy of Science Part B: Studies in History and Philosophy of Modern Physics 34*: 135–142.

Andersen, A., J. Madsen, C. Reichelt, S.R. Ahl, B. Lautrup, C. Ellegaard, M.T. Levinsen, and T. Bohr. Double-slit experiment with single wave-driven particles and its relation to quantum mechanics. *Physical Review E 92*: 013006.

Arbelaiz, J., A.U. Oza, and J.W.M. Bush. 2018. Promenading pairs of walking droplets: dynamics and stability. *Physical Review Fluids 3*: 013604.

Aspect, A., P. Grangier, and G. Roger. 1982a. Experimental realization of Einstein-Podolsky-Rosen-Bohm *Gedankenexperiment*: A new violation of Bell's inequalities. *Physical Review Letters 49*: 91.

Aspect, A., J. Dalibard, and G. Roger. Experimental test of Bell's inequalities using time-varying analyzers. *Physical Review Letters 49*: 1804.

Bassi, A., and G. Ghirardi. 2003. Dynamical reduction models. *Physics Reports 379*(5–6): 257–426.

Bassi, A., K. Lochan, S. Satin, T.P. Singh, and H. Ulbricht. 2013. Models of wave-function collapse, underlying theories, and experimental tests. *Reviews of Modern Physics 85*: 471–527. Publisher: American Physical Society.

Batchelor, G.K. 2008. *The life and legacy of G.I. Taylor*. Cambridge: Cambridge University Press.

Bell, J.S. 1964. On the Einstein Podolsky Rosen paradox. *Physics, 1*(3): 195–200.
Bell, J.S. 1987. *Speakable and Unspeakable in Quantum Mechanics*. Cambridge: Cambridge University Press.
Bohm, D. 1951. *Quantum Theory*. Englewood Cliffs: Prentice-Hall.
Bohm, D. 1952a. A suggested interpretation of the quantum theory in terms of hidden variables, i. *Physical Review 85*: 66–179.
Bohm, D. 1952b. A suggested interpretation of the quantum theory in terms of hidden variables, ii. *Physical Review 85*: 180–193.
Bohr, N. 1935. Can quantum-mechanical description of physical reality be considered complete? *Physical Review 48*: 696. American Physical Society.
Borghesi, C., J. Moukhtar, M. Labousse, A. Eddi, E. Fort, and Y. Couder. 2014. Interaction of two walkers: Wave-mediated energy and force. *Physical Review E 90*: 063017.
Bricmont, J. 2016. *Making Sense of Quantum Mechanics*. Berlin: Springer.
Bricmont, J. 2017. History of quantum mechanics or the comedy of errors. *International Journal of Quantum Foundation 3*: 31–64.
Brunner, N., D. Cavalcanti, S. Pironio, V. Scarani, and S. Wehner. 2014. Bell nonlocality. *Reviews of Modern Physics 86*: 419–478.
Bush, J.W.M. 2015. Pilot-wave hydrodynamics. *Annual Review of Fluid Mechanics 47*: 269–292.
Bush, J.W.M., and A.U. Oza. 2020. Hydrodynamic quantum analogs. *Reports on Progress in Physics 84*(017001): 1–41.
Catillon, P., N. Cue, M.J. Gaillard, R. Genre, M. Gouanère, R.G. Kirsch, J.C. Poizat, J. Remillieux, L. Roussel, and M. Spighel. 2008. A search for the de Broglie particle internal clock by means of electron channeling. *Foundations of Physics 38*(7): 659–664.
Clauser, J.F., and M.A. Horne. 1974. Experimental consequences of objective local theories. *Physical Review D 10*: 526–535.
Clauser, J.F., M.A. Horne, A. Shimony, and R.A. Holt. Proposed experiment to test local hidden-variable theories. *Physical Review Letters 23*: 880–884.
Colin, S., T. Durt, and R. Willox. 2017. L. de Broglie's double solution program: 90 years later. *Annales de la Fondation Louis de Broglie 42*: 19–71.
Couchman, M.M.P., and J.W.M. Bush. 2020. Free rings of bouncing droplets: stability and dynamics. *Journal of Fluid Mechanics 903*: A49.
Couder, Y., and E. Fort. 2006. Single particle diffraction and interference at a macroscopic scale. *Physical Review Letters 97*: 154101.
Couder, Y., S. Protière, E. Fort, and A. Boudaoud. 2005. Walking and orbiting droplets. *Nature 437*: 208.
Cristea-Platon, T., P.J. Sáenz, and J.W.M. Bush. 2018. Walking droplets in a circular corral: Quantisation and chaos. *Chaos 28*: 096116.
Crommie, M.F., C.P. Lutz, and D.M. Eigler. 1993a. Imaging standing waves in a two-dimensional electron gas. *Nature 363*: 524–527.
Crommie, M.F., C.P. Lutz, and D.M. Eigler. 1993b. Confinement of electrons to quantum corrals on a metal surface. *Science 262*: 218–220.
Dagan, Y., and J.W.M. Bush. 2020. Hydrodynamic quantum field theory: The free particle. *Comptes Rendus Mécanique 348*(6–7): 552–570.
Damiano, A.P., P.T. Brun, D.M. Harris, C.A. Galeano-Rios, and J.W.M. Bush. 2016. Surface topography measurements of the bouncing droplet experiment. *Experiments in Fluids 57*: 163.
Davies, P.C.W., and J.R. Brown. 1993. *The Ghost in the Atom: An Interview with John Bell*. Cambridge: Cambridge University Press.
de Broglie, L. 1926. *Ondes et mouvements*. Paris: Gautier Villars.
de Broglie, L. 1970. The reinterpretation of wave mechanics. *Foundations of Physics 1*(1): 5–15.
de Broglie, L. 1987. Interpretation of quantum mechanics by the double solution theory. *Annales de la Fondation Louis de Broglie 12*: 1–23.
de la Peña, L., A.M. Cetto, and T.A. Brody. 1972. On hidden-variable theories and Bell's inequality. *Lettere al Nuovo Cimento 5*: 2.

de la Peña, L., A.M. Cetto, and A. Valdés-Hernández. 2015. *The Emerging Quantum: The Physics Behind Quantum Mechanics*. Cham: Springer.

Dirac, P. A. M. 1958. *Principles of Quantum Mechanics Quantum*. London: Oxford University Press.

Durey, M., and J.W.M. Bush. 2020. Hydrodynamic quantum field theory: the onset of particle motion and the form of the pilot wave. *Frontiers in Physics 8*: 300.

Durey, M., and J.W.M. Bush. 2021. Classical pilot-wave dynamics: the free particle. *Chaos 31*(033136): 1–11.

Eddi, A., D. Terwagne, E. Fort, Y. Couder. 2008. Wave propelled ratchets and drifting rafts. *Europhysics Letters 82*: 44001.

Eddi, A., E. Fort, F. Moisy, and Y. Couder. 2009a. Unpredictable tunneling of a classical wave-particle association. *Physical Review Letters 102*: 240401.

Eddi, A., A. Decelle, E. Fort, and Y. Couder. 2009b. Archimedean lattices in the bound states of wave interacting particles. *Europhysics Letters 87*: 56002.

Eddi, A., E. Sultan, J. Moukhtar, E. Fort, M. Rossi, and Y. Couder. 2011. Information stored in Faraday waves: the origin of a path memory. *Journal of Fluid Mechanics 674*: 433–463.

DeVoe, R.G., and R.G. Brewer. 1996. Observation of superradiant and subradiant spontaneous emission of two trapped ions. *Physical Review Letters 76*: 2049–2052.

Durey, M., P.A. Milewski, and J.W.M. Bush. 2018. Dynamics, emergent statistics and the mean-pilot-wave potential of walking droplets. *Chaos 28*: 096108.

Durey, M., P.A. Milewski, and Z. Wang. 2020. Faraday pilot-wave dynamics in a circular corral. *Journal of Fluid Mechanics 891*: A3.

Eddington, S.A.S. 1927. *The Nature of the Physical World: Gifford Lectures*. Cambridge: Cambridge University Press.

Einstein, A., B. Podolsky, and N. Rosen. 1935. Can quantum-mechanical description of physical reality be considered complete? *Physical Review 47*: 777–780.

Ellegaard, C., and M.T. Levinsen. 2020. Interaction of wave-driven particles with slit structures. *Physical Review E 102*: 023115.

Englert, B.G., M.O. Scully, G. Süssmann, and H. Walther. 1992. Surrealistic Bohm Trajectories. *Zeitschrift für Naturforschung A 47*(12): 1175–1186.

Euve, L.P., F. Michel, R. Parentani, T.G. Philbin, and G. Rousseaux. 2016. Observation of noise correlated by the Hawking effect in a water tank. *Physical Review Letters 117*: 121301.

Everett, H. 1957. Relative state formulation of quantum mechanics. *Reviews of Modern Physics 29*: 454–462.

Feynman, R.P. 1964. *The Feynman Lectures on Physics*. Reading: Addison-Wesley.

Fiete, G.A., and E.J. Heller. 2003. Theory of quantum corrals and quantum mirages. *Reviews of Modern Physics 75*: 933–948.

Folse, H.J. 1985. *The Philosophy of Niels Bohr: The Framework of Complementarity*. Amsterdam: Elsevier.

Fort, E., A. Eddi, J. Moukhtar, A. Boudaoud, and Y. Couder. 2010. Path-memory induced quantization of classical orbits. *Proceedings of the National Academy of Sciences 107*(41): 17515–17520.

Frumkin, V., D. Darrow, W. Struyve, and J.W.M. Bush. 2022. Real surreal trajectories in pilot-wave hydrodynamics. *Physical Review A 106*: L010203.

Frumkin, V., K. Papatryfonos, and J.W.M. Bush. 2023. Superradiant droplet emission from parametrically excited cavities. *Physical Review Letters 130*: 064002.

Galeano-Rios, C.A., M.M.P. Couchman, P. Caldairou, and J.W.M. Bush. 2018. Ratcheting droplet pairs. *Chaos 28*: 096112.

Griffiths, D.J. 2014. *Introduction to Quantum Mechanics: Second Edition*. London: Pearson.

Hance, J.R., and S. Hossenfelder. 2022. What does it take to solve the measurement problem? *Journal of Physical Communication 6*: 102001. Publisher: IOP Publishing.

Harris, D.M., and J.W.M. Bush. 2014. Drops walking in a rotating frame: From quantized orbits to multimodal statistics. *Journal of Fluid Mechanics 739*: 444–464.

Harris, D.M., J. Moukhtar, E. Fort, Y. Couder, and J.W.M. Bush. 2013. Wavelike statistics from pilot-wave dynamics in a circular corral. *Physical Review E 88*(011001): 1–5.
Harris, D.M., P.T. Brun, A. Damiano, L. Faria, and J.W.M. Bush. 2018. The interaction of a walking droplet and a submerged pillar: From scattering to the logarithmic spiral. *Chaos 28*: 096106.
Hasan, M.A., L. Calderin, T. Lata, P. Lucas, K. Runge, and P.A. Deymier. 2019. The sound of Bell states. *Communications Physics 2*: 106.
Hatifi, M., R. Willox, S. Colin, and T. Durt. 2018. Bouncing oil droplets, de Broglie's quantum thermostat, and convergence to equilibrium. *Entropy 20*: 780.
Healey, R. (2017). *The Quantum Revolution in Philosophy*. London: Oxford University Press.
Heisenberg, W. 1930. *The Physical Principles of the Quantum Theory*. Chicago: University of Chicago Press.
Hestenes, D. 1990. The Zitterbewegung interpretation of quantum mechanics. *Foundations of Physics 20*(10): 1213–1232.
Holland, P.R. 1995. *The Quantum Theory of Motion: An Account of the de Broglie-Bohm Causal Interpretation of Quantum Mechanics*. Cambridge: Cambridge University Press.
Home, D., and R. Nair. 1994. Wave function collapse as a nonlocal quantum effect. *Physics Letters A 187*: 224–226.
Horodecki, R., P. Horodecki, M. Horodecki, and K. Horodecki. 2009. Quantum entanglement. *Reviews of Modern Physics 81*: 865.
Karimi, E., and R.W. Boyd. 2015. Classical entanglement? *Science, 350*: 1172–1173. Publisher: American Association for the Advancement of Science Section: Perspective.
Karnieli, A., N. Rivera, A. Arie, and I. Kaminer. 2021. Superradiance and subradiance due to quantum interference of entangled free electrons. *Physical Review Letters 127*: 060403.
Kupczynski, M. 2020. Is the moon there if nobody looks: Bell inequalities and physical reality. *Frontiers in Physics 8:* 273.
Kutz, J.N., Nachbin, A., Baddoo, P.J. and Bush, J.W.M. 2023. Pilot-wave dynamics: Using dynamic mode decomposition to characterize bifurcations, routes to chaos and emergent statistics. *Physical Review E 108*: 034213.
Labousse, M., S. Perrard, Y. Couder, and E. Fort. 2014a. Build-up of macroscopic eigenstates in a memory-based constrained system. *New Journal of Physics 16*: 113027.
Mahler, D.H., L. Rozema, K. Fisher, L. Vermeyden, K.J. Resch, H.M. Wiseman, and A. Steinberg. 2016. Experimental nonlocal and surreal Bohmian trajectories. *Science Advances 2:* e11501466.
Makarov, A.A., and V. Letokhov. 2004. Metastable entangled states of atomic systems in macroscale: Radiation dynamics and spectrum. In *International Workshop on Quantum Optics 2003 (Proceedings of the SPIE)*, 54–64.
Manoharan, H., C. Lutz, and D. Eigler. 2000. Quantum mirages formed by coherent projection of electronic structure. *Nature 403*: 512–515.
Maudlin, T. 2014. What Bell did. *Journal of Physics A: Mathematical and Theoretical 47*(42): 424010.
Maxwell, J.C. 1873. *A Treatise on Electricity and Magnetism*. Oxford: Clarendon Press.
McKagan, S.B., W. Handley, K.K. Perkins, and C.E. Wieman. 2009. A research-based curriculum for teaching the photoelectric effect. *American Journal of Physics 77*: 87–94.
Miller, D.A.B. 1991. Huygens's wave propagation principle corrected. *Optics Letters 16*: 1370–1372.
Moon, C.R., C.P. Lutz, and H.C. Manoharan. 2008. Single-atom gating of quantum-state superpositions. *Nature Physics 4*: 454–458.
Nachbin, A. 2018. Walking droplets correlated at a distance. *Chaos 28*: 096110.
Nachbin, A. 2022. Effect of isolation on two-particle correlations in pilot-wave hydrodynamics. *Physical Review Fluids 7*: 093604.
Nachbin, A., P.A. Milewski, and J.W.M. Bush. 2017. Tunneling with a hydrodynamic pilot-wave model. *Physical Review Fluids 2*: 034801.
Newton, I. 1704. *Opticks: Or, a Treatise of the Reflections, Refractions, Inflections and Colours of Light*. Oxford: Oxford University.

Oza, A.U., D.M. Harris, R.R. Rosales, and J.W.M. Bush. 2014a. Pilot-wave dynamics in a rotating frame: On the emergence of orbital quantization. *Journal of Fluid Mechanics 744*: 404–429.

Oza, A.U., O. Wind-Willassen, D.M. Harris, R.R. Rosales, and J.W.M. Bush. 2014b. Pilot-wave hydrodynamics in a rotating frame: exotic orbits. *Physics of Fluids 26*: 082101.

Oza, A.U., E. Siéfert, D.M. Harris, J. Moláček, and J.W.M. Bush. 2017. Orbiting pairs of walking droplets: dynamics and stability. *Physical Review Fluids 2*: 053601.

Oza, A.U., R.R. Rosales, and J.W.M. Bush. 2018. Hydrodynamic spin states. *Chaos 28*: 096106.

Papatryfonos, K., G. Rodary, C. David, F. Lelarge, A. Ramdane, and J.C. Girard. 2015. One-dimensional nature of inas/inp quantum dashes revealed by scanning tunneling spectroscopy. *Nano Letters 15*(7): 4488–4497.

Papatryfonos, K., M. Ruelle, C. Bourdiol, A. Nachbin, J.W.M. Bush, and M. Labousse. 2022a. Hydrodynamic superradiance in wave-mediated cooperative tunneling. *Communications Physics 5*: 142.

Papatryfonos, K., L. Vervoort, A. Nachbin, M. Labousse, and J.W.M. Bush. 2022b. Bell test in a classical pilot-wave system. arXiv:2208.08940.

Perrard, S., M. Labousse, M. Miskin, E. Fort, and Y. Couder. 2014. Self-organization into quantized eigenstates of a classical wave-driven particle. *Nature Communications 5*: 3219.

Pinheiro, A.R.C., C.E.R. Souza, D.P. Caetano, J.A.O. Huguenin, A.G.M. Schmidt, and A.Z. Khoury. 2013. Vector vortex implementation of a quantum game. *Journal of the Optical Society of America B 30*: 3210–3214.

Protière, S., S. Bohn, and Y. Couder. 2008. Exotic orbits of two interacting wave sources. *Physical Review E 78*: 036204.

Pucci, G., D.M. Harris, L.M. Faria, and J.W.M. Bush. 2018. Walking droplets interacting with single and double slits. *Journal of Fluid Mechanics 835*: 1136–1156.

Qian, X.F., B. Little, J.C. Howell, and J.H. Eberly. 2015. Shifting the quantum-classical boundary: theory and experiment for statistically classical optical fields. *Optica 2*: 611–615.

Rousseaux, G., and H. Kellay. 2020. Classical hydrodynamics for analogue space–times: Open channel flows and thin films. *Philosophical Transactions of the Royal Society A 378*: 20190233.

Sáenz, P.J., T. Cristea-Platon, and J.W.M. Bush. 2018. Statistical projection effects in a hydrodynamic pilot-wave system. *Nature Physics 14*: 315–319.

Sáenz, P.J., T. Cristea-Platon, and J.W.M. Bush. 2020. A hydrodynamic analog of Friedel oscillations. *Science Advances 6*: 20.

Sáenz, P.J., G. Pucci, S.E. Turton, A. Goujon, R.R. Rosales, J. Dunkel, and J.W.M. Bush. 2021. Emergent order in hydrodynamic spin lattices. *Nature 596*: 58–62.

Santos, E.A., F. Castro, and R. Torres. 2018. Huygens-fresnel principle: analyzing consistency at the photon level. *Physical Review A 97*: 043853.

Sautoy, M.D. 2017. *The Great Unknown: Seven Journeys to the Frontiers of Science*. New York: Viking.

Schmid, C., A.P. Flitney, W. Wieczorek, N. Kiesel, H. Weinfurter, and L.C.L. Hollenberg. 2010. Experimental implementation of a four-player quantum game. *New Journal of Physics 12*: 063031.

Schrödinger, E. 1926. Quantisierung als Eigenwertproblem. *Annals of Physics 384*(4): 361–376.

Schrödinger, E. 1935. Die gegenwärtige Situation in der Quantenmechanik. *Naturwissenschaften 23*: 807–812.

Schützhold, R., and W.G. Unruh. 2002. Gravity wave analogues of black holes. *Physical Review D 66*: 044019.

Simon, B.N., S. Simon, F. Gori, M. Santarsiero, R. Borghi, N. Mukunda, R. Simon. 2010. Nonquantum entanglement resolves a basic issue in polarization optics. *Physical Review Letters 104*: 023901.

Spreeuw, R.J.C. 1998. A classical analogy of entanglement. *Foundations of Physics 28*: 361–374.

Stapp, H.P. 1975. Bell's theorem and world process. *Il Nuovo Cimento B 29*: 270–276. Springer.

Tadrist, L., T. Gilet, P. Schlagheck, and J.W.M. Bush. 2020. Predictability in a hydrodynamic pilot-wave system: Resolution of walker tunneling. *Physical Review E 102*: 013104.

Tanji-Suzuki, H., I.D. Leroux, M.H. Schleier-Smith, M. Cetina, A.T. Grier, J. Simon, and V. Vuletić. 2011. Chapter 4 - interaction between atomic ensembles and optical resonators: Classical description. In *Advances in Atomic, Molecular, and Optical Physics*, ed. E. Arimondo, P. Berman, and C. Lin. *Advances in Atomic, Molecular, and Optical Physics*, vol. 60, 201–237. Cambridge: Academic Press.

Taylor, G.I. 1909. Interference fringes with feeble light. *Proceedings of the Cambridge Philosophical Society 15*: 114–115.

Tonomura, A., J. Endo, T. Matsuda, T. Kawasaki, and H. Ezawa. 1989. Demonstration of single electron buildup of an interference pattern. *American Journal of Physics*: 57.

Töppel, F., A. Aiello, C. Marquardt, E. Giacobino, and G. Leuchs, Classical entanglement in polarization metrology. *New Journal of Physics 16*: 073019.

Turton, S.E., M.M.P. Couchman, and J.W.M. Bush, A review of the theoretical modeling of walking droplets: towards a generalized pilot-wave framework. *Chaos 28*: 096111.

Vervoort, L. 2018. Are hidden-variable theories for pilot-wave systems possible? *Foundations of Physics 48*: 803–826.

Unruh, W.G. 1981. Experimental black-hole evaporation? *Physical Review Letters 46*: 1351–1353.

Weihs, G., T. Jennewein, C. Simon, H. Weinfurter, and A. Zeilinger. 1998. Violation of bell's inequality under strict einstein locality conditions. *Physical Review Letters 81*: 5039–5043.

Wigner, E.P. 1995. Remarks on the Mind-Body Question. In *Philosophical Reflections and Syntheses*, ed. J. Mehra, The Collected Works of Eugene Paul Wigner, 247–260. Berlin: Springer.

Yin, J., Y. Cao, Y.H. Li, S.K. Liao, L. Zhang, J.G. Ren, W.Q. Cai, W.Y. Liu, B. Li, H. Dai, G.B. Li, Q.M. Lu, Y.H. Gong, Y. Xu, S.L. Li, F.Z. Li, Y.Y. Yin, Z.Q. Jiang, M. Li, J.J. Jia, G. Ren, D. He, Y.L. Zhou, X.X. Zhang, N. Wang, X. Chang, Z.C. Zhu, N.L. Liu, Y.A. Chen, C.Y. Lu, R. Shu, C.Z. Peng, J.Y. Wang, and J.W. Pan. 2017. Satellite-based entanglement distribution over 1200 kilometers. *Science 356*: 1140–1144.

Zurek, W.H. 2003. Decoherence, einselection, and the quantum origins of the classical. *Reviews of Modern Physics 75*: 715–775. Publisher: American Physical Society.

Chapter 3
Establishing Long-Range Pilot-Wave Interactions

André Nachbin

Abstract Long-range intrinsically related correlations between two particles is our topic of interest in the context of the hydrodynamic pilot-wave. In particular how particles, at a distance, can share the same statistics in a non-separable fashion. Namely, the bipartite system has to be studied as a whole and not through its sub-systems. We report different bipartite computational simulations where these properties are observed with a classical bouncing droplet model. In particular we describe how numerical Faraday instability tests are instrumental in designing multi-cavity structures where long-range pilot-wave interactions can be enhanced.

Keywords Wave-mediated correlations · Coupled oscillators · Spontaneous synchronization · Hydrodynamic pilot-wave · Bouncing droplets

3.1 Background and Introductory Discussion

The groundbreaking discovery by Couder et al. (2005) and Couder and Fort (2006) led to many works on pilot-wave hydrodynamic (Bush 2015; Bush and Oza 2021). Pilot-wave hydrodynamics originates from bouncing droplets of silicon oil that can levitate over a vibrating bath of the same fluid and be self-propelled by the Faraday waves it generates. This wave-droplet association is viewed as a hydrodynamic quantum analogue of a wave-particle system. Our main goal here is to report on our quest for long-range correlations which can be established when two walking droplets are trapped in cavities placed at a distance. This is naturally motivated by the following facts from Quantum Mechanics.

In their review article, Horodecki et al. (2009) discuss properties of bipartite systems, in particular cases where the particle's individual statistics are *intrinsically related.* They mention the "existence of global states of a composite system

A. Nachbin (✉)
Department of Mathematical Sciences, Worcester Polytechnic Institute, Worcester, MA, USA
e-mail: anachbin@wpi.edu

P. Castro et al. (eds.), *Advances in Pilot Wave Theory*, Boston Studies in the Philosophy and History of Science 344,
https://doi.org/10.1007/978-3-031-49861-9_3

which cannot be written as a product of the states of individual subsystems". In this case the subsystems are referred to as *non-separable* (or *non-factorizable*) statistical states (Horodecki et al. 2009). Schrödinger called "entanglement" the intrinsic statistical relations between particles of a compound quantum system which involves *nonclassical correlations* between subsystems (Horodecki et al. 2009).

Our recent goal has been to identify intrinsically related correlations between largely separated droplets. In recent articles we have studied computationally a classical composite system with two walking droplets (walkers) guided by their pilot-waves. We have designed different scenarios with droplets trapped in remote cavities. Each walker represents a particle oscillating in a given potential well. As will be described, we found potential well configurations where intrinsically related droplet statistics were observed. The respective potential well configurations are created through different cavity-structures and their aspect ratios. Examples, to be discussed, are displayed schematically in Fig. 3.1.

Figure 3.1a displays 3 cavities separated by 2 barriers. The two-dimensional (2D) domain is filled with silicon oil and vibrated vertically at (say) 80 Hz. A bouncing droplet is placed on the vibrating surface of each side-cavity. Due to the high barriers these droplets oscillate laterally while trapped within their respective cavity (potential well). The mid-cavity is deeper which helps establishing a long-range interaction between the two particles. The role of the mid-cavity will be further

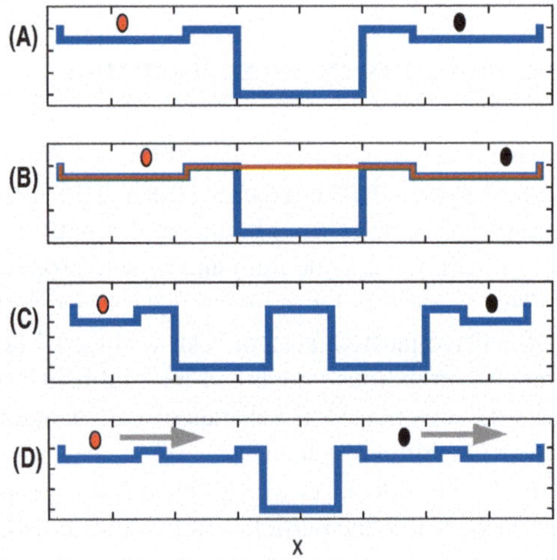

Fig. 3.1 A schematic description of the domains considered. (**a**) Droplets oscillating laterally while trapped within a cavity. The intervening mid-cavity can enhance the droplets' interaction at a distance. (**b**) The intervening mid-cavity suddenly disappears isolating the oscillators after a preliminary interaction. (**c**) Droplets oscillating laterally while trapped within a cavity. Two intervening mid-cavities are used to further explore the droplets' interaction at a greater distance. (**d**) Droplets can tunnel within their two-cavity sub-system. The aspect ratio of the mid-cavity can enhance the tunneling probability of both droplets

discussed in the present manuscript. Nachbin (2018) showed that the two oscillators could spontaneously synchronize after an initial period of adjustments. Inspired by the celebrated Kuramoto model (Acebrón et al. 2005), orbits in phase-space eventually synchronized for both droplets on the same limit-cycle (Nachbin 2018). The droplets were then further separated (Nachbin 2018). The cavity configuration is that shown in Fig. 3.1c. The droplets are now oscillators tapped within the shallower extreme-cavities. Through the underlying wave field, the two deeper mid-cavities allow for a long-range interaction between the two oscillators. The spontaneous Kuramoto-like synchronization is lost. Nevertheless the probability density function (PDF) of both droplets is identical. These PDFs are characterized through histograms computed in phase space. The PDFs are identical and indicate the *statistical indistinguishability* between the two oscillators. Coherence is not by phase-locking, as in the Kuramoto model, but in a more general statistical fashion which is akin to quantum modeling.

The intrinsic statistical relation of a bipartite system was then further explored by Papatryfonos et al. (2022). As depicted in Fig. 3.1d, the domain comprises 5 cavities. Along the sides, each pair of shallow cavities form a subsystem for each droplet. Through the intervening deep cavity we are able to design a two-energy-level subsystem for each particle. In order to trap a droplet in its respective subsystem we have used our experience from Nachbin et al. (2017) where a single droplet could tunnel between two identical cavities separated by a high barrier. In Papatryfonos et al. (2022) the droplet can tunnel only within its respective two-level subsystem. Each cavity of the subsystem has a different energy-level (Papatryfonos et al. 2022). As depicted in Fig. 3.2 the topography is designed so that each of the two-energy-level cavities represent a metastable double-well potential for each particle. The time of residence in the higher-energy well is much shorter than that in the ground state, namely the cavity where the droplet is trapped for longer time intervals before tunneling. This is clearly observed in Fig. 3.2 where the ground state refers to the outer most cavities. The long-range influence of the intervening cavity is clearly seen by the wave pattern in Fig. 3.2. Note that every time one of the droplets tunnels a whiter pattern is observed in the mid-cavity. In Fig. 3.3 one of these regions is detailed. We can see a pilot-wave propagating in the direction in which the droplet tunnels. When tunneling does not take place, the wave pattern is that of standing waves with a small lateral oscillation, which we refer to as *wave sloshing*.

Through the set-up described above, long-range cooperative tunneling events were demonstrated by Papatryfonos et al. (2022). The long-range interactions enhanced the probability of tunneling, namely through the presence of the other droplet. These findings led to an analogy with super- or subradiance in ion-ion interactions (DeVoe and Brewer 1996). The observation of superradiance, and the related nonseparability of the subsystems, hint the plausibility of developing hydrodynamic analog Bell-tests and hopefully the establishment of the first hydrodynamic analog of quantum entanglement.

Furthermore, the well-known (1935) Einstein-Podolsky-Rosen *gedankenexperiment* motivated a new series of hydrodynamic pilot-wave simulations. One may inquire what happens to the established long-range correlations (Nachbin 2018;

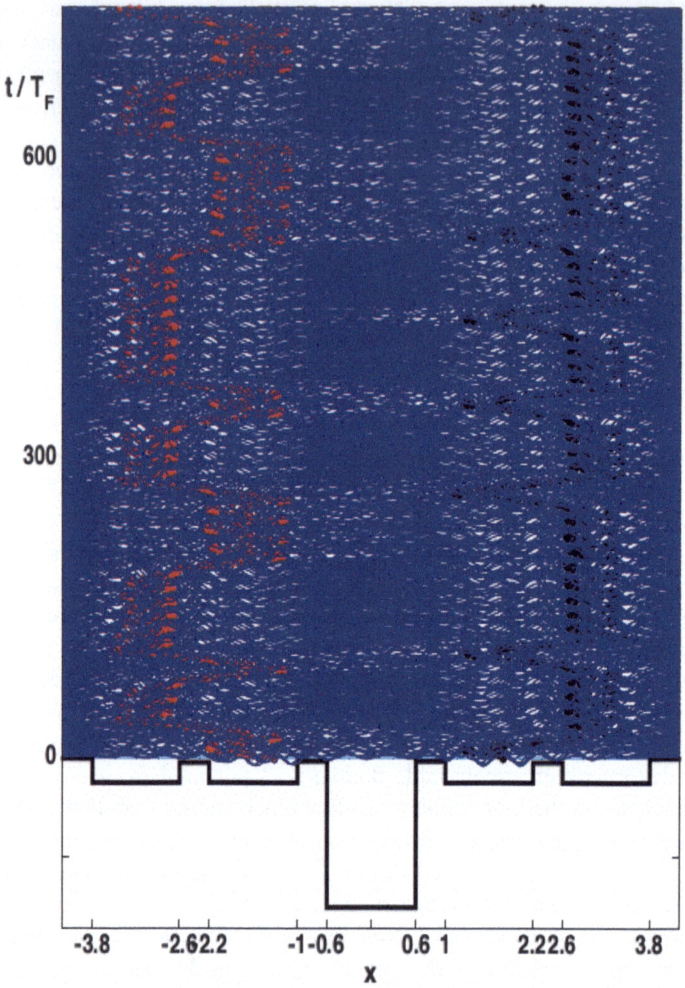

Fig. 3.2 Cooperative tunneling in HQA. Each droplet is trapped to a two-cavity subsystem. They randomly tunnel between their ground state (outer most cavity) and their excited state (inner most cavity). Their long-range interaction is wave mediated and enhanced by the aspect ratio of the intervening mid-cavity

Papatryfonos et al. 2022) after the droplets are set into isolation. Quantum mechanics predicts correlated measurements on the two particles (Aspect et al. 1982) after they have been separated. In Nachbin (2022) we have performed the first computational simulations in this direction. This is depicted in Fig. 3.1b. In order to isolate the droplets after a preliminary interaction, a wide barrier suddenly appears in between them at a prescribed time. This is indicated by the red line that covers the mid-cavity joining the two barriers into one very wide barrier. The intrinsically related correlations and non-separability, as defined in Nachbin (2018),

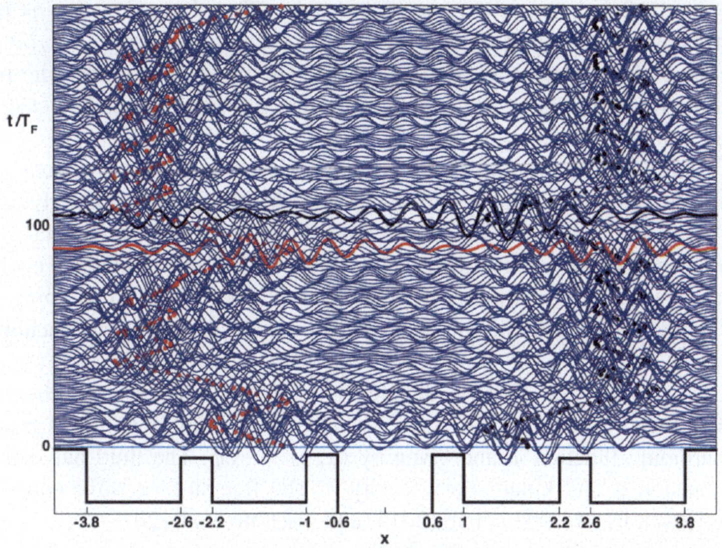

Fig. 3.3 A detail of Fig. 3.2. During a tunneling event the wave pattern in the mid-cavity is altered, as depicted by the clear areas in Fig. 3.2. Here one such clear area is observed in more detail. As highlighted in red and black, note that the respective pilot-waves have more freedom to propagate and therefore guide the droplet over the barrier. At other times standing waves sloshing (laterally) in the mid-cavity prevent tunneling

persist even after the bipartite system is brought into isolation through the wide barrier. Upon isolation, the bipartite system quickly migrates to new identical PDFs. These new PDFs are different from those had the droplet been isolated at the onset. Pre-established intrinsically related correlations persist and the bipartite system's non-separability remains even after the domain has been altered.

In the next section we briefly describe the mathematical model used in the simulations described above. In the following section we summarize some of the steps considered in designing a cavity structure for long-range pilot-wave interactions.

3.2 Remarks on the Mathematical Model

A brief description is given on our one-dimensional (1D) pilot-wave hydrodynamic model which allows for a complex cavity system. Some simulations have been performed using one droplet while others use two droplets. Each droplet's horizontal trajectory is described through

$$m\frac{d^2X_j}{dt^2} + 0.01\, G(t)\frac{dX_j}{dt} = -G(t)\frac{\partial \eta}{\partial x}(X_j(t), t), \tag{3.1}$$

where $j = 1, 2$ indicates the respective droplet. The above equation is namely Newton's Law. The droplet's one dimensional position is given by $X_j(t)$, while $\eta(x, t)$ stands for the wave elevation in time. The pilot-wave guides the particle through the forcing term and takes place through the wave-slope. The steeper the wave the faster the droplet moves.

The coupling between particles is implicit since X_2 does not appear in X_1's equation and vice-versa. The coupling is established through the underlying wave field. The magnitude of this propulsive force is denoted by $G(t)$ and takes place during contact time, here prescribed as $Tc = T_F/4$. This time-dependent coefficient depends on surface tension and also appears in the (skin-friction) damping coefficient. Further modeling details are given in Milewski et al. (2015) and Nachbin et al. (2017).

Let the velocity potential be denoted by $\phi(x, z, t)$, where z is the vertical coordinate pointing upwards. The velocity field in the bulk of the fluid is associated to the harmonic function ϕ and given by $(u, v) = \nabla\phi$. The fluid has density ρ, surface tension σ and kinematic viscosity ν. The free surface wave equations at $z = 0$ are given by Milewski et al. (2015) and Nachbin et al. (2017):

$$\frac{\partial\phi}{\partial t} = -g(t)\eta + \frac{\sigma}{\rho}\eta_{xx} + 2\nu\,\phi_{xx} - \frac{1}{\rho}\sum_{j=1,2} P_d(x - X_j(t)), \tag{3.2}$$

$$\frac{\partial\eta}{\partial t} = DtN[\phi] + 2\nu\,\eta_{xx}. \tag{3.3}$$

The Bernoulli law displays the presence of each droplet through the pressure terms P_d. This is a pressure-based wave-making term, which acts over a spatial interval equal to the droplet's diameter and centered at $X_j(t)$. Similarly to what happens with the droplet's ordinary differential equations (ODEs), these pressure terms are activated periodically only for a quarter of the Faraday period T_F. The kinematic condition (3.3) contains the Dirichlet-to-Neumann operator (DtN) which is a Fourier integral operator mapping the Dirichlet data onto the Neumann data as $DtN[\phi(x, 0, t)] = \partial\phi/\partial z(x, 0, t)$. The mathematical use of properties of the Laplace operator permits an analytical reduction of the 2D fluid dynamics problem into a simpler 1D system along the undisturbed free surface (Nachbin et al. 2017; Nachbin 2018). The above equations are solved by a Fourier spectral method in space and a Verlet method in time.

3.3 Establishing Long-Range Interactions

We start by discussing a laboratory experiment conducted my Miles Couchman (MIT) and reported at the 2016 Division of Fluid Dynamics/APS meeting (Nachbin et al. 2016). Consider an elongated 3-cavity domain designed to reproduce numeri-

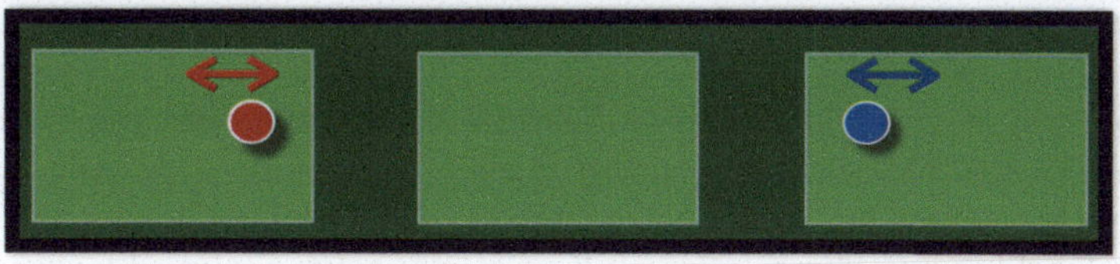

Fig. 3.4 Top view of an elongated domain with three cavities and two droplets. All cavities have the same depth (0.5 cm) and approximately 2 Faraday wavelengths long, in the direction of the horizontal oscillation. The transversal direction is smaller with the pilot-wave mildly touching the sides and therefore locking in a one-dimensional oscillation along the main axis

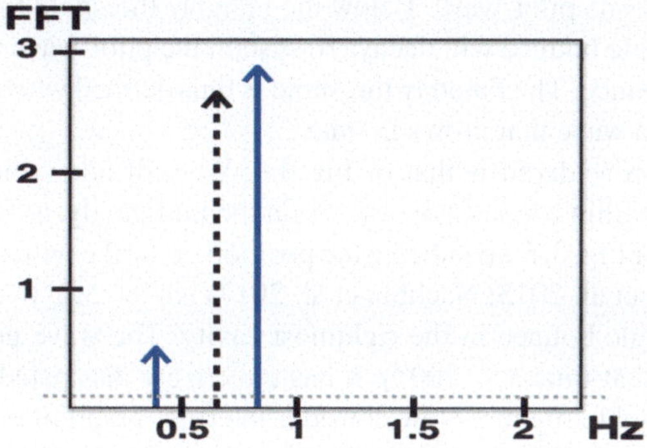

Fig. 3.5 The main Fourier content for the right droplet as an oscillator moving laterally according to Fig. 3.4. The horizontal axis displays the frequency in Hz while the vertical axis has the Fourier mode's amplitude

cal results from the 1D mathematical model described above. The aspect ratio of the cavities is such that the pilot-wave feels the sides of the cavity. This lateral boundary guides the droplet to walk in a one-dimensional (horizontal) fashion along the center of the cavity. The cavity is a 1D waveguide for the pilot-wave. In the elongated direction the cavity is about 2 Faraday wavelengths long. The cavity is short but has enough room for both droplets to oscillate laterally as indicated by the arrows in Fig. 3.4. All cavities have the same depth.

It was observed that the droplets oscillated in synchrony moving inwards and outwards at the same time. The frequency of the droplet-oscillator was calculated using an FFT (Fast Fourier Transform). A result from Nachbin et al. (2016) is presented in the simplified schematic Fig. 3.5, considering only the right droplet. The oscillator has a dominant mode, represented by the tall solid (blue) arrow and a second harmonic depicted through the short solid (blue) arrow. Eventually in the experiment the left droplet coalesced into the fluid leaving the right droplet by itself. The dominant Fourier mode then shifted to the dashed line and the second harmonic disappeared. This is a clear indication that the left droplet was affecting the right oscillator at a distance, even with an empty cavity in between them. Not only it generated a phase shift in the dominant mode but also a second harmonic. Further numerical evidence of forcing at a distance was presented by Nachbin (2020), where even an empty cavity at a distance could force a solitary oscillating droplet. The action of the forcing would depend on the barrier heights and other parameters.

In Fig. 3.1 the mid-cavity was always deeper. As mentioned this enhances the long-range interaction between droplets. We now explain how this was chosen. The intuition for this nonuniform cavity configuration is better conveyed by examining some computational Faraday instability tests.

The Faraday threshold for these complex cavity systems has to be computed numerically. Below the threshold we know that Faraday waves do not emerge. But when a droplet does one bounce it triggers the most unstable mode locally, which eventually becomes its pilot-wave. Below the Faraday threshold the wave elevation generated by a single bounce will decay. To sustain the pilot-wave the droplet needs to continuously bounce. The Faraday threshold is thus defined when a single droplet-bounce generates a wave that grows in time.

The geometry considered is that of Fig. 3.6. We will allow the droplet to take its single bounce within a side cavity or within the mid-cavity, as shown in Fig. 3.6. Along the x-axis of Fig. 3.7 we indicate the position of the 3 cavities. The vibrational forcing (Milewski et al. 2015; Nachbin et al. 2017) is $\Gamma = 5.2$. In the left graph the droplet does a single bounce in the rightmost cavity. The wave generated steadily decays in time and at time $t = 200T_F$ it has effectively dissipated. This means we are in a Faraday-stable regime. Now consider the right graph in Fig. 3.7. The same vibrational forcing is used but the single bounce takes place in the deeper mid-cavity. This case is Faraday-stable as before, but the decay is slower than before. Hence we see that a cavity depth can change the dissipation rates locally. The same happens for a regime above the Faraday threshold, where we have an instability. The growth rate also depends on the local depth.

In Fig. 3.8 the forcing has been slightly increased to $\Gamma = 5.3$.

The wave, due to the droplet's single bounce in the rightmost cavity, decays for a while but then grows indefinitely indicating that this regime is Faraday-unstable, thus defining the Faraday threshold. To better highlight the instability and the local growth rates, the vibrational forcing is increased to $\Gamma = 5.45$. Figure 3.9 displays the wave elevation for a single bounce in both the right and center cavity. The

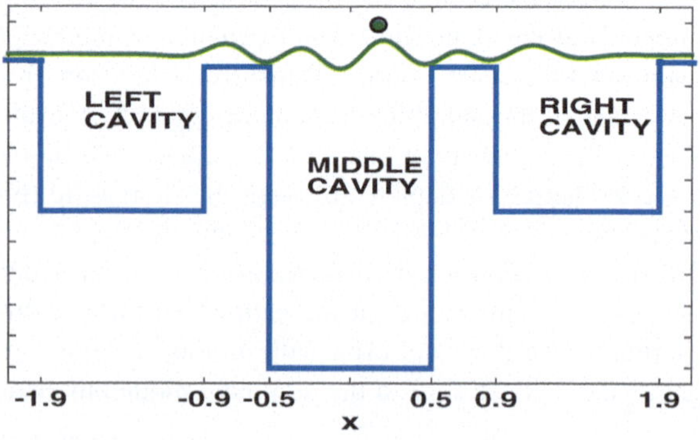

Fig. 3.6 The three-cavity setup for the numerical Faraday instability tests

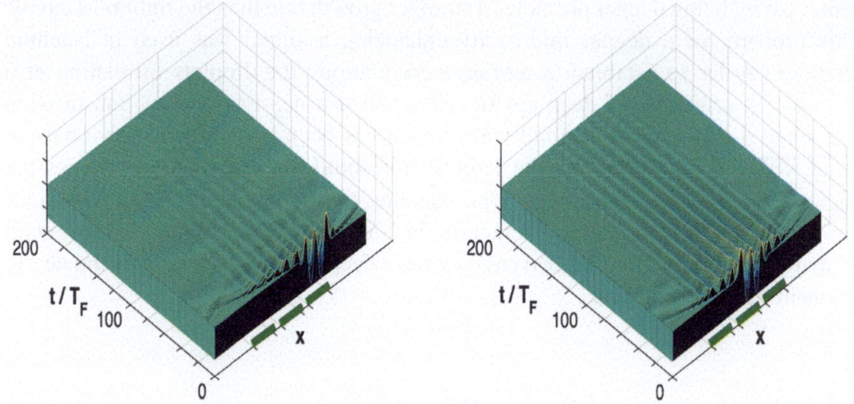

Fig. 3.7 The numerical Faraday instability test for a single bounce in the shallower cavity (to the right) and for a single bounce in the deeper cavity (mid-cavity). Both indicate a Faraday-stable regime but the decay pattern is different

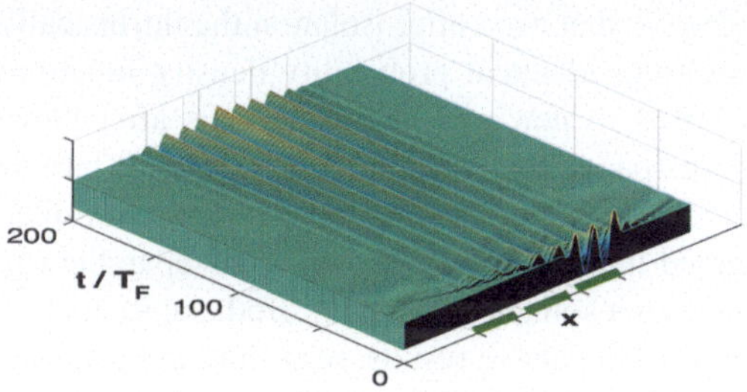

Fig. 3.8 The numerical Faraday instability test for a single bounce in the shallower cavity (to the right). At first there is a mild decay but eventually the wave height grows indefinitely indicating a Faraday-unstable regime

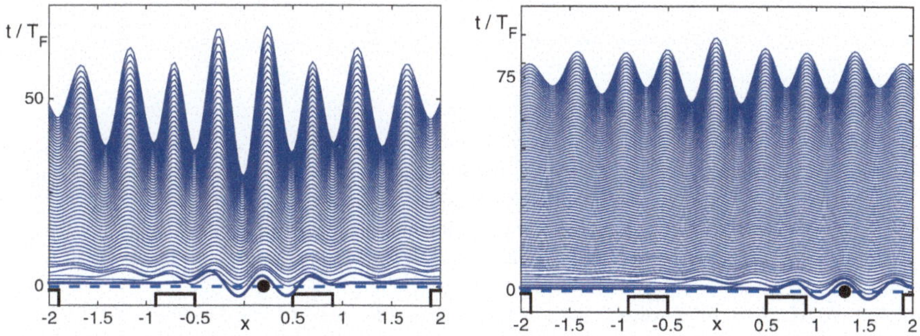

Fig. 3.9 Faraday instability at $\Gamma = 5.45$. At the right we a have a droplet performing its single bounce in the right cavity. At the left we a have a droplet performing its single bounce in the deeper mid-cavity. The cavity barriers are indicated at the bottom. The growth rate is stronger in the mid-cavity

center cavity being deeper promotes a stronger growth rate than the rightmost cavity. This property of a deeper mid-cavity enhancing a signal was used in Nachbin (2018) in order to establish long-range correlations for droplets oscillating at a distance. A sequence of deep cavities was called a nearly resonant transmission line since it can amplify a small wave which has been transmitted over a barrier. Even with the forcing below the Faraday threshold the mid-cavity enhances the background vibrating field, namely propagating ripples emanating from both sides and establishing a long-range interaction. In Nachbin (2018) we had one case with 4 cavities, where the middle two cavities were deeper and thus acting as the nearly resonant transmission line.

3.4 Conclusion and Future Directions

Recently Nachbin (2022) allowed two oscillating droplets to establish long-range intrinsically related correlations, as in Nachbin (2018), followed by the sudden appearance of a barrier covering the mid-cavity in order to isolate the oscillators. It was striking to observe that even after isolation the intrinsically-related statistics persisted. The preexisting identical probability density functions (PDF) for each droplet quickly adjusted to new and identical PDFs after they were isolated. It was shown that phase-space orbits of both particles divert into displaying the new statistics faster than any signal emanating from the other droplet (Nachbin 2022). The statistics presented in all cases above were non-separable since they could not be identified by studying a solitary subsystem. Both subsystems were needed to be accounted for. Therefore the above results, regarding the persistence of long-range correlations, have hinted the plausibility of performing hydrodynamic analog of Bell tests. This is an ongoing effort with collaborators.

Acknowledgments The author was supported by CNPq under (PQ1D) 307078/2021-3 and FAPERJ *Cientistas do Nosso Estado* project E-26/201.156/2021.

References

Acebrón, J., L. Bonilla, V. Perez, F. Ritort, and R. Spigler. 2005. The Kuramoto model: a simple paradigm for synchronization phenomena. *Reviews of Modern Physics* 77: 137–185.

Aspect, A., P. Grangier, and G. Roger. 1982. Experimental realization of Einstein-Podolsky-Rosen-Bohm gedankenexperiment: a new violation of Bell's inequalities. *Physical Review Letters* 49: 91.

Bush, J. W. M. 2015. Pilot-wave hydrodynamics. *Annual Review of Fluid Mechanics* 47: 269.

Bush, J. W. M., and A. Oza. 2021. Hydrodynamic quantum analogs. *Reports on Progress in Physics* 84: 017001.

Couder, Y., and E. Fort. 2006. Single-particle diffraction and interference at a macroscopic scale. *Physical Review Letters* 97: 154101.

Couder, Y., S. Protière, E. Fort, and A. Boudaoud. 2005. Walking and orbiting droplets. *Nature* 437: 208.

DeVoe, R.G., and R. Brewer. 1996. Observation of superradiant and subradiant spontaneous emission of two trapped ions. *Physical Review Letters* 76: 2049–2052.

Horodecki, R., P. Horodecki, M. Horodecki, and K. Horodecki. 2009. Quantum entanglement. *Reviews of Modern Physics* 81: 865–942.

Milewski, P.A., C.A. Galeano-Rios, A. Nachbin, and J.W.M. Bush. 2015. Faraday pilot-wave dynamics: modelling and computation. *Journal of Fluid Mechanics* 778: 361.

Nachbin, A. 2018. Walking droplets correlated at a distance. *Chaos* 28: 096110.

Nachbin, A. 2020. Kuramoto-like synchronization mediated through Faraday surface waves. *Fluids* 5 (4): 226.

Nachbin, A., M.M.P. Couchman, and J.W.M. Bush. 2016. Non-local features of a hydrodynamic pilot-wave system. 43rd Annual Meeting of the Division of Fluid Dynamics, American Physical Society 61.

Nachbin, A., P. Milewski, and J. Bush. 2017. Tunneling with a hydrodynamic pilot-wave model. *Physical Review Fluids* 2: 034801.

Nachbin, A. 2022. Effect of isolation on two-particle correlations in pilot-wave hydrodynamics. *Physical Review Fluids* 7:093604

Papatryfonos, K., M. Ruelle, C. Bourdiol, A. Nachbin, J. W. M. Bush, and M. Labousse. 2022. Hydrodynamic superradiance in wave-mediated cooperative tunneling. *Nature Communications Physics* 5: 142.

Chapter 4
Hydrodynamically Inspired Pilot-Wave Theory: An Ensemble Interpretation

Yuval Dagan

Dedicated to the memory of my father, Dr. Mordechai Dagan

Abstract This chapter explores a deterministic hydrodynamically-inspired ensemble interpretation for free relativistic particles, following the original pilot wave theory conceptualized by de Broglie in 1924 and recent advances in hydrodynamic quantum analogs. We couple a one-dimensional periodically forced Klein-Gordon wave equation and a relativistic particle equation of motion, and simulate an ensemble of multiple uncorrelated particle trajectories. The simulations reveal a chaotic particle dynamic behavior, highly sensitive to the initial random condition. Although particles in the simulated ensemble seem to fill out the entire spatiotemporal domain, we find coherent spatiotemporal structures in which particles are less likely to cross. These structures are characterized by de Broglie's wavelength and the relativistic modulation frequency kc. Markedly, the probability density function of the particle ensemble correlates to the square of the absolute wave field, solved here analytically, suggesting a classical deterministic interpretation of de Broglie's matter waves and Born's rule.

Keywords Hydrodynamic quantum analog · de Broglie waves · Ensemble interpretation · Klein-Gordon equation

4.1 Introduction

In 1924, de Broglie proposed that particles may be associated with an intrinsic clock oscillating at the Compton frequency. De Broglie envisaged a particle as a localized wavefield guided by a pilot wave, exchanging rest-mass energy with field

Y. Dagan (✉)
Faculty of Aerospace Engineering, Technion - Israel Institute of Technology, Haifa, Israel
e-mail: yuvalda@technion.ac.il

P. Castro et al. (eds.), *Advances in Pilot Wave Theory*, Boston Studies in the Philosophy and History of Science 344,
https://doi.org/10.1007/978-3-031-49861-9_4

energy (De Broglie 1924). Interactions between the particle generating the wave field and, in turn, the wavefield guiding the particle constituted de Broglie's realistic picture of matter (De Broglie 2012).

The particles in de Broglie's theory were conceptualized as a localized yet infinite, spatially decaying field guided by relativistic phase waves. Relativistic considerations, and in particular the notion of *Harmony of Phases* associating the particle to its guiding wave, were imperative in de Broglie's theory:

> I think the theory of relativity plays a major role much greater than most people usually think in the basic ideas of wave mechanics and that if one wants to really understand its origins, one has to come back to relativistic considerations...—Louis de Broglie (1967).

This note from an interview (translated here from French) reveals de Broglie's view on the fundamental importance of relativistic motion in the dynamics of particles. Thus, relativistic considerations could also play an important role in the hydrodynamic pilot-wave analogy (Dagan 2023). In de Broglie's studies, matter waves were generally realized as monochromatic plane waves or at least quasi-monochromatic, as appeared in his later notes (De Broglie 1970). However, de Broglie neither defined a specific guiding wavefield in his theory nor suggested a mechanism for the generation of such waves (although he did suggest several candidates for such waves, including the Klein-Gordon equation and the Dirac equation (De Broglie 1970)).

The astounding success of Born and Heisenberg in mathematically describing the statistical nature of quantum mechanics, and perhaps de Broglie's inability to finish his incomplete program, eventually led to generally discarding the realistic pilot-wave view of quantum particle dynamics. In retrospect, in an attempt to find a realistic picture of matter, de Broglie's assumptions may have seemed too simplified and thus incomplete. Whether the pilot wave theory conceptualized by de Broglie describes a realistic picture of matter or not, there is a reason to believe that it should constitute more complex two-way coupled particle-wave interactions that were not realized in de Broglie's program.

Nonlinear particle-wave interactions, similar to ideas conceptualized by de Broglie's later publications, are frequently encountered in fluid dynamics. Since the introduction of the Madelung transformation (Madelung 1926) of the Schrödinger equation to a fluid mechanical system of equations, multifold attempts to realize quantum particle dynamics relying on fluid mechanics principles have been proposed. De Broglie-Bohm theory (Bohm 1952), Nelson's theory (Nelson 1966), and the Stochastic Electrodynamics (SED) (Boyer 1975; De la Peña et al. 2015) have raised the possibility that underlying unknown statistical mechanisms—the so-called 'hidden variables'—govern the dynamics of quantum particles giving rise to the expected quantum statistical signature.

However, de Broglie's original picture of matter was neither a hidden variable approach nor statistical but rather a classical deterministic one. It is, therefore, instructive to follow de Broglie's deterministic views and review some of the recent advances in deterministic hydrodynamic quantum mechanical analogs. One of the most successful analogies to quantum mechanics was found by Couder and Fort, who experimentally observed millimetric oil droplets bouncing over a vibrating bath that remarkably feature the statistical behavior of many quantum mechanical systems (Couder et al. 2005; Couder and Fort 2006; Bush and Oza 2020). In this hydrodynamic quantum analogy (HQA), droplets interact in resonance with a quasi-monochromatic wavefield they generate and exhibit a self-propelling mechanism. This analog has extended the range of classical physics to include many features previously thought to be exclusively quantum, including tunneling (Eddi et al. 2009; Hubert et al. 2017; Nachbin et al. 2017; Tadrist et al. 2020), Landau levels (Fort et al. 2010; Harris and Bush 2014; Oza et al. 2014), quantum harmonic oscillator (Perrard et al. 2014; Durey and Bush 2020), the quantum corral (Harris et al. 2013; Gilet 2014, 2016; Sáenz et al. 2018; Cristea-Platon et al. 2018), the quantum mirage (Sáenz et al. 2018), and Friedel oscillations (Sáenz et al. 2020). Remarkably, Couder was able to demonstrate in his hydrodynamic analogy a mechanism for single-particle diffraction (Couder and Fort 2006).

The ability of a classical dynamic system to resolve quantum mechanical problems raises the question: Are we able to conceptualize a similar deterministic mechanism on the microscopic (quantum mechanical) scale? In this chapter, we shall address this question theoretically using hydrodynamic considerations in an attempt to realize deterministic particle dynamics on the quantum scale.

Several authors have attempted to formulate a form of quantum dynamics based on insights from HQA (Andersen et al. 2015; Borghesi 2017; Drezet et al. 2020). Andersen et al. (2015) explored the behavior of a dynamical system in which a particle locally excites a waveform satisfying Schrödinger's equation, which then moves in response to gradients of the phase of that field. Orbital quantization was shown to arise along with an analog to the Bohr-Sommerfeld quantization rule; Borghesi (2017) proposed an elastic pilot-wave model wherein a point particle moves within a non-dissipative elastic substrate. The coupled dynamics of the particle and elastic medium in Borghesi's study are governed by a modified Klein-Gordon equation. Shinbrot (2019) examined the Klein-Gordon equation with an oscillatory potential as a model for a quantum particle emitting and absorbing pilot waves. He concluded that bound state solutions with half-integer spin exist provided the particle rest mass oscillates in time, which aligns with de Broglie's physical picture (De Broglie 1970).

Recently, we developed a hydrodynamically-inspired quantum theory (Dagan and Bush 2020), a theoretical model of relativistic quantum dynamics inspired by de Broglie's pilot wave theory. In this framework, the particle is assumed as a localized—yet infinite—oscillating disturbance, externally forcing a Klein-Gordon wave equation. A relativistic dynamic equation couples the motion of the localized particle to the wave. Using this deterministic framework, several features of quantum mechanics are revealed, associated with inline oscillations corresponding to the relation $p = \hbar k$, realized through interactions with the wave field. Notably, particle speed modulations are averaged at the de Broglie wavelength and modulated by the relativistic frequency kc. Although the nonlinear system of free particles in this framework is chaotic in nature, and inline oscillations may be characterized by multiple oscillation modes, de Broglie wavelength is the most pronounced and may be realized as *quasi-monochromatic* modulation of the particle motion, as also suggested by de Broglie.

Excitation of motion and the waveform of the same framework at non-relativistic speeds were examined by Durey and Bush (2020), who revealed the wave generation and self-propelling mechanism for the coupled wave-particle system and provided analytical validity to our subsequent work on the hydrodynamic pilot-wave theory.

To further explore the extent to which a classical hydrodynamic theory may be realized as a viable interpretation of de Broglie's theory and relativistic quantum dynamics, and in particular, the possibility of such a theory to account for de Broglie wavelength, a fully classical non-relativistic dynamic system was also considered (Dagan 2023). This is in contrast to our previous study (Dagan and Bush 2020), where a relativistic dynamic equation is introduced to properly satisfy a Lorentz covariant formulation. However, this formulation closely correlates to the hydrodynamic analog and allows the isolation of the role of classical wave mechanics in producing relativistic quantum signatures.

In the present study, we extend the discussion on the relativistic pilot wave framework developed by Dagan and Bush (2020) by conducting multiple uncorrelated particle simulations and analyzing the statistics emerging from the ensemble of particle trajectories.

4.2 Hydrodynamically-Inspired Quantum Theory

Several options for realizing particle-wave interactions may be postulated. We follow our previous study (Dagan and Bush 2020), where the pilot waves are generated by the oscillating particle and evolve according to a one-dimensional KG equation:

$$\frac{\partial^2 \phi}{\partial t^2} - c^2 \frac{\partial^2 \phi}{\partial x^2} + \omega_c^2 \phi = \epsilon_p f(t) \delta_a (x - x_p) , \tag{4.1}$$

Here, ϕ is a real wave, ω_c is the Compton frequency, and c is the speed of light. The term on the RHS of Eq. (4.1) represents an external forcing of the wavefield localized about the particle location x_p, where $f(t) = \sin(2\omega_c t)$, ϵ_p is a constant and $\delta_a = \frac{1}{|a|\sqrt{\pi}} e^{-(x/a)^2}$ is a modified delta function that serves to localize the driving oscillation to the vicinity of the particle location. The parameter a defines the width of the modified delta function and the extent of the particle's influence on the wave field. Here, as in Dagan and Bush (2020), $a = \lambda_c/2$, where $\lambda_c = h/(mc)$ is the Compton wavelength. Hence, the Gaussian perturbation has a characteristic width of λ_c. ϵ_p is the forcing amplitude, and $f(t)$ is chosen as a harmonic function with twice the Compton frequency, $f(t) = sin(2\omega_c t)$. Inspired by HQA and de Broglie's guidance equation, a trajectory equation can be written in which the momentum is determined by the gradients of the waves such that

$$\gamma \dot{x}_p = -\alpha \frac{\partial \phi}{\partial x'}_{x=x_p}, \tag{4.2}$$

where $\dot{x}_p$ is the particle velocity, and x' denotes the particle location after proper Lorentz translation from the particle frame of reference to the stationary one. Note that this translation ensures the equation of motion is relativistically covariant yet not invariant as the Klein-Gordon equation. α is a free coupling parameter between the particle and its waves, and γ is the Lorentz boost factor. The LHS represents the relativistic momentum, divided by the rest mass m_0, which is absorbed into the coupling parameter, α. Notably, when a classical particle is coupled to periodic and oscillating flows, clustering of particles may occur, which may be viewed as quantization of their probability distribution (Dagan et al. 2017a,b).

We shall now solve the coupled equations numerically for a single free particle. The coupled Klein–Gordon equation and the equation of motion are discretized using finite differences. An explicit finite-difference method was derived to solve the Klein–Gordon wave equation and a Runge–Kutta scheme was employed to advance the nonlinear guidance equation in time. More details on the numerical method and numerical parameters may be found in Dagan and Bush (2020).

Figure 4.1 demonstrates the typical spatiotemporal evolution of the wavefield and particle dynamics for two different cases, which differ only by the initial condition imposed on the particle; in both cases, an initial stage of motion may be observed, at which the particle seems stagnant during approximately the first ten Compton periods. The particle oscillates at twice the Compton frequency, exciting localized standing waves. In order to break the symmetry and initiate motion, an initial random wave perturbation is applied, approximately four orders of magnitude smaller than the characteristic wave amplitude generated by the stagnant particle. This perturbation initially causes small oscillations about its initial position.

Since the particle is initially deflected randomly, inline oscillations about $x/\lambda_C = 0$ occur but are too small to be observed in the scale of the figures. However, symmetry breaking in both the wavefield and particle motion is apparent between $t/\tau_c = 10$ and $t/\tau_c = 20$. In Fig. 4.1a, the particle locks into a quasi-steady motion, as also reported in our previous study (Dagan and Bush 2020). In Fig. 4.1b, however, a more chaotic motion is apparent, for which we cannot determine a characteristic quasi-steady speed, although the same slope, i.e., speed of $\beta = 0.25$, may be observed locally. In this case, as in any other simulated trajectory, the particle will eventually lock to the quasi-steady speed if the simulation is integrated for a long enough time.

This peculiar behavior reveals the complexities that may emerge from the highly nonlinear coupled system comprising our fully deterministic model. On the one hand, the evolution of the wavefield and, in turn, the particle motion is highly sensitive to initial conditions and reflects chaotic dynamics. On the other hand, the particle locks into inline oscillations, where coherent waveform may be observed.

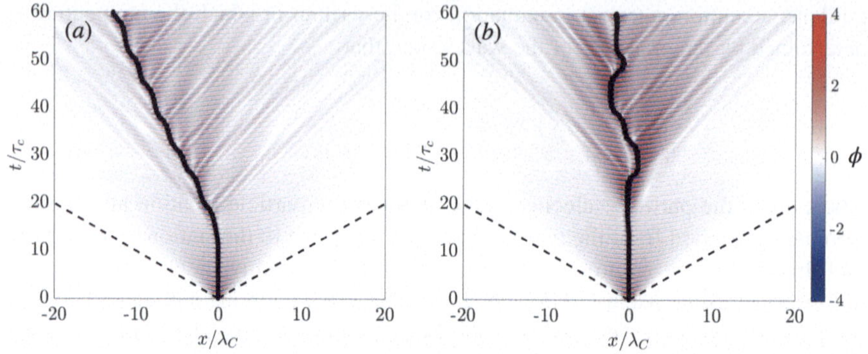

Fig. 4.1 Pilot-wave dynamics of two free particles demonstrated in (**a**) and (**b**) for two different initial conditions. The spatiotemporal map illustrates the evolution of the normalized pilot-wave field, ϕ (color map), generated by the trajectory of a free particle (indicated in black). The coupling constant is set to $\alpha = 0.045$, corresponding to a quasi-steady speed of approximately $\beta = 0.25$, as calculated from the wavelength in the particle vicinity. The dashed black line indicates the particle's light cone with slope $\beta = 1$

Notably, as previously reported (Dagan and Bush 2020), the instantaneous momentum of the particle at any time corresponds to $p = \hbar k$, where k is measured locally in the vicinity of the particle. Farther away from the particle location, the dispersion of waves and superposition with waves reflected from previous changes of motion affects the wavelength, and the correlation to the particle momentum is less significant. This was confirmed for several low and high speeds along the trajectory shown in Fig. 4.1b.

The complex dynamics show promising features for a deterministic hydrodynamic interpretation of quantum mechanics. These include a mechanism for motion excitation, rendering the system unstable, and quantized momentum at the de Broglie wavelength. To further investigate the ability of this model to account for quantum statistics, multiple trajectories are simulated and analyzed in the following section.

4.3 Ensemble of Particle Trajectories

We proceed by computing multiple particle trajectories. An ensemble of 1000 overlapping trajectories is presented in Fig. 4.2a, and a close-up view in Fig. 4.2b under the same conditions presented in Fig. 4.1. Here, each line represents a distinct simulation, which differs from all others only by applying an initial condition of random perturbation.

The three highlighted trajectories of Fig. 4.2b illustrate different scenarios for the spatiotemporal particle evolution. Dashed lines mark the light cone with respect to the initial position of all simulations at $x/\lambda_c = 0$.

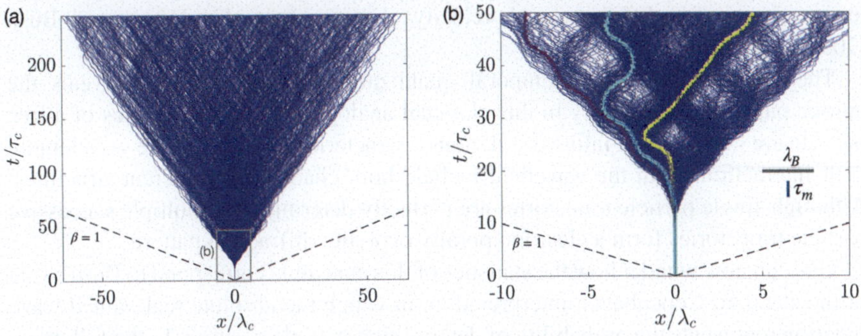

Fig. 4.2 (**a**) Ensemble of 1000 particle trajectories simulated at the same settings as in Fig. 4.1. (**b**) A close-up view of the initial stages of evolution. Blue lines show all trajectories; Three particular trajectories highlighted in colors illustrate different scenarios for the particle spatiotemporal path. Dashed lines mark the speed of light cone of $\beta = 1$. Reference scales for de Broglie wavelength λ_B and relativistic modulation τ_m are shown here in horizontal grey and vertical blue lines, respectively

In the first stages of evolution, all trajectories start at the exact spatial location at $x/\lambda_c = 0$. This location is sustained for some time until symmetry braking is apparent. Trajectories seem to diverge after about ten Compton periods, and particles fill the space.

As in Fig. 4.1, each particle is characterized by inline oscillations corresponding to the local wave field and de Broglie wavelength. When viewed at a length scale around $\lambda_B/\lambda_c \approx 100$, particle trajectories seem randomly scattered, and as time proceeds, we observe a more spatially uniform distribution of particles. However, a closer look at the first stages of the ensemble evolution (Fig. 4.2b) reveals a clear spatial preference for the spatiotemporal path of the particle. Coherent structures of the ensemble appear through frequent changes in the mean particle direction, resulting in spatiotemporal voids in which particles are less likely to cross. Notably, these voids have a typical length scale comparable to de Broglie wavelength λ_B, and time scale τ_m (scales shown for reference in Fig. 4.2b).

It is interesting to note three distinct trajectories (highlighted): The purple trajectory attains the quasi-steady speed of $\beta = 0.25$ and follows a path similar to the envelope of the ensemble; the trajectory in cyan appears to change directions after approximately 30, and then 40 Compton periods, while the yellow trajectory appears to lock into a quasi-steady mean speed after abruptly changing directions around 25 Compton periods. Note that the purple and cyan trajectories start roughly the same path until diverting into two distinct routes.

Thus, particle-wave interactions seem to create spatiotemporal unstable nodes in which a bifurcation appears only when observed as an ensemble. In fact, all trajectories are characterized by the same quasi-steady speed observed in Fig. 4.1. We may also assume they all interact with the local wave field and follow de Broglie's relation $p = \hbar k$ (this was confirmed for multiple points in the Spatiotemporal map for different local velocity changes—see also Dagan and Bush (2020)).

The emergence of spatiotemporal quantization of the ensemble reveals the unique particle-wave duality in this classical analog. While the dynamics of a free particle are governed by inline oscillations characterized by de Broglie wavelength, their interactions with the wavefield include both chaotic and coherent structures. Although single particle trajectories are perfectly deterministic, multiple successive particle trajectories form a clear temporally evolving diffraction pattern.

One can now wonder how the statistics of this ensemble correspond to Born's rule of the standard Copenhagen interpretation, in which the absolute real-valued wave function represents the probability of determining particle position. In the following sections, we shall show more quantitatively that our deterministic framework captures, in general, the correct statistical form of the standard quantum mechanical representation.

4.4 Quantum Statistics of a Classical Ensemble

We shall now investigate the ensemble's statistics to better understand its evolution and viability in reproducing quantum statistics.

Due to a relatively large number of simulated trajectories, we may extract the probability density function (PDF) of the particle location at each time step. The time-evolving PDF is demonstrated in Fig. 4.3a. Initially, the PDF takes the form of a Gaussian (due to its normal distribution kernel), representing all particles localized at $x/\lambda_c = 0$. One may notice that it then evolves into a spatial structure similar to that of a wave field responding to an initial perturbation. To test this hypothesis, we explore the response of the Klein-Gordon wave field to an initial Gaussian perturbation of the same characteristics as determined from the simulated particle PDF.

Wave Response to a Gaussian Disturbance

In order to analytically resolve the underlying wave field, we consider the response of the unsteady Klein-Gordon equation in one dimension to an initial spatial disturbance $\psi_0(x)$

$$\frac{\partial^2 \psi}{\partial t^2} - c^2 \frac{\partial^2 \psi}{\partial x^2} + \omega_c^2 \psi = 0\,, \tag{4.3}$$

$$\psi(x, 0) = \psi_0(x)$$

We shall now define separation parameters and assume

$$\psi = X(x)T(t)\,, \tag{4.4}$$

which transforms Eq. (4.3) into

$$X(x)\ddot{T}(t) - c^2 T(t)X''(x) + \omega_c^2 X(x)T(t) = 0 \tag{4.5}$$

and can be separated as

$$\frac{\omega_c^2 T(t) + \ddot{T}(t)}{c^2 T(t)} = \frac{X''(x)}{X(x)} = q\,. \tag{4.6}$$

Here, since $X'(x) = ikX(x)$, $q = -k^2$ is the separation constant, and we may solve the set of separated equations:

$$\begin{aligned} X_n''(x) + k_n^2 X_n(x) &= 0 \\ \ddot{T}_n(t) + (\omega_c^2 + c^2 k_n^2)T_n(t) &= 0 \end{aligned} \tag{4.7}$$

that can be solved for the complex mode $k = k_r + ik_i$ and written as

$$\begin{aligned} X_n &= C_{1n} e^{i2n\pi x} e^{ik_i} + C_{2n} e^{-i2n\pi x} e^{-ik_i} \\ T_n &= C_{1n} \cos\left(\sqrt{c^2 k_n^2 + \omega_c^2 t}\right) + C_{2n} \sin\left(\sqrt{c^2 k_n^2 + \omega_c^2 t}\right)\,, \end{aligned} \tag{4.8}$$

where $k_n = 2\pi n;\ \ n = 0, \pm 1, \pm 2, \ldots$. If we assume that $\frac{\partial \psi}{\partial t} = 0$, we can write a solution for ψ as the following superposition of modes,

$$\psi(x,t) = \sum_{n=-N/2}^{N/2} C_n e^{ik_n x} \cos\left(\sqrt{c^2 k_n^2 + \omega_c^2 t}\right)\,, \tag{4.9}$$

(absorbing all the constants in C_{1n} and C_{2n}). At $t = 0$

$$\psi(x,0) = \sum_{n=-N/2}^{N/2} C_n e^{ik_n x}\,, \tag{4.10}$$

and the coefficients can be calculated using a Fourier transform

$$C_n = \int_{-L}^{L} \psi_0(x) e^{-ik_n x} dx = \int_{-L}^{L} \psi_0(x) e^{-i2\pi n x} dx\,. \tag{4.11}$$

For the initial disturbance, we assume a Gaussian effective width of twice the Compton wavelength centered at the particle location, x_p, as follows. This parameter is chosen here to fit the initial particle PDF solution of our ensemble simulations. The initial disturbance takes the form

$$\psi_0(x) = \beta e^{-\left(\frac{x-x_p}{a}\right)^2}, \tag{4.12}$$

from which the coefficients,

$$C_n = \beta \int_{-L}^{L} e^{-\left(\frac{x-x_p}{a}\right)^2} e^{-i2\pi n x} dx \tag{4.13}$$

may be calculated. Finally, by substituting the solution for Eq. (4.13) in Eq. (4.9), we can express the spatiotemporal evolution of the wave to an initial Gaussian disturbance as

$$\psi(x,t) = \beta \sum_{n=-N/2}^{N/2} \left\{ e^{i2\pi n x} \cos\left(\sqrt{c^2 k_n^2 + \omega_c^2 t}\right) \frac{\sqrt{\pi} a}{2} e^{-\pi^2 a^2 n^2 - i2\pi n x_p} \times \left[erf\left(\frac{L+x_p}{a} - i\pi n a\right) + erf\left(\frac{L-x_p}{a} + i\pi n a\right)\right]\right\}. \tag{4.14}$$

The temporal evolution of the Klein-Gordon wavefield can now be plotted and compared to the evolution of the particle PDF in our analogy. For the interpretation with Born's rule, the squared absolute value of the wavefield, $|\psi|^2$, is shown in Fig. 4.3b, and compared to the temporal evolution of the particle PDF extracted from the ensemble simulations (Fig. 4.3a).

One may observe a general similarity between the two plots, although they are not the same. Initially, the Gaussian distribution is identical between the KG response and the particle PDF. However, as expected, as time proceeds, the main discrepancies between the two plots emerge at points where the wavefield nullifies. These locations are not allowed in the standard statistical quantum mechanical interpretation. Clearly, this is not the case for the particle PDF in our simulations since all trajectories are generally possible. Nevertheless, the apparent quantum statistics emerge in our hydrodynamic interpretation through the emergence of coherent structures within the spatially evolving wavefield. More specifically, the regions in which particles are less likely to cross (see Fig. 4.2b) are responsible for the apparent troughs in the PDF of Fig. 4.3a.

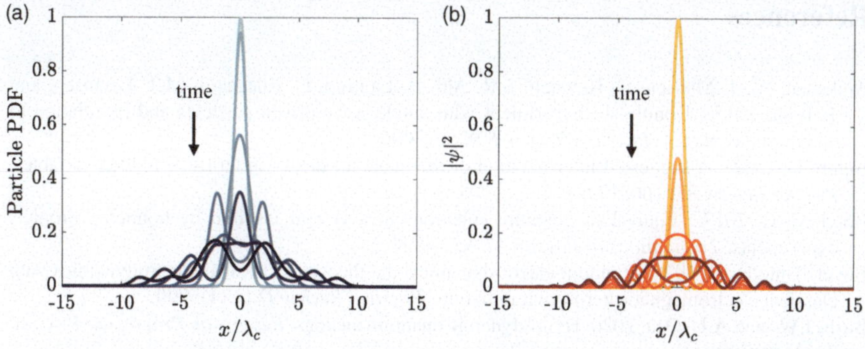

Fig. 4.3 (**a**) Temporal evolution of particle probability distribution function collected from 1000 simulations (illustrated here in Fig. 4.2). (**b**) Normalized analytical solutions for the Gaussian wave response of the time-dependent Klein-Gordon equation. The absolute square value is shown for particle statistics

4.5 Concluding Remarks

Following our previous work on a relativistic pilot-wave theory, we introduced a new deterministic ensemble interpretation based on a hydrodynamically-inspired pilot-wave framework. This analysis proposes a plausible mechanism for quantum statistics emerging from the ensemble of multiple uncorrelated deterministic particle trajectories. Each trajectory is characterized by a quasi-steady speed modulated by inline oscillations at the de Broglie wavelength, which, due to nonlinear particle-wave interactions, occasionally changes its direction of motion.

We show that through the ensemble of particle trajectories, clear, coherent spatiotemporal structures appear at which particles are less likely to cross, similar to previous works on hydrodynamic analogies. However, the coherent structures in the current framework are comparable to the de Broglie wavelength, suggesting a deterministic mechanism for single-particle diffraction on the quantum mechanical scale.

Finally, we find that the evolution of the particle PDF of this interpretation closely follows the evolution of the square of the absolute-valued Klein-Gordon wave field, suggesting a deterministic approach to Born's rule. This is in contrast to the standard Copenhagen interpretation and many-worlds interpretation, in which particles exist nonlocally in all possible locations predicted by quantum statistics.

Although simplified, we hope this model and its findings may provoke a renewed discussion on realistic pilot waves and the possible resolution of the long sought-after 'hidden variables' in the dynamics of quantum particles.

Acknowledgments The author would like to thank Professor John W.M. Bush for the encouragement and timeless conversations that greatly improved this chapter.

References

Andersen, A., J. Madsen, C. Reichelt, S.R. Ahl, B. Lautrup, C. Ellegaard, M.T. Levinsen, and T. Bohr. 2015. Double-slit experiment with single wave-driven particles and its relation to quantum mechanics. *Physical Review E* 92: 013006.

Bohm, D. 1952. A suggested interpretation of the quantum theory in terms of hidden variables. *Physics Review* 85: 166–179.

Borghesi, C. 2017. Equivalent quantum equations in a system inspired by bouncing droplets experiments. *Foundations of Physics* 47: 933–958.

Boyer, Timothy H. 1975. Random electrodynamics: the theory of classical electrodynamics with classical electromagnetic zero-point radiation. *Physical Review D* 11 (4): 790.

Bush, J.W., and A.U. Oza. 2020. Hydrodynamic quantum analogs. *Reports on Progress in Physics*, 84 (1): 017001.

Couder, Yves, and Emmanuel Fort. 2006. Single-particle diffraction and interference at a macroscopic scale. *Physical Review Letters* 97 (15): 154101.

Couder, Y., E. Fort, C.H. Gautier, and A. Boudaoud. 2005. From bouncing to floating: non-coalescence of drops on a fluid bath. *Physical Review Letters* 94: 177801.

Cristea-Platon, Tudor, Pedro J. Sáenz, and John W.M. Bush. 2018. Walking droplets in a circular corral: quantisation and chaos. *Chaos: An Interdisciplinary Journal of Nonlinear Science* 28 (9): 096116.

Dagan, Yuval. 2023. Relativistic hydrodynamic interpretation of de Broglie matter waves. *Foundations of Physics* 53 (1): 20.

Dagan, Yuval, and John W.M. Bush. 2020. Hydrodynamic quantum field theory: the free particle. *Comptes Rendus. Mécanique* 348 (6–7): 555–571.

Dagan, Y., J.B. Greenberg, and D. Katoshevski. 2017a. Similarity solutions for the evolution of polydisperse droplets in vortex flows. *International Journal of Multiphase Flow* 97: 1–9.

Dagan, Yuval, David Katoshevski, and J. Barry Greenberg. 2017b. Particle and droplet clustering in oscillatory vortical flows. *Atomization and Sprays* 27 (7): 629.

De Broglie, Louis. 1924. Recherches sur la théorie des quanta. PhD thesis, Migration-université en cours d'affectation.

de Broglie, Louis. 1970. The reinterpretation of wave mechanics. *Foundations of Physics* 1 (1): 5–15.

De Broglie, Louis. 2012. *Heisenberg's Uncertainties and the Probabilistic Interpretation of Wave Mechanics: With Critical Notes of the Author*, vol. 40. Berlin: Springer. Springer Science & Business Media.

De la Peña, Luis, Ana Marıa Cetto, and Andrea Valdés-Hernández. 2015. *The Emerging Quantum. The Physics Behind Quantum Mechanics*. Cham: Springer International Publishing.

Drezet, A., P. Jamet, D. Bertschy, A. Ralko, and C. Poulain. 2020. Mechanical analog of quantum Bradyons and Tachyons. *Physical Review E* 102 (052206): 1–10.

Durey, Matthew, and John W.M. Bush. 2020. Hydrodynamic quantum field theory: the onset of particle motion and the form of the pilot wave. *Frontiers in Physics* 8: 300.

Eddi, A., Emmanuel Fort, F. Moisy, and Yves Couder. 2009. Unpredictable tunneling of a classical wave-particle association. *Physical Review Letters* 102 (24): 240401.

Fort, Emmanuel, Antonin Eddi, Arezki Boudaoud, Julien Moukhtar, and Yves Couder. 2010. Path-memory induced quantization of classical orbits. *Proceedings of the National Academy of Sciences* 107 (41): 17515–17520.

Gilet, Tristan. 2014. Dynamics and statistics of wave-particle interactions in a confined geometry. *Physical Review E* 90 (5): 052917.

Gilet, Tristan. 2016. Quantumlike statistics of deterministic wave-particle interactions in a circular cavity. *Physical Review E* 93 (4): 042202.

Harris, Daniel M., and John W.M. Bush. 2014. Droplets walking in a rotating frame: from quantized orbits to multimodal statistics. *Journal of Fluid Mechanics* 739: 444–464.

Harris, Daniel M., Julien Moukhtar, Emmanuel Fort, Yves Couder, and John W.M. Bush. 2013. Wavelike statistics from pilot-wave dynamics in a circular corral. *Physical Review E* 88 (1): 011001.

Hubert, Maxime, Matthieu Labousse, and Stéphane Perrard. 2017. Self-propulsion and crossing statistics under random initial conditions. *Physical Review E* 95 (6): 062607.

Madelung, E. 1926. Quantentheorie in hydrodynamischen form. *Zeitschrift für Physik* 40: 322–326.

Nachbin, André, Paul A. Milewski, and John W.M. Bush. 2017. Tunneling with a hydrodynamic pilot-wave model. *Physical Review Fluids* 2 (3): 034801.

Nelson, Edward. 1966. Derivation of the Schrödinger equation from Newtonian mechanics. *Physics Review* 150: 1079–1085.

Oza, Anand U., Daniel M. Harris, Rodolfo R. Rosales, and John W.M. Bush. 2014. Pilot-wave dynamics in a rotating frame: on the emergence of orbital quantization. *Journal of Fluid Mechanics* 744: 404–429.

Perrard, Stéphane, Matthieu Labousse, Marc Miskin, Emmanuel Fort, and Yves Couder. 2014. Self-organization into quantized eigenstates of a classical wave-driven particle. *Nature Communications* 5 (1): 1–8.

Sáenz, Pedro J., Tudor Cristea-Platon, and John W.M. Bush. 2018. Statistical projection effects in a hydrodynamic pilot-wave system. *Nature Physics* 14 (3): 315.

Sáenz, Pedro J., Tudor Cristea-Platon, and John W.M. Bush. 2020. A hydrodynamic analog of Friedel oscillations. *Science Advances* 6 (20): eaay9234.

Shinbrot, T. 2019. Dynamic pilot-wave bound states. *Chaos* 29: 113124.

Tadrist, Loïc, Tristan Gilet, Peter Schlagheck, and John W.M. Bush. 2020. Predictability in a hydrodynamic pilot-wave system: resolution of walker tunneling. *Physical Review E* 102 (1): 013104.

Chapter 5
On Multi-Time Correlations in Stochastic Mechanics

Maaneli Derakhshani and Guido Bacciagaluppi

Abstract We address a long-standing criticism of the stochastic mechanics approach to quantum theory by one of its pioneers, Edward Nelson: multi-time correlations in stochastic mechanics differ from those in textbook quantum theory. We elaborate upon an answer to this criticism by Blanchard et al. (Phys Rev D 34(12):3732–3738, 1986), who showed that if the (derived) wave function in stochastic mechanics is assumed to collapse to a delta function in a position measurement, the collapse will change the stochastic process for the particles (because the stochastic process depends on derivatives of the wave function), and the resulting multi-time correlations will agree with those in textbook quantum theory. We show that this assumption can be made rigorous through the tool of 'effective collapse' familiar from pilot-wave theories, and we illustrate this with an example involving the double-slit experiment. Hence one of the major lingering objections to stochastic mechanics is dissolved. We finally show that in the case of multi-time correlations between multiple particles, effective collapse implies nonlocal influences between particles.

Keywords Stochastic mechanics · Multi-time correlations · Effective collapse

5.1 Introduction

Stochastic mechanics is an approach to quantum theory that aims to recover it from an underlying stochastic process in configuration space. As developed by Nelson (1966, 1985), and briefly described in Sect. 5.2 for the N-particle case,

M. Derakhshani
Department of Mathematics, Rutgers University, New Brunswick, NJ, USA
e-mail: md1485@math.rutgers.edu

G. Bacciagaluppi (✉)
Department of Mathematics and Descartes Centre, Utrecht University, Utrecht, Netherlands
e-mail: g.bacciagaluppi@uu.nl

P. Castro et al. (eds.), *Advances in Pilot Wave Theory*, Boston Studies in the Philosophy and History of Science 344,
https://doi.org/10.1007/978-3-031-49861-9_5

such a recovery starts from a time-reversible description of a diffusion process in configuration space. By then imposing a number of (time-symmetric) dynamical conditions on the process, most elegantly the variational principle due to Yasue (1981), one obtains the Madelung equations for two real functions R and S, which can be combined into a Schrödinger equation for $\psi = e^{\frac{1}{\hbar}(R+iS)}$.

The resulting theory has a number of similarities with de Broglie and Bohm's pilot-wave theory. Indeed, particle trajectories in Nelson's stochastic mechanics can be intuitively thought of as de Broglie–Bohm trajectories with a superimposed white noise (although in pilot-wave theory ψ is interpreted as a fundamental quantity). This means, of course, that stochastic mechanics shares the same non-local features as pilot-wave theory (one of the reasons why Nelson himself eventually lost interest in his theory). As discussed elsewhere (Bacciagaluppi 2005), however, stochastic mechanics has a number of potential advantages over pilot-wave theory. The main reason why it is not as popular as de Broglie–Bohm theory (or other fundamental approaches to quantum theory such as spontaneous collapse or many-worlds), we surmise, is that it is perceived as suffering from two main technical problems not shared by other better-known approaches.

These are:

(a) The problem, raised by Wallstrom (1994), that the Madelung equations are only equivalent to the Schrödinger equation if one imposes a seemingly ad hoc quantisation condition on their solutions.
(b) The problem, raised by Nelson himself (1986), that the theory appears to fail to recover the multi-time correlations predicted by quantum theory.

For different reasons, however, neither of these problems are decisive. Wallstrom's problem proved recalcitrant for many years, but after a number of arguably only partially convincing attempts,[1] it has now found what appears to us to be a natural resolution, based on combining Nelson's strategy with de Broglie's original strategy for deriving the quantisation conditions (Derakhshani 2017). And, as we argue in this paper, Nelson's problem of multi-time correlations is a red herring.

The problem of multi-time correlations (which we describe in more detail in Sect. 5.3) can be expressed very intuitively in terms of series of measurements. In standard quantum theory, in order to calculate the correlations between the position of a particle at multiple times, one considers a series of position measurements, which localise the particle at the given times. Straightforward application of the Born rule provides the desired result. In Nelson's mechanics, the particle has a well-defined position at all times, and the stochastic process also naturally determines the correlations in position at multiple times. The quantum mechanical calculation and the stochastic mechanical calculation, however, lead to different results, as spelled out by Nelson (1986).

At the root of the problem lies the fact that in the quantum mechanical calculation at each measurement one substitutes a new collapsed wave function for the original

[1] The most interesting probably being Schmelzer (2011).

unitarily evolved wave function, while the wave function that is derived in Nelson's mechanics from the underlying stochastic process always follows the unitary Schrödinger evolution. Indeed, as already pointed out by Blanchard et al. (1986) (and summarised below in Sect. 5.4), if the corresponding substitution is made also in the stochastic mechanical calculation, the results will coincide. The obvious question, however, is *why on earth* should one feel licensed to invoke collapse in Nelson's theory where the wave function is not even a fundamental object?

The answer, as we shall argue in detail in Sect. 5.5, lies just below the surface. Since the current velocity $\mathbf{v}$ in stochastic mechanics has the same form as the particle velocity in de Broglie–Bohm theory, many of the results and techniques developed in the context of pilot-wave theory can be straightforwardly imported into stochastic mechanics, even though in the latter theory the wave function is not considered a fundamental quantity. One example, transferred to stochastic mechanics in Bacciagaluppi (2012), is the notion of ('sub-quantum') disequilibrium and consequences thereof, i.e. the idea that frequencies in an actual sample could differ from the theoretical probability distribution. The pilot-wave notion we focus on in this paper, however, is that of *effective collapse*, which is the tool developed by Bohm (1952) to describe in pilot-wave theory measurements of quantities other than position. The key idea is that one must include the apparatus into the pilot-wave theoretic description. Since the trajectory of one system depends on the actual position of other systems with which it is entangled, including information about the position of such other systems changes the way one calculates the trajectory of a system. And in measurement situations, including information about the measurement result means that one can calculate trajectories as if the wave function of the system had collapsed. As we shall discuss, such effective collapse can be transferred also to the setting of stochastic mechanics, and as we have suggested previously (Bacciagaluppi 2012, p. 5, fn. 8), it is the missing ingredient that allows one to solve Nelson's problem of multi-time correlations.[2]

We conclude that the common perception of stochastic mechanics as suffering from major technical problems is mistaken.

---

[2] The notion of effective collapse has been discussed in stochastic mechanics by Goldstein (1987) as well as by Blanchard et al. (1991), but does not seem to have found wide application. Peruzzi and Rimini (1996) give it a unified treatment in pilot-wave theory, stochastic mechanics and variants thereof, and Bacciagaluppi (2003) makes essential use of it in discussing identical particles in pilot-wave theory and stochastic mechanics. Blanchard et al. (1991, p. 162) clearly realise its relevance to the problem of multi-time correlations (since they reference Blanchard et al. (1986)), but they do not spell out the connection explicitly. Nelson, to the best of our knowledge, never discussed or recognised the notion of effective collapse in his writings on stochastic mechanics; if he was unaware of the notion, this could help explain why he thought there was a fundamental discrepancy between stochastic mechanics and quantum mechanics for multi-time correlations.

5.2 Review of Nelson–Yasue Stochastic Mechanics

The N-particle version of Nelson–Yasue stochastic mechanics (NYSM) begins with the hypothesis that 3-D space is pervaded by a homogeneous and isotropic ether with classical stochastic fluctuations that impart a frictionless, conservative diffusion process to a point particle of mass m and charge e immersed within the ether. Accordingly, for N point particles of masses m_i and charges e_i immersed in the ether, each particle will in general have its position 3-vector $\mathbf{q}_i(t)$ constantly undergoing diffusive motion with drift, as modelled by the first-order forward stochastic differential equations

$$d\mathbf{q}_i(t) = \mathbf{b}_i(q(t), t)dt + d\mathbf{W}_i(t). \tag{5.1}$$

Here $q(t) := \{\mathbf{q}_1(t), \mathbf{q}_2(t), ..., \mathbf{q}_N(t)\} \in \mathbb{R}^{3N}$, $\mathbf{b}_i(q(t), t)$ is the deterministic mean forward drift velocity of the ith particle (which in general may be a function of the positions of all the other particles, such as in the case of particles interacting with each other gravitationally and/or electrostatically), and $\mathbf{W}_i(t)$ is the Wiener process modeling the ith particle's interaction with the ether fluctuations. (The 'mean' here refers to averaging over the Wiener process, in the sense of the conditional expectation at time t.)

The Wiener increments $d\mathbf{W}_i(t)$ are assumed to be Gaussian with zero mean, independent of $d\mathbf{q}_i(s)$ for $s \le t$, and with variance

$$\mathrm{E}_t\left[dW_{in}(t)dW_{im}(t)\right] = 2\nu_i\delta_{nm}dt, \tag{5.2}$$

where E_t denotes the conditional expectation at time t. We then hypothesise that the magnitudes of the diffusion coefficients ν_i are given by

$$\nu_i = \frac{\hbar}{2m_i}. \tag{5.3}$$

In addition to (5.1), we also have the backward stochastic differential equations

$$d\mathbf{q}_i(t) = \mathbf{b}_{i*}(q(t), t)dt + d\mathbf{W}_{i*}(t), \tag{5.4}$$

where $\mathbf{b}_{i*}(q(t), t)$ are the mean backward drift velocities, and $d\mathbf{W}_{i*}(t)$ are the backward Wiener processes. The $d\mathbf{W}_{i*}(t)$ have all the properties of $d\mathbf{W}_i(t)$ except that they are independent of the $d\mathbf{q}_i(s)$ for $s \ge t$. With these conditions on $d\mathbf{W}_i(t)$ and $d\mathbf{W}_{i*}(t)$, Eqs. (5.1) and (5.4) respectively define forward and backward Markov processes for N particles on $\mathbb{R}^3$ (or, mathematically equivalently, for a single particle on $\mathbb{R}^{3N}$).

Associated to the trajectories $\mathbf{q}_i(t)$ is the N-particle probability density $\rho(q, t) := n(q, t)/N$ where $n(q, t)$ is the number of particles per unit volume. Corresponding to (5.1) and (5.4), then, are the N-particle forward and backward

Fokker-Planck equations

$$\frac{\partial \rho(q,t)}{\partial t} = -\sum_{i=1}^{N} \nabla_i \cdot [\mathbf{b}_i(q,t)\rho(q,t)] + \sum_{i=1}^{N} \frac{\hbar}{2m_i} \nabla_i^2 \rho(q,t), \tag{5.5}$$

and

$$\frac{\partial \rho(q,t)}{\partial t} = -\sum_{i=1}^{N} \nabla_i \cdot [\mathbf{b}_{i*}(q,t)\rho(q,t)] - \sum_{i=1}^{N} \frac{\hbar}{2m_i} \nabla_i^2 \rho(q,t), \tag{5.6}$$

where we assume that the solutions $\rho(q,t)$ in each time direction satisfy the normalisation condition

$$\int_{\mathbb{R}^{3N}} \rho_0(q) d^{3N} q = 1. \tag{5.7}$$

Up to this point, (5.5) and (5.6) correspond to independent diffusion processes in opposite time directions.[3] To fix the diffusion process uniquely for both time directions, we must constrain the diffusion process to simultaneous solutions of (5.5) and (5.6).

Note that the sum of (5.5) and (5.6) yields the N-particle continuity equation

$$\frac{\partial \rho(q,t)}{\partial t} = -\sum_{i=1}^{N} \nabla_i \cdot [\mathbf{v}_i(q,t)\rho(q,t)], \tag{5.8}$$

where

$$\mathbf{v}_i(q,t) := \frac{1}{2}(\mathbf{b}_i(q,t) + \mathbf{b}_{i*}(q,t)) \tag{5.9}$$

is the current velocity field of the ith particle. We shall also require that $\mathbf{v}_i(q,t)$ is equal to the gradient of a scalar potential $S(q,t)$ (since, if we allowed $\mathbf{v}_i(q,t)$ a non-zero curl, then the time-reversal operation would change the orientation of the curl, thus distinguishing time directions).[4] And for particles classically interacting with an external vector potential $\mathbf{A}_i^{ext} := \mathbf{A}^{ext}(\mathbf{q}_i, t)$, the current velocities get modified

[3] In fact, given all possible solutions to (5.1), one can define as many forward processes as there are possible initial distributions satisfying (5.5); likewise, given all possible solutions to (5.4), one can define as many backward processes as there are possible 'initial' distributions satisfying (5.6). Consequently, the forward and backward processes are both underdetermined, and neither (5.1) nor (5.4) has a well-defined time-reversal.

[4] Alternatively, assuming that $\mathbf{v}_i$ is a gradient ensures satisfaction of the symmetry condition $D\mathbf{b}_i = D_*\mathbf{b}_{i*}$ (cf. (5.15) and (5.16) below) (de la Peña and Cetto 1982, eq. (16); Bacciagaluppi 2012, Appendix).

by the usual expression

$$\mathbf{v}_i(q,t) = \frac{\nabla_i S(q,t)}{m_i} - \frac{e_i}{m_i c}\mathbf{A}_i^{ext}. \tag{5.10}$$

So (5.8) becomes

$$\frac{\partial \rho(q,t)}{\partial t} = -\sum_{i=1}^{N} \nabla_i \cdot \left[\left(\frac{\nabla_i S(q,t)}{m_i} - \frac{e_i}{m_i c}\mathbf{A}_i^{ext} \right) \rho(q,t) \right], \tag{5.11}$$

which is now a time-reversal invariant evolution equation for ρ.

The function S is an N-particle velocity potential, defined here as a field over the possible positions of the particles (hence the dependence of S on the generalised coordinates $\mathbf{q}_i$), and generates different possible initial irrotational velocities for the particles via (5.10). No assumptions are made at this level as to whether or not S can be written as a sum of single-particle velocity potentials. Rather, this will depend on the initial conditions and constraints specified for a system of N Nelsonian particles, as well as the dynamics we obtain for S.

Note also that subtracting (5.5) from (5.6) yields the equality on the right-hand side of

$$\mathbf{u}_i(q,t) := \frac{1}{2}\left[\mathbf{b}_i(q,t) - \mathbf{b}_{i*}(q,t)\right] = \frac{\hbar}{2m_i}\frac{\nabla_i \rho(q,t)}{\rho(q,t)}, \tag{5.12}$$

where $\mathbf{u}_i(q,t)$ is the osmotic velocity field of the ith particle. From (5.10) and (5.12), we then have $\mathbf{b}_i = \mathbf{v}_i + \mathbf{u}_i$ and $\mathbf{b}_{i*} = \mathbf{v}_i - \mathbf{u}_i$, which when inserted back into (5.5) and (5.6), respectively, return (5.11). Thus ρ is fixed as the unique, single-time, 'quantum equilibrium' distribution for the solutions of (5.1) and (5.4), and evolves by (5.11).

To give physical meaning to the osmotic velocities, we postulate the presence of an external 'osmotic' potential which couples to the ith particle as $R(q(t),t) = \mu U(q(t),t)$ and imparts to it a momentum $\nabla_i R(q,t)|_{q=q(t)}$. This momentum then gets counter-balanced by the ether fluid's osmotic impulse pressure, $(\hbar/2m_i)\nabla_i \ln[n(q,t)]|_{q=q(t)}$, leading to the equilibrium condition $\nabla_i R/m_i = (\hbar/2m_i)\nabla_i \rho/\rho$ (using $\rho = n/N$), hence $\rho = e^{2R/\hbar}$ for all times.

In order to formulate the second-order dynamics of the particles, we need to generalise to N particles Nelson's (1966) construction of mean forward and backward derivatives. This generalisation is straightforwardly given by

$$D\mathbf{q}_i(t) = \lim_{\Delta t \to 0^+} \mathrm{E}_t \left[\frac{q_i(t+\Delta t) - q_i(t)}{\Delta t} \right], \tag{5.13}$$

and

$$D_* \mathbf{q}_i(t) = \lim_{\Delta t \to 0^+} \mathrm{E}_t \left[\frac{q_i(t) - q_i(t-\Delta t)}{\Delta t} \right]. \tag{5.14}$$

By the Gaussianity of $d\mathbf{W}_i(t)$ and $d\mathbf{W}_{i*}(t)$, we obtain $D\mathbf{q}_i(t) = \mathbf{b}_i(q(t), t)$ and $D_*\mathbf{q}_i(t) = \mathbf{b}_{i*}(q(t), t)$. We note that $\mathbf{v}_i(q(t), t) = \frac{1}{2}(D + D_*)\,\mathbf{q}_i(t)$. The computation of $D\mathbf{b}_i(q(t), t)$ and $D_*\mathbf{b}_i(q(t), t))$ is given in the appendix and yields

$$D\mathbf{b}_i(q(t), t) = \left[\frac{\partial}{\partial t} + \sum_{i=1}^{N} \mathbf{b}_i(q(t), t) \cdot \nabla_i + \sum_{i=1}^{N} \frac{\hbar}{2m_i} \nabla_i^2 \right] \mathbf{b}_i(q(t), t), \tag{5.15}$$

$$D_*\mathbf{b}_{i*}(q(t), t) = \left[\frac{\partial}{\partial t} + \sum_{i=1}^{N} \mathbf{b}_{i*}(q(t), t) \cdot \nabla_i - \sum_{i=1}^{N} \frac{\hbar}{2m_i} \nabla_i^2 \right] \mathbf{b}_{i*}(q(t), t). \tag{5.16}$$

We now generalise the construction given by Yasue (1981) in the single-particle case. Using (5.15)–(5.16), and assuming that particle i also couples to an external electric potential, $\Phi_i^{ext} := \Phi^{ext}(\mathbf{q}_i(t), t)$, as well as to the other particles by the Coulomb interaction potential $\Phi_i^{int} := \frac{1}{2}\sum_{j=1}^{N(j \neq i)} \frac{e_j}{|\mathbf{q}_i(t) - \mathbf{q}_j(t)|}$, we can construct the N-particle generalisation of Yasue's ensemble-averaged, time-symmetric mean action:

$$\begin{aligned} J = \mathrm{E}\Bigg[\int_{t_I}^{t_F} \sum_{i=1}^{N} \Bigg\{ & \frac{1}{2}\left[\frac{1}{2} m_i \mathbf{b}_i^2 + \frac{1}{2} m_i \mathbf{b}_{i*}^2 \right] + \frac{e_i}{c} \mathbf{A}_i^{ext} \cdot \frac{1}{2}(D + D_*)\,\mathbf{q}_i(t) \\ & - e_i \left[\Phi_i^{ext} + \Phi_i^{int} \right] \Bigg\} dt \Bigg] \\ = \mathrm{E}\Bigg[\int_{t_I}^{t_F} \sum_{i=1}^{N} \Bigg\{ & \frac{1}{2} m_i \mathbf{v}_i^2 + \frac{1}{2} m_i \mathbf{u}_i^2 + \frac{e_i}{c} \mathbf{A}_i^{ext} \cdot \mathbf{v}_i - e_i \left[\Phi_i^{ext} + \Phi_i^{int} \right] \Bigg\} dt \Bigg], \end{aligned} \tag{5.17}$$

where E [...] denotes the absolute expectation.[5]

Upon imposing the conservative diffusion constraint through the N-particle generalisation of Yasue's variational principle

$$J = extremal, \tag{5.18}$$

[5] $\mathrm{E}[...] := \int \rho(q,t)\,[...]\,d^{3N}q$.

a straightforward computation (Derakhshani 2017, Appendix 7.1) shows that (5.18) implies

$$\sum_{i=1}^{N} \frac{m_i}{2} \left[D_* D + D D_*\right] \mathbf{q}_i(t)$$
$$= \sum_{i=1}^{N} e_i \left[-\frac{1}{c} \partial_t \mathbf{A}_i^{ext} - \nabla_i \left(\Phi_i^{ext} + \Phi_i^{int} \right) + \frac{\mathbf{v}_i}{c} \times \left(\nabla_i \times \mathbf{A}_i^{ext} \right) \right] |_{q=q(t)}. \tag{5.19}$$

and the stochastic accelerations

$$\begin{aligned} m_i \mathbf{a}_i(q(t), t) &= \frac{m_i}{2} \left[D_* D + D D_*\right] \mathbf{q}_i(t) \\ &= \left[-\frac{e_i}{c} \partial_t \mathbf{A}_i^{ext} - e_i \nabla_i \left(\Phi_i^{ext} + \Phi_i^{int} \right) \right. \\ &\quad \left. + \frac{e_i}{c} \mathbf{v}_i \times \left(\nabla_i \times \mathbf{A}_i^{ext} \right) \right] |_{q=q(t)}, \end{aligned} \tag{5.20}$$

for $i = 1, ..., N$. Applying the mean derivatives in (5.20), using that $\mathbf{b}_i = \mathbf{v}_i + \mathbf{u}_i$ and $\mathbf{b}_{i*} = \mathbf{v}_i - \mathbf{u}_i$, and replacing $q(t)$ with q in the functions on both sides, straightforward manipulations show that (5.19) turns into

$$\begin{aligned} &\sum_{i=1}^{N} m_i \left[\partial_t \mathbf{v}_i + \mathbf{v}_i \cdot \nabla_i \mathbf{v}_i - \mathbf{u}_i \cdot \nabla_i \mathbf{u}_i - \frac{\hbar}{2m_i} \nabla_i^2 \mathbf{u}_i \right] \\ &= \sum_{i=1}^{N} \left[-\frac{e_i}{c} \partial_t \mathbf{A}_i^{ext} - e_i \nabla_i \left(\Phi_i^{ext} + \Phi_i^{int} \right) + \frac{e_i}{c} \mathbf{v}_i \times \left(\nabla_i \times \mathbf{A}_i^{ext} \right) \right]. \end{aligned} \tag{5.21}$$

Using (5.10) and (5.12), integrating both sides of (5.21), and setting the arbitrary integration constants equal to zero, we then obtain the N-particle quantum Hamilton–Jacobi equation

$$\begin{aligned} -\partial_t S(q, t) &= \sum_{i=1}^{N} \frac{\left[\nabla_i S(q, t) - \frac{e_i}{c} \mathbf{A}_i^{ext}\right]^2}{2m_i} \\ &\quad + \sum_{i=1}^{N} e_i \left[\Phi_i^{ext} + \Phi_i^{int} \right] - \sum_{i=1}^{N} \frac{\hbar^2}{2m_i} \frac{\nabla_i^2 \sqrt{\rho(q, t)}}{\sqrt{\rho(q, t)}}, \end{aligned} \tag{5.22}$$

which describes the total energy of the possible mean trajectories of the particles, and, upon evaluation at $q = q(t)$, the total energy of the actual particles along their mean trajectories. So (5.11) and (5.22) together define the N-particle HJM equations.

Let us now combine (5.11) and (5.22) into an N-particle Schrödinger equation and write down the most general form of the N-particle wave function. To do this, we first need to impose the N-particle quantisation condition

$$\sum_{i=1}^{N} \oint_L \nabla_i S(q,t) \cdot d\mathbf{q}_i = nh. \tag{5.23}$$

(For a detailed justification of condition (5.23), cf. Derakhshani (2017).) Then we can combine (5.11) and (5.22) into

$$i\hbar \frac{\partial \psi(q,t)}{\partial t} = \sum_{i=1}^{N} \left[\frac{\left[-i\hbar\nabla_i - \frac{e_i}{c}\mathbf{A}_i^{ext}\right]^2}{2m_i} + e_i\left(\Phi_i^{ext} + \Phi_i^{int}\right)\right]\psi(q,t), \tag{5.24}$$

where $\psi = \sqrt{\rho}e^{iS/\hbar} = e^{\frac{1}{\hbar}(R+iS)}$. Our resolution to the problem of multi-time correlations will depend on being able to calculate Nelson trajectories from a quantum-mechanical wave function, so it is essential for us that (5.23) is imposed.

5.3 Nelson's Argument

Here we describe Nelson's argument for the case of a one-dimensional harmonic oscillator, as discussed by Blanchard et al. (1986). Although Nelson (1986) considered the case of two non-interacting one-dimensional harmonic oscillators with the same frequency, the one-dimensional oscillator case illustrates the same point more simply.[6]

Suppose that the one-dimensional oscillator is in the ground state with wave function

$$\psi(x,t) = \frac{1}{(2\pi\sigma^2)^{1/4}} e^{\left[-\frac{1}{2}\left(i\omega_0 t + \frac{x^2}{2\sigma^2}\right)\right]}, \tag{5.25}$$

[6] Nelson's example is interesting because it brings in considerations of nonlocality. We return to these in our concluding Sect. 5.6.

where $\sigma := \hbar/2m\omega_0$. For the oscillator in the ground state, the forward stochastic differential equation is (Nelson 1967)

$$\begin{aligned} dX(t) &= \left(\frac{1}{m}\frac{dS(x,t)}{dx} + \frac{1}{m}\frac{dR(x,t)}{dx}\right)|_{x=X(t)}dt + dW(t) \\ &= -\omega_0 X(t)dt + dW(t), \end{aligned} \tag{5.26}$$

where $S = \frac{\hbar}{2i}\ln(\frac{\psi}{\psi^*})$ and $R = \hbar \ln|\psi|$. Solving (5.26) by quadrature yields

$$X(t) = e^{-\omega_0 t}\left[X(0) + \int_0^t e^{\omega_0 t'} dW(t')\right], \tag{5.27}$$

where we assume $\omega_0 > 0$ and $m > 0$. The two-time correlation function obtained using (5.27) is thus

$$\mathrm{E}\left[X(0)X(t)\right] = \sigma^2 e^{-\omega_0 t} \tag{5.28}$$

for $t \geq 0$.

Now we calculate the two-time correlation from the Heisenberg picture of textbook quantum mechanics. For the oscillator Hamiltonian

$$\hat{H} = -\frac{\hbar}{2m}\partial_x^2 + \frac{1}{2}m\omega_0^2\hat{x}^2 \tag{5.29}$$

the Heisenberg position operator at time t is defined as

$$\begin{aligned} \hat{x}(t) &= e^{\frac{i}{\hbar}\hat{H}t}\hat{x}(0)e^{-\frac{i}{\hbar}\hat{H}t} \\ &= \hat{x}(0)\cos(\omega_0 t) + \frac{\sin(\omega_0 t)}{m\omega_0}(-i\hbar\partial_x). \end{aligned} \tag{5.30}$$

Hence the commutator

$$\left[\hat{x}(0), \hat{x}(t)\right] = i\hbar\frac{\sin(\omega_0 t)}{m\omega_0}, \tag{5.31}$$

which vanishes for $t = n\pi/\omega_0$, where $n \in \mathbb{Z}$. For these times, and assuming the ground state wave function (5.25), the two-time correlation function becomes

$$\langle\psi|\,\hat{x}(0)\hat{x}(t)\,|\psi\rangle = (-1)^n\,\sigma^2, \tag{5.32}$$

which clearly differs from (5.28).[7] In particular, the latter vanishes as $t \to \infty$.

In Nelson's example, since the two harmonic oscillators are non-interacting their position operators (at the same or different times) commute and the above analysis carries over straightforwardly, i.e., one obtains the same disagreement in results. Another worked example, involving a particle with an initially Gaussian wave function scattering through a potential, is given by Blanchard et al. (1986).

5.4 Blanchard et al.'s Resolution

The Nelsonian expression (5.27) has a perfectly coherent meaning as the expectation value of the product of the position of the oscillator at $X(0)$ and $X(t)$ when the oscillator's position is *not* measured, and for this case the two-time correlation is indeed the one with exponential decay, (5.28). On the other hand, as Blanchard et al. (1986) note for Nelson's theory, the collapse of a subsystem's wave function in a position measurement (in particular the first measurement at $t = 0$) changes the mean forward drift velocity in (5.26) or (5.1), which in turn changes the particle's stochastic trajectory. The reason is that the ith drift velocity $\mathbf{b}_i(q(t), t)$ depends on a wave function $\psi = e^{\frac{1}{\hbar}(R+iS)}$ satisfying the Schrödinger equation (5.25) via

$$\mathbf{b}_i(q(t), t) = \frac{\hbar}{m_i}\,(\mathrm{Re} + \mathrm{Im})\,\nabla_i \ln\psi(q, t)|_{q=q(t)}, \tag{5.33}$$

hence the drift velocity after (say) an ideal position measurement of a Nelsonian particle will differ from before the measurement, assuming that the wave function collapses as a consequence of the measurement. With this observation in hand, it is not at all surprising that (5.27) for the unmeasured oscillator's position should differ from the quantum mechanical correlation (5.31), the latter referring to measurements of the oscillator's position.

In the case of the oscillator, Blanchard et al. take its collapsed wave function at $t = 0$, when the ideal position measurement is made with result X_0, to be a delta function

$$\lim_{t\to 0^+} \psi^{X_0}(x, t) = \delta(x - X_0) \tag{5.34}$$

[7] Blanchard et al. (1986) give the general expression for the two-time correlation function when the operators do not commute, see their equation (34). Our Eq. (5.32) is valid for the special times when the operators commute.

which thereafter evolves via the oscillator's Hamiltonian (5.29) to some excited state at finite time $t > 0$. They note that since Schrödinger's equation with the Hamiltonian (5.29) has the propagator (Sakurai and Napolitano 1993, pp. 116–119)

$$K\left(x, t; x'; 0\right) = \sqrt{\frac{m\omega_0}{2\pi i\hbar \sin\left(\omega_0 t\right)}} \times \exp\left[\left(\frac{im\omega_0}{2\hbar}\right) \times \frac{\left(x^2 + x'^2\right)\cos\left(\omega_0 t\right) - 2xx'}{\sin\left(\omega_0 t\right)}\right], \tag{5.35}$$

with the property that $\lim_{t\to 0^+} K\left(x, t; x'; 0\right) = \delta\left(x - x'\right)$, the time-evolved wave function for the oscillator at $t > 0$, with initial condition (5.34), is just

$$\psi^{X_0}(x, t) := \int_{-\infty}^{\infty} K(x, t; x'; 0)\psi^{X_0}(x', 0)dx' = K(x, t; X_0, 0). \tag{5.36}$$

(Our form for the propagator in (5.35) differs from the form given in Blanchard et al. (1986); however it can be readily shown, using the definition of the propagator (Sakurai and Napolitano 1993, pp. 116–119), that the two forms are equivalent. Our form for the propagator makes the calculation of the drift velocity more transparent.) Using (5.36) in (5.33), we can calculate the drift velocity for the oscillator at time t and obtain

$$b^{X_0}\left(x, t\right) = \omega_0 \left[\frac{x}{\tan\left(\omega_0 t\right)} - \frac{X_0}{\sin\left(\omega_0 t\right)}\right]. \tag{5.37}$$

Inserting this drift velocity in the stochastic differential equation

$$dX(t) = b^{X_0}\left(X(t), t\right)dt + dW(t) \tag{5.38}$$

and solving by quadrature results in a periodic oscillator trajectory for $t > 0$:

$$X(t) = \left[\cos\left(\omega_0 t\right) - \sin\left(\omega_0 t\right)\cot\left(\omega_0 s\right)\right] X_0 + \frac{\sin\left(\omega_0 t\right)}{\sin\left(\omega_0 s\right)} X(s) + \sin\left(\omega_0 t\right)\int_s^t \frac{dW(t')}{\sin\left(\omega_0 t'\right)}, \tag{5.39}$$

for $0 < s \leq t$. Clearly, for $t = n\pi/\omega_0$, the trajectory $X(t)$ is just given by $(-1)^n X_0$. So the stochastic mechanical two-time correlation $\mathrm{E}[X(0)X(t)]$, for $t = n\pi/\omega_0$ and using again the probability density $|\psi(x, t)|^2$ to compute the absolute expectation, where $\psi(x,t)$ is the ground-state wave function (5.25) prior to the position measurement, reads[8]

$$\mathrm{E}[X(0)X(t)] = (-1)^n \sigma^2, \tag{5.40}$$

which agrees with the Heisenberg picture result (5.32).

We note that the drift velocity (5.34) for a wave function corresponding to a delta function is undefined. Of course, the assumption that the wave function collapses to a delta function in an ideal position measurement is itself an idealisation. A more realistic assumption is that the wave function collapses to a Gaussian of some narrow width w. For such a function, the drift velocity exists and is well-behaved. If we make this more realistic assumption, i.e., that $\lim_{t\to 0^+} \psi^{X_0}(x,t) = g(x-\widetilde{x})$, where $g(x-\widetilde{x})$ is a normalised Gaussian function of width w centered on some point $\widetilde{x}$ at $t=0$, we need only multiply the propagator $K\left(x,t;x';0\right)$ with an initial wave function corresponding to $g\left(x'-\widetilde{x}\right)$ and integrate over all space (that, is over x') to obtain the oscillator's wave function at $t>0$. It is easy to confirm that, for such a wave function, the resulting drift velocity and oscillator trajectory for $t>0$ will be periodic. It is therefore reasonable to expect that the quantum mechanical two-time correlation will also be reproduced.

The obvious question is whether one can justify using collapsed wave functions in stochastic mechanics. At first sight this seems incoherent: the wave function in stochastic mechanics is a derived notion, and one may not introduce separate postulates for its evolution. As we have seen, the evolution of the wave function in stochastic mechanics is given by the Schrödinger equation (5.24).

On the other hand, the stochastic mechanics calculation in the previous section does not model measurements using the Schrödinger equation. Indeed, the calculation assumes that the oscillator is following the free Schrödinger evolution throughout. Measurement is apparently taken into account by conditioning on $X(0)$, but that means one is assuming that a measurement simply reveals the oscillator's position without affecting in any way its subsequent dynamics. This assumption is surely not licensed. One needs to model the measurement explicitly and check whether the resulting drift equals that given by the collapsed wave function of textbook quantum mechanics.

Blanchard et al. (1986) make one possible suggestion in this regard. Precisely since the wave function in stochastic mechanics is a derived notion, if as a consequence of the measurement one updates the distribution ρ, this will lead to a different wave function also in stochastic mechanics. Such 'non-equilibrium' constraints on the distribution of the process can indeed be modelled using a new wave function, but require introducing also a non-linearity in the resulting Schrödinger equation (Bacciagaluppi 2012).

[8] We note that this expression also agrees with the more general expression for the two-time correlation given by Blanchard et al. (1986), see equation (33) therein.

In the next section, we shall provide a justification for the use of effectively collapsed wave functions fashioned after that used in de Broglie and Bohm's pilot-wave theory, explicitly modelling measurements by inclusion of appropriate degrees of freedom of the measuring apparatus.

5.5 Effective Collapse Resolution

How does pilot-wave theory treat measurements and recover the appearance of collapse, even though there is no 'true' collapse in the theory?

We shall start by treating a very simple case, that of performing a position measurement by letting a particle go through a slit. Assume the particle, initially guided by a wave $\psi(t)$, goes through the slit at time $t = 0$. At times larger than 0, the pilot wave has developed one 'transmitted' and one 'reflected' component:

$$\psi(t) = \psi_T(t) + \psi_R(t). \tag{5.41}$$

It is important to note that these two components do not overlap. As a consequence, if the particle indeed goes through at $t = 0$, for times $t > 0$ the only component of the pilot wave that is relevant for the motion of the particle is the transmitted component. By the same token, if the particle fails to go through, the only component of the pilot wave that is relevant for the future motion of the particle is the reflected component. For all effects and purposes (which in pilot-wave theory means: for the purpose of guiding the motion of the particle), the passage or failed passage through the slit has collapsed the wave.

If we translate this to the framework of stochastic mechanics, we see that through the simple inclusion of the external potential, conditioning on $\mathbf{x}(0)$ in effect picks out a collapsed wave function. If we compare the predictions of stochastic mechanics with those of quantum mechanics, the predictions of the two theories are identical. But in this simple case it is also unsurprising, because the quantum mechanical predictions for later position measurements are insensitive to whether or not we actually detect the particle at the slit and collapse the wave function. Thus, it is actually immaterial whether stochastic mechanics in this case is able to recover the notion of collapse.

This simple model, simply using an external potential, is essentially how de Broglie (1928) could treat diffraction phenomena. But it fails already for slightly more complex measurements.

Imagine a double-slit experiment. Simply including the external potential covers the case in which both slits are open. But now it makes all the difference whether or not we detect *which* open slit the particle goes through. In order to model this in pilot-wave theory (and by extension in stochastic mechanics), we need to include also *apparatus degrees of freedom*.

Let us do so, and let the particle interact with an 'apparatus' with initial wave function $\phi(\mathbf{y}, t)$ for measuring the particle's position to within the aperture of the

U(pper) or L(ower) slit. We schematically model the interaction as:

$$\psi(\mathbf{x}, t)\phi(\mathbf{y}, t) \rightarrow a\,\psi_U(\mathbf{x}, t)\phi_U(\mathbf{y}, t) + b\,\psi_L(\mathbf{x}, t)\phi_L(\mathbf{y}, t). \tag{5.42}$$

Note again that the two components of the wave do not overlap. So, analogously as above, if the particle goes through the upper or lower slit its motion is guided, respectively, by the effectively collapsed wave functions $\psi_U(\mathbf{x}, t)$ or $\psi_L(\mathbf{x}, t)$. This time, however, this is crucially due to the explicit inclusion of the apparatus degrees of freedom: $\psi_U(\mathbf{x}, t)$ and $\psi_L(\mathbf{x}, t)$ will spread out from the slits and overlap, but the macroscopically different $\phi_U(\mathbf{y}, t)$ and $\phi_L(\mathbf{y}, t)$ do not (or negligibly so), thus ensuring the non-overlap of the two components in (5.42).

Again, translating to the case of stochastic mechanics, conditioning on $\mathbf{x}(0)$ effectively picks out the collapsed wave function $\psi_U(\mathbf{x}, t)$ or $\psi_L(\mathbf{x}, t)$, and the predictions of stochastic mechanics coincide with those of standard quantum mechanics with collapse. This example generalises to arbitrary sequences of position measurements, thereby solving Nelson's problem of multi-time correlations: whenever position is actually measured in such a way that the measurement result is encoded in the degrees of freedom of some macroscopic object, the multi-time correlations calculated in stochastic mechanics will coincide with those calculated in standard quantum mechanics.

For completeness, let us consider also measurements of observables other than position. Take for instance a measurement not of the position but of the component of the particle's momentum parallel to the slits (maybe using a suspended double slit). This might be modelled as:

$$\psi(\mathbf{x}, t)\phi(\mathbf{y}, t) \rightarrow c\,\psi_\uparrow(\mathbf{x}, t)\phi_\uparrow(\mathbf{y}, t) + d\,\psi_\downarrow(\mathbf{x}, t)\phi_\downarrow(\mathbf{y}, t). \tag{5.43}$$

As in the case of (5.42), even though of course $\psi_\uparrow(\mathbf{x}, t)$ and $\psi_\downarrow(\mathbf{x}, t)$ overlap, the two components of (5.43) do not overlap because $\phi_\uparrow(\mathbf{y}, t)$ and $\phi_\downarrow(\mathbf{y}, t)$ do not. Therefore, after the measurement the particle will be guided by either $\psi_\uparrow(\mathbf{x}, t)$ or $\psi_\downarrow(\mathbf{x}, t)$, depending on the result of the measurement as encoded in $\phi_\uparrow(\mathbf{y}, t)$ and $\phi_\downarrow(\mathbf{y}, t)$, respectively.

In the case of stochastic mechanics, conditioning on $\mathbf{x}(0)$ no longer picks out a collapsed wave function, but the measurement was not a measurement of position, so $\mathbf{x}(0)$ does not encode the result of the measurement. It is $\mathbf{y}(0)$ that does, and conditioning on $\mathbf{y}(0)$ accordingly picks out the collapsed wave function $\psi_\uparrow(\mathbf{x}, t)$ or $\psi_\downarrow(\mathbf{x}, t)$.

The treatment of measurements and of 'effective collapse' we have just sketched was Bohm's (1952) crucial contribution to pilot-wave theory. It allows the theory to recover the predictions of quantum mechanics in all cases of measurement, as long as the measurement results are encoded in some macroscopic positions (whether the positions of pointers of or ink drops on a printout).

In stochastic mechanics, it explains how the drift velocity $\mathbf{b}$ after a measurement is the one corresponding to the usual collapsed wave function of standard quantum mechanics. The problem of multi-time correlations disappears.

5.6 Conclusion

Of the two traditional problems besetting stochastic mechanics, Wallstrom's (1994) of the apparent ad hoc character of the quantisation of the wave function and Nelson's (1986) of multi-time correlations, Nelson himself considered the latter to be the most pressing.[9] We believe both of these problems have been addressed in a satisfactory way. In particular we have pointed out in this paper how the explicit inclusion of measurements in stochastic mechanics resolves the problem of multi-time correlations. Nevertheless, we presume Nelson would not have been entirely happy, because the resolution of this problem comes at the price of nonlocality.

In both Nelson (1986) and Nelson (2005) the problem of multi-time correlations is presented as a Bell-type dilemma between reproducing the quantum mechanical predictions and locality. Instead of taking a single one-dimensional oscillator, as we did in Sect. 5.3, Nelson takes two non-interacting one-dimensional oscillators of the same frequency ω_0 that are entangled in such a way that at $t = 0$ their positions are, say, perfectly correlated. The two oscillators may, however, be at arbitrary separation along some other dimension.

Since $\hat{x}_2(t)$ is periodic, so is the quantum mechanical correlation function $\langle\psi_0|\, \hat{x}_1(0)\hat{x}_2(t)\, |\psi_0\rangle$. But, as with (5.28), the Nelsonian correlation function $\mathrm{E}\,[X_1(0)X_2(t)]$ will have an exponential decay in time. Thus, in general,

$$\mathrm{E}[X_1(0)X_2(t)] \neq \langle\psi_0|\, \hat{x}_1(0)\hat{x}_2(t)\, |\psi_0\rangle\,. \tag{5.44}$$

And the only way of getting equality is if the measurement on the first oscillator at $t = 0$ somehow nonlocally affects the later trajectory of the second oscillator in some appropriate way.

As we have seen, including the measurement apparatus in the stochastic mechanical description automatically yields the same predictions as the standard textbook collapse. In the case of two entangled systems, it will reproduce the collapse at a distance (as is well-known from pilot-wave theory). Thus, a measurement on one oscillator does have an effect on the drift velocity of the other oscillator. We note, incidentally, that the same mechanism is what enables stochastic mechanics to violate the Bell inequalities.[10] In conclusion, stochastic mechanics is nonlocal and the dilemma is resolved in the way Nelson would have liked least.

[9] Personal communication to one of us (GB) at a conference in Vienna in November 2011.

[10] This also addresses the objection to stochastic mechanics raised by Kiukas and Werner (2010), who discuss violations of the Bell inequalities based on measurements of position at different times and suggest that the existence of joint multi-time distributions in stochastic mechanics would prevent it from violating these Bell inequalities.

Appendix: Computation of Forward and Backward Stochastic Derivatives

To compute $D\mathbf{b}_i(q(t), t)$ (or $D_*\mathbf{b}_i(q(t), t)$) expand $\mathbf{b}_i$ in a Taylor series up to terms of order two in $d\mathbf{q}_i(t)$:

$$d\mathbf{b}_i(q(t), t) = \frac{\partial \mathbf{b}_i(q(t), t)}{\partial t}dt + \sum_{i=1}^{N} d\mathbf{q}_i(t) \cdot \nabla_i \mathbf{b}_i(q, t)|_{q=q(t)}$$
$$+ \sum_{i=1}^{N} \frac{1}{2} \sum_{n,m} dq_{in}(t) dq_{im}(t) \frac{\partial^2 \mathbf{b}_i(q, t)}{\partial q_{in} \partial q_{im}}|_{q=q(t)} + \dots . \tag{5.45}$$

From (5.1), we can replace $d\mathbf{q}_i(t)$ by $d\mathbf{W}_i(t)$ in the last term, and when taking the conditional expectation at time t in (5.13), we can replace $d\mathbf{q}_i(t) \cdot \nabla_i \mathbf{b}_i|_{q=q(t)}$ by $\mathbf{b}_i(\mathbf{q}(t), t) \cdot \nabla_i \mathbf{b}_i|_{q=q(t)}$ since $d\mathbf{W}_i(t)$ is independent of $\mathbf{q}_i(t)$ and has mean 0. From (5.2), we then obtain

$$D\mathbf{b}_i(q(t), t) = \left[\frac{\partial}{\partial t} + \sum_{i=1}^{N} \mathbf{b}_i(q(t), t) \cdot \nabla_i + \sum_{i=1}^{N} \frac{\hbar}{2m_i} \nabla_i^2\right] \mathbf{b}_i(q(t), t), \tag{5.46}$$

and likewise

$$D_*\mathbf{b}_{i*}(q(t), t) = \left[\frac{\partial}{\partial t} + \sum_{i=1}^{N} \mathbf{b}_{i*}(q(t), t) \cdot \nabla_i - \sum_{i=1}^{N} \frac{\hbar}{2m_i} \nabla_i^2\right] \mathbf{b}_{i*}(q(t), t). \tag{5.47}$$

References

Bacciagaluppi, G. 2003. Derivation of the symmetry postulates for identical particles from pilot-wave theories. https://arxiv.org/pdf/quant-ph/0302099.pdf
Bacciagaluppi, G. 2005. *Endophysics, Time, Quantum and the Subjective*, 367–388. Singapore: World Scientific. Revised version: http://philsci-archive.pitt.edu/8853/
Bacciagaluppi, G. 2012. *Journal of Physics: Conference Series* 361: 012017, 1–12.
Blanchard, P., S. Golin, and M. Serva. 1986. *Physical Review D* 34 (12): 3732–3738.
Blanchard, P., M. Cini, and M. Serva. 1991. *BiBoS–463/91*, 149–171 Bielefeld: BiBoS.
Bohm, D. 1952. *Physical Review* 85 (2): 166–179 and 180–193.
de Broglie, L. 1928. *Electrons et photons*, 105–141. Paris: Gauthier-Villars.
de la Peña, L., and A.M. Cetto. 1982. *Foundations of Physics* 12: 1017–1037.
Derakhshani, M. 2017. PhD Thesis (Utrecht). https://arxiv.org/abs/1804.01394
Goldstein, S. 1987. *Journal of Statistical Physics* 47 (5/6): 645–667.
Kiukas, J., and R.F. Werner. 2010. *Journal of Mathematical Physics* 51 (7): 072105, 1–16.
Nelson, E. 1966. *Physical Review* 150 (4): 1079–1085.
Nelson, E. 1967. *Dynamical Theories of Brownian Motion*. Princeton: PUP.

Nelson, E. 1985. *Quantum Fluctuations*. Princeton: PUP.
Nelson, E. 1986. Field theory and the future of stochastic mechanics. https://doi.org/10.1007/3540171665_87
Nelson, E. 2005. The mystery of stochastic mechanics. https://web.math.princeton.edu/~nelson/papers/talk.pdf
Peruzzi, G., and A. Rimini. 1996. *Foundations of Physics Letters* 9 (6): 505–519.
Sakurai, J.J., and J. Napolitano. 1993. *Modern Quantum Mechanics*. Revised ed. Boston: Addison-Wesley.
Schmelzer, I. 2011. An answer to the Wallstrom objection against Nelsonian stochastics. https://arxiv.org/abs/1101.5774v3
Wallstrom, T. 1994. *Physical Review A* 49 (3): 1613–1617.
Yasue, K. 1981. *Journal of Mathematical Physics* 22 (5): 1010–1020.

Chapter 6
A Version of de Broglie's Double Solution Theory Reproducing Landau's Quantization in a Uniform Magnetic Field

Pierre Jamet and Aurélien Drezet

Abstract Following the recent development of a mechanical atomic model based on Louis de Broglie's double solution theory, we show how it is possible to obtain the famous Landau quantization rules for a purely classical electron coupled to a wave. We present both a qualitative but intuitive explanation of our model of wave-particle duality, as well as a more rigorous derivation of the solutions. This work completes a set of three articles, each using a different combination of external electric and magnetic forces on the same base model. Despite this similarity with previous approaches, it also uncovers some unique properties of this system. We will see that, in order to correctly recover some useful non-relativistic limits, it is necessary to use a scalar field whose mass is the same as an effective mass for the particle that naturally appears in our model. This could be a crucial step to better characterize the necessary ingredients of a double solution theory.

Keywords Mechanical quantum analogs · De Broglie's double solution · Pilot wave · Bohr's atom · Landau's levels · Wave-particle duality

6.1 Introduction

Quantization of various physical quantities like momentum or energy is the hallmark of modern quantum mechanics. In fact, it even gave it its name, due to the importance of this aspect in the development of the old quantum theory as developed originally by Planck, Einstein, Bohr and many others in the first decades of the twentieth century. However, it is well known today—and it was already at the time—that quantization is actually a property of waves in a broad sense, rather than a

P. Jamet · A. Drezet (✉)
Institut Néel, UPR 2940, CNRS-Université Grenoble Alpes, Grenoble, France
e-mail: pierre.jamet@protonmail.com; aurelien.drezet@neel.cnrs.fr

P. Castro et al. (eds.), *Advances in Pilot Wave Theory*, Boston Studies in the Philosophy and History of Science 344,
https://doi.org/10.1007/978-3-031-49861-9_6

specificity of the microscopic world. In that context the paradigmatic discovery was the wave particle duality associated with the name of Albert Einstein for the photon and Louis de Broglie for the electron and other particles (de Broglie 1925). The idea of de Broglie to associate a wave to a particle soon evolved into the family of so called pilot wave theories (de Broglie 1927; Bacciagaluppi and Valentini 2009). More precisely, the goal of such models was to attach to each particle a wave, which would guide its dynamics so that some aspects of undulatory mechanics could then be applied to it. The idea was inspired by works made by Fermat and Hamilton showing some analogies between optics and classical mechanics of point-like objects. For instance, if we consider the case of an electron orbiting around the nucleus of an atom, its associated wave would have to satisfy a periodicity condition along the perimeter of the trajectory, which would then translate to a quantization of its physical properties, e.g. wavelength, energy, frequency and so on. The electron, being so closely related to the wave, would then acquire the same restrictions on its own properties, and a quantization of its angular momentum, energy, radius and velocity would result. One way to achieve this duality between a wave and a particle was devised by de Broglie, who postulated the existence of an internal periodic phenomenon inside each particle, a clock of sorts, which would have to be synchronized with the oscillations of the wave. If, then, the frequency of the wave had to be quantized due to some boundary condition, the particle would also need to constrain its own dynamics in order to maintain the synchronization during the motion. Remarkably, the two ideas first introduced by de Broglie almost a century ago as the basis of pilot wave theories, namely the existence of a phase wave and its synchronization with the oscillation of some property of the particle itself, still remain in some form or another in current approaches.

In fact in the past two decades, macroscopic hydrodynamic experiments have been performed that manage to reproduce a good number of the properties of quantum objects (Couder and Fort 2006; Bush 2015; Fort et al. 2010). In light of these recent developments, one can wonder how we could translate these results to a microscopic model, i.e. applied to electrons and atoms rather than oil droplets for example. Compared to other quantum theories and interpretations (e.g., the Copenhagen interpretation or the Many Worlds interpretation), this approach has some unique constraints, the main one being that this model needs to remain a mechanical one all throughout the development. It means that we need to have "hidden" variables and, of course, that the field and interactions must propagate locally at all times. On the other hand, it will be necessary to deviate from the usual macroscopic descriptions as we will see throughout this document. The model should also differ from the so called 'Bohmian mechanics' (Bohm and Hiley 1993) which is a logical development of the original de Broglie pilot wave approach (realized after the rediscovery of the theory by Bohm in 1952). In Bohmian mechanics a wave guides a particle but this is done in the configuration space and not in the 3D physical space. Also the physical nature of this guiding wave is ambiguous and mysterious since it acts on the particle but there is no back-reaction on the wave. De Broglie didn't consider the pilot wave approach as a good starting point. He hoped (somehow in agreement with Einstein) to find a mechanical explanation (the double solution (de Broglie 1960)) justifying the existence of the coupling between a localized object and an extended wave.

In the following, we will present one such theoretical mechanical model developed in recent years (Drezet et al. 2020; Jamet and Drezet 2021, 2022) following de Broglie's intuition about wave-particle duality, and show that it is indeed able to predict the Landau quantization of an electron orbiting in a homogeneous magnetic field (Landau 1930). This result is to be put in comparison with an experimental demonstration of the Landau levels, achieved in a system consisting of a droplet bouncing on a vibrating bath (Fort et al. 2010). Compared to macroscopic experiments however, some different ingredients are needed as mentioned. For instance, it is required to work under a relativistic framework, but also to make use of complex quantities—in particular a complex field—in order to take into account the electric charge. While forcing us to deviate from what we are used to in hydrodynamical descriptions, these different aspects have the benefit of giving a really robust and self-consistent theory, so that we can for example use a completely covariant formalism in an elegant way.

The model that is presented here is an extension of one that has been developed in the past three years (motivated by a previous work (Borghesi 2017)). As we will see, it makes use of the fact that if we couple a scalar field with a point-like particle, there exists a regime where the equations of motion for the various objects decouple from each other. In that regime, called transparency, the initially complex and chaotic dynamics can become linear and stable on top of being able to reproduce results of the quantum theory. This result is of course very promising as an attempt to give a completely mechanical description of quantum objects, and in particular as it relates to the hydrodynamics experiments mentioned earlier. It is also a satisfying illustration of wave-particle duality, since in that specific regime the two objects no longer affect each other and are so closely linked that they behave as one, so that to distinguish them becomes extremely challenging. We will first see in Sect. 6.2 how the model itself is built and how it relates to de Broglie's original ideas on wave-particle duality as well as pilot wave theories. In Sect. 6.3, we will obtain the expressions for the different quantities defining the dynamics of the particle. Finally in Sect. 6.4, we will derive the complete analytical solutions for the field, and conclude by comparing these results with the well-known Landau effect.

6.2 Description of the System

Inspired by the ideas of de Broglie, and more specifically his double solution theory, our goal is to build a simple mechanical model (Drezet et al. 2020; Jamet and Drezet 2021, 2022) able to reproduce some of the quantum properties of a particle, for example here an electron. We start off with some basic ingredients in order to construct a pilot wave model, namely a complex scalar field $u(x) = u(t, \mathbf{x}) \in \mathbb{C}$ (here $x^\mu := [t, \mathbf{x}] \in \mathbb{R}^4$ denotes the relativistic position four-vector) acting as the guiding wave which we couple to a point-like particle of mass m_p located at the position $\mathbf{x}_\mathrm{p}(t) \in \mathbb{R}^3$. This particle will be constrained to live on the field at all times, meaning that we introduce another degree of freedom $z_\mathrm{p}(t) \in \mathbb{C}$ which one can think of as an internal oscillation analogous to de Broglie's internal clock, later synchronized with the oscillations of the field. To account for this natural periodic phenomenon for the particle, one needs to add a term akin to a harmonic force acting on this quantity z_p, with a proper frequency Ω_p, or equivalently a stiffness $m_\mathrm{p}\Omega_\mathrm{p}^2$. Our system[1] is thus characterized by the action

$$
\begin{aligned}
I = &-\int \left[m_\mathrm{p} - \frac{1}{2} m_\mathrm{p}\sigma \left(|\dot{z}(\tau)|^2 - \Omega_\mathrm{p}^2 |z(\tau)|^2\right)\right] \mathrm{d}\tau - e\int A(x_\mathrm{p}(\tau))\dot{x}_\mathrm{p}(\tau)\mathrm{d}\tau \\
&+ \int \left\{\mathcal{N}(\tau)\left[z(\tau) - u(x_\mathrm{p}(\tau))\right]^* + \mathcal{N}^*(\tau)\left[z(\tau) - u(x_\mathrm{p}(\tau))\right]\right\} \mathrm{d}\tau \\
&+ T\int \left[(Du)(Du)^* - \omega_0^2 u u^*\right] \mathrm{d}^4 x
\end{aligned}
\tag{6.1}
$$

which is split into three components:

(i) a Lagrangian for the particle integrated over the proper time τ comprised of a usual relativistic kinetic term, the internal degree of freedom, as well as the local interaction with a four-potential $A_\mu(x)$ through a coupling constant $e = -|e| < 0$ (e.g. an electric charge for the 'electron').

(ii) a Lagrangian density integrated over space-time for the field u, with the covariant derivative $D_\mu = \partial_\mu + ieA_\mu$ using the same four-potential as it appears for the particle.

(iii) an interaction Lagrangian between the field and the particle integrated over the proper time, making use of a Lagrange multiplier $\mathcal{N}(\tau)$ interpreted as the reaction force of the field on the particle.

We stress that we here use the proper time τ as a parameter to label the position of the particle $x(\tau)$ along its trajectory. Minimizing this action $\delta I = 0$ soon leads to a system of coupled Euler-Lagrange differential equations, from which we can extract a regime called transparency where the force $\mathcal{N}(\tau)$ vanishes, meaning that the particle and the field are no longer affecting each other. This specific regime is an interesting example of a wave-particle duality, in the sense that the two objects behave as one and share most of their respective properties, as we will see after obtaining the detailed solutions.

[1] In the following we work with covariant notations, and in particular we have the contravariant space-time four-vector $x^\mu := [t, \mathbf{x}]$, the four-potential $A^\mu := [0, \mathbf{A}]$ and four-gradient $\partial^\mu := [\partial_t, -\nabla]$, in a metric $g_{\mu\nu} = \mathrm{diag}(1, -1, -1, -1)$.

In the case of the transparency, the equations decouple, apart from the holonomic constraint

$$z_{\mathrm{p}}(\tau) - u(\tau, x_{\mathrm{p}}(\tau)) = 0. \tag{6.2}$$

The dynamics of $z_{\mathrm{p}}(\tau)$ then becomes a simple harmonic motion at the proper frequency Ω_{p}

$$z_{\mathrm{p}}(\tau) = z_0 e^{-i\Omega_{\mathrm{p}}\tau}, \tag{6.3}$$

while the particle and the field follow the equations

$$m_{\mathrm{eff.}} \ddot{x}_{\mathrm{p}\mu}(\tau) = eF_{\mu\nu}(x_{\mathrm{p}}(\tau))\dot{x}_{\mathrm{p}}^{\nu} \tag{6.4}$$

$$\left(D^2 + \omega_0^2\right) u(t, x) = 0 \tag{6.5}$$

with $m_{\mathrm{eff.}} = m_{\mathrm{p}}\left(1 + \sigma\Omega_{\mathrm{p}}^2 |z_0|^2\right)$ an effective mass that takes into account both the usual inertia m_{p} and the oscillatory motion along z, $F_{\mu\nu} = \partial_\mu A_\nu - \partial_\nu A_\mu$ the electromagnetic tensor, $D_\mu = \partial_\mu + ieA_\mu$ the covariant derivative and finally ω_0 the mass of the scalar field. We chose to consider a massive scalar field to keep the model as general as possible. We can however also decide to use a massless one where $\omega_0 \to 0$ as we did in other works at a later time. The advantage of using a Klein-Gordon equation for u is to be able to give the correct non-relativistic limit to the field as we will see at the end of Sect. 6.4, which a simple d'Alembertian would not allow due to the nature of the problem at hand, and especially the fact that we consider a pure magnetic field as our external force.

At this point we can introduce another of de Broglie's ideas in order to give a qualitative description of the solutions. Indeed, if we consider that the particle must see, in its proper frame $\mathcal{R}'$, a stationary field $u'(t', \varphi') = f(t')g(\varphi')$, it follows quickly that along the path followed by the particle, the field must be written as a sum of two counterpropagating plane waves

$$u(t, \varphi) = u_+ + u_- = \frac{u_0}{2}\left(e^{i(\omega_+ t - k_+ \rho_{\mathrm{p}}\varphi)} + e^{i(\omega_- t - k_- \rho_{\mathrm{p}}\varphi)}\right), \tag{6.6}$$

or alternatively, a product of a phase and group wave

$$u(t, x) = u_0 e^{i(k\rho_{\mathrm{p}}\varphi - \omega t)} \cos\left[\frac{k_+ + k_-}{2}\rho_{\mathrm{p}}\varphi - \frac{\omega_+ - \omega_-}{2}t\right], \tag{6.7}$$

with

$$k = \frac{k_+ - k_-}{2} \quad \text{and} \quad \omega = \frac{\omega_+ + \omega_-}{2}. \tag{6.8}$$

This way, we are able to recover the famous result of de Broglie that, through the holonomic constraint (6.2), the frequencies $\Omega_{\rm p}$ of the particle appearing in the oscillation $z(\tau)$ and ω of the field must match according to the relation

$$\omega = \Omega_{\rm p}. \tag{6.9}$$

On top of that, we see that the group velocity for the field must also match the particle velocity

$$v_g = v_{\rm p}. \tag{6.10}$$

More specifically, the relation between $k_\pm$ and $\omega_\pm$

$$k_\pm = \omega_\pm + \varepsilon_\pm \tag{6.11}$$

lets us write the field

$$u = u_0 e^{i(k\rho_{\rm p}\varphi - \omega t)} \cos\left[(\omega + \varepsilon)\rho_{\rm p}\varphi - \frac{k-\eta}{\omega+\varepsilon}t\right], \tag{6.12}$$

with

$$\varepsilon = \frac{\varepsilon_+ + \varepsilon_-}{2} \quad \text{and} \quad \eta = \frac{\varepsilon_+ - \varepsilon_-}{2}, \tag{6.13}$$

so that Eq. (6.10) gives us

$$v_g = \frac{k-\eta}{\omega+\varepsilon} \equiv \frac{P_{\rm p} - eA_\varphi}{E_{\rm p}} = v_{\rm p} \tag{6.14}$$

from which we immediately see that $\varepsilon = 0$ and $\eta = eA_\varphi$. This equivalence also obviously implies the relation $(\hbar\omega, \hbar\mathbf{k}) = (E_{\rm p}, \mathbf{P}_{\rm p})$ where we reintroduced the constant $\hbar$ explicitely, meaning that we recover the famous de Broglie hypothesis. The relation (6.14) is fundamental since it justifies the guidance of the particle by the wave. Looking at the field, we have in the end two components that we can separate and try to interpret in different ways:

(i) a phase wave following the vertical dynamics of the particle, i.e. its oscillation with frequency $\Omega_{\rm p}$, displaying the properties of a monochromatic plane wave and able to introduce some phenomena like quantization rules under specific boundary conditions.
(ii) a group wave following the horizontal dynamics of the particle, i.e. its uniform motion at velocity $v_{\rm p}$, displaying the properties of a localized energy packet.

This is of course a useful description to understand qualitatively what happens along the particle's trajectory, as de Broglie did, but we will need a more complete and robust mathematical description in order to obtain the field at all space-time points

x and derive from it the specific values of its parameters, such as its amplitude and frequencies. Before we tackle the complex field equations however, we will benefit from doing the same for the particle's dynamics, which will be the topic of the next section.

6.3 The Particle Dynamics

Assuming a circular motion in the (x, y) plane with acceleration $\mathbf{a}_{\mathrm{p}} = \frac{\mathrm{d}^2\mathbf{x}_{\mathrm{p}}}{\mathrm{d}t^2} = -\frac{v_{\mathrm{p}}^2}{\rho}\hat{\mathbf{e}}_{\rho}$ as shown on Fig. 6.1, let us write Eq. (6.4) again with the specific form of the four-potential $A^{\mu} = (0, \frac{1}{2}B\rho\hat{\mathbf{e}}_{\varphi})$, i.e., corresponding to a uniform magnetic field $\mathbf{B} = B\hat{\mathbf{e}}_z$ along z. The relativistic dynamical equation for the particle reads

$$- m_{\mathrm{eff.}}\gamma\frac{v_{\mathrm{p}}^2}{\rho} = ev_{\mathrm{p}}B \tag{6.15}$$

where we used the form of the Lorentz force $e\mathbf{v}_{\mathrm{p}}\times\mathbf{B} = ev_{\mathrm{p}}B\hat{\mathbf{e}}_{\rho}$ (note that the electric charge is here negative) and $\gamma = (1 - v_{\mathrm{p}}^2)^{-\frac{1}{2}}$. We also define a Bohr-Sommerfeld integral of motion linked to the angular momentum:

$$\oint P_{\mathrm{p}}\mathrm{d}l = 2\pi\rho\left(\gamma m_{\mathrm{eff.}}v_{\mathrm{p}} + \frac{1}{2}eB\rho\right) = 2\pi n, \tag{6.16}$$

with P_{p} the linear momentum of the particle, and where n is for now only a quantity translating the conservation of the angular momentum rather than its quantization,

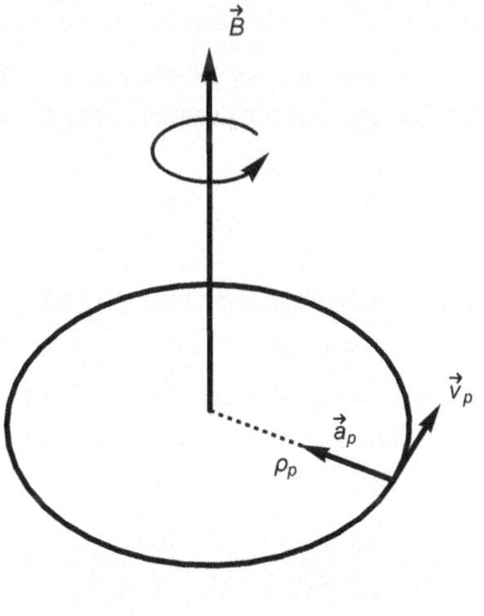

Fig. 6.1 Representation of the particle's motion

meaning that we only assume that it is a real constant. The quantization will of course appear later as we reintroduce the field, but is not necessary at that stage.

From these two equations we derive the velocity

$$v_n = \frac{1}{\sqrt{1 + \frac{m_{\text{eff.}}}{2n\omega_L}}}, \tag{6.17}$$

the radius

$$\rho_n = \sqrt{\frac{2n}{m_{\text{eff.}}\omega_L}}, \tag{6.18}$$

and finally the energy

$$E_n = m_{\text{eff.}}\sqrt{1 + 2\frac{n\omega_L}{m_{\text{eff.}}}}, \tag{6.19}$$

where we labeled each of the quantities with the index n to indicate that they correspond to a specific orbit of the particle, and where we introduced a frequency

$$\omega_L = -\frac{eB}{m_{\text{eff.}}} \tag{6.20}$$

which we will identify as the usual Landau frequency (we stress that since $e < 0$ we have indeed $\omega_L > 0$).

We see that there is a dependency on ω_L in each of the quantities, so that in particular we get for the energy E_n in the non-relativistic limit

$$E_n \approx m_{\text{eff.}} + n\omega_L. \tag{6.21}$$

This means that as we increase or decrease the intensity of the magnetic field B, the energy will vary linearly with it as is expected in these configurations (Landau 1930).

6.4 Solutions of the Field Equation

In order to derive the solutions for the complex massive scalar field u, we start from the Klein-Gordon equation in cylindrical coordinates with covariant derivatives $D^\mu := [\partial_t, -\nabla + \frac{1}{2}ieB\rho\hat{\mathbf{e}}_\varphi]$

$$-\partial_t^2 u + \frac{1}{\rho}\partial_\rho\left(\rho\partial_\rho u\right) + \frac{1}{\rho^2}\partial_\varphi^2 u + \partial_z^2 u - ieB\partial_\varphi u - \frac{1}{4}e^2B^2\rho^2 u = \omega_0^2 u, \tag{6.22}$$

giving us a general solution with separate variables

$$u(t, \rho, \varphi, z) = R(\rho)e^{i(k_z z + m\varphi - \omega t)}, \quad m \in \mathbb{Z}. \tag{6.23}$$

Introducing a new variable $\xi = -\frac{1}{2}eB\rho^2$, we can extract the equation on $R(\xi)$

$$\xi R''(\xi) + R'(\xi) + R(\xi)\left(\beta - \frac{\xi}{4} - \frac{m^2}{4\xi}\right) = 0 \tag{6.24}$$

with $\beta = -\frac{\omega^2 - \omega_0^2 - k_z^2}{2eB} - \frac{m}{2}$ a constant. This equation admits solutions of the form

$$R(\xi) = Ce^{-\frac{\xi}{2}}\xi^{\frac{|m|}{2}}\,{}_1F_1\left(-\left(\beta - \frac{|m|+1}{2}\right), |m|+1, \xi\right) \tag{6.25}$$

with C a normalization constant and ${}_1F_1(a, b, z)$ the confluent hypergeometric function solution of

$$zf''(z) + (b - z)f'(z) + af(z) = 0. \tag{6.26}$$

Alternatively, this solution is often written in terms of the equivalent Whittaker function $\mathcal{M}_{\kappa,\mu}(z)$, giving us the radial part

$$R(\xi) = C\sqrt{\xi}\mathcal{M}_{\beta, \frac{|m|}{2}}(\xi). \tag{6.27}$$

As usual we will get the total field by superposing two counterpropagating solutions

$$u_\pm(t, \rho, \varphi, z) = C_\pm R \pm (\xi)e^{i(\pm m_\pm + k_{z\pm} z - \omega_\pm t)}. \tag{6.28}$$

If we impose that the radial function be finite everywhere, the quantity $\beta - (|m| + 1)/2$ must equal an integer $n_\rho \in \mathbb{N}$ and, knowing the expression of β, we deduce the value of the energy ω

$$\omega = \sqrt{\omega_0^2 - 2eB\left(n_\rho + \frac{|m| + m + 1}{2}\right) + k_z^2}. \tag{6.29}$$

We will now restrict ourselves to cases where $k_z = 0$, meaning to planar motions rather than helicoidal ones. Knowing also that we defined m_+ and m_- to be respectively positive and negative integers, the total energy of the field may be defined with the energies of the two counterpropagating waves $\omega_{\pm}$ as

$$\begin{aligned}\omega &= \frac{\omega_+ + \omega_-}{2} \\ &= \frac{1}{2}\omega_0\left[\sqrt{1 - 2\frac{eB}{\omega_0^2}\left(n_{\rho+} + m_+ + \frac{1}{2}\right)} + \sqrt{1 - 2\frac{eB}{\omega_0^2}\left(n_{\rho-} + \frac{1}{2}\right)}\right]\end{aligned} \tag{6.30}$$

which in the limit $-eB/\omega_0^2 \ll 1$ approximates to

$$\omega \approx \omega_0 - \frac{eB}{\omega_0}\frac{n_{\rho+} + n_{\rho-} + m_+ + 1}{2}. \tag{6.31}$$

This is equivalent to Landau's formula for the energy $\omega = \omega_0 + n\omega_L$ with the quantum number n

$$n := \frac{m_+ - m_-}{2} = \frac{n_{\rho+} + n_{\rho-} + m_+ + 1}{2} \tag{6.32}$$

on the condition that the mass of the field and the particle are equal

$$\omega_0 = m_{\text{eff.}}. \tag{6.33}$$

This last point is especially interesting since it highlights the very constrained nature of this model. Indeed, to correctly recover the weak field limit, it is for one necessary to have a massive scalar field with $\omega_0 \neq 0$, but also that this mass be equal to the mass of the particle, and finally that the particle quantum number n of Sect. 6.3 matches $\frac{m_+ - m_-}{2}$ of Eq. (6.32). This does not mean that a massless field is impossible to work with, as was demonstrated when we introduced a central Coulomb potential in a previous article, but it emphasizes once again the dual nature of our two objects—that is, the field and particle—which constrain one another and their respective dynamics.

6.5 Conclusion and Perspectives

We developed a quantum mechanical analog of the Landau quantization of the cyclotron orbits of a charged particle in a uniform magnetic field (Landau 1930). The model is based on a version of the double solution theory that we already presented in Drezet et al. (2020) and Jamet and Drezet (2021, 2022). Our approach starts with a relativistically covariant theory in which a particle is interacting with a wave $u(x)$. The model therefore circumvents the usual objection made against Bohmian mechanics concerning action/reaction. Here however, we have a so called transparency regime which is a particular case of a much more complicated dynamics. In this regime the wave and the particle are not interacting. It could be interesting to consider the more general (chaotic) regime perhaps associated with transitions between orbits. We mention that the model is rather peculiar since it relies on an extended wave guiding the particle but the physical reason for this wave is not explained in our model (causality and the boundary conditions are imposed by hand). We remind that in the double solution theory of de Broglie (1960) the idea was to find a localized wave (like a soliton) for defining the particle guided by a base extended wave. But here, we still have a kind of dualistic dynamics with a particle and an extended wave. In a subsequent work we are going to show how to modify the present model in order to agree with the core idea of the double solution theory by introducing a localized wave.

References

Bacciagaluppi, G., and A. Valentini. 2009. Quantum Theory at the Crossroads: Reconsidering the 1927 Solvay Conference. Cambridge: Cambridge University Press.

Bohm, D., and B.J. Hiley. 1993. *The Undivided Universe*. London: Routledge.

Borghesi, C. 2017. Dualité onde-corpuscule formée par une masselotte oscillante dans un milieu élastique: étude théorique et similitudes quantiques. *Annales de la Fondation Louis de Broglie* 42: 161–196.

Bush, J.W.M. 2015. The new wave of pilot-wave theory. *Physics Today* 68: 47–53.

Couder, Y., and E. Fort. 2006. Single-particle diffraction and interference at a macro-scopic scale. *Physical Review Letters* 97: 154101.

de Broglie, L. 1925. Recherches sur la théorie des quanta. *Annals of Physics* 10: 22–128.

de Broglie, L. 1927. La mécanique ondulatoire et la structure atomique de la matière et du rayonnement. *Journal de Physique et Le Radium* 8: 225–241.

de Broglie, L. 1960. *Nonlinear Wave Mechanics: A Causal Interpretation*. Amsterdam: Elsevier.

Drezet, A., P. Jamet, D. Bertschy, A. Ralko, and C. Poulain. 2020. Mechanical analog of quantum bradyons and tachyons. *Physical Review E* 102: 052206.

Fort, E., A. Eddi, A. Boudaoud, J. Moukhtar, and Y. Couder. 2010. Path-memory induced quantization of classical orbits. *Proceedings of the National Academy of Sciences of the United States of America* 107: 17515.

Jamet, P., and A. Drezet. 2021. A mechanical analog of Bohr's atom based on de Broglie's double-solution approach. *Chaos* 31: 103120.

Jamet, P., and A. Drezet. 2022. A classical analog of the quantum Zeeman effect. *Chaos* 32: 033101.

Landau, L. 1930. Diamagnetismus der metalle. *Zeitschrift für Physik* 64: 629–637.

Chapter 7
Convergence to Quantum Equilibrium: Deterministic vs Stochastic Pilot Wave Dynamics

Mohamed Hatifi, Ralph Willox, and Thomas Durt

Abstract We compare the onset of quantum equilibrium in the (deterministic) de Broglie-Bohm (dBB) formalism with its counterpart in the (stochastic) formalism of Bohm, Hiley, and Nelson (BHN). We bring to light the role played by microstructures in the position distributions regarding the approach to quantum equilibrium. We show that they provide a marker which makes it possible to distinguish deterministic chaos *à la* dBB from stochastic dynamics *à la* BHN. We also introduce a measure of the degree of mixing of two ensembles initially separated in space; this measure is inspired by Bernoulli processes.

Keywords de Broglie-Bohm · Quantum equilibrium · Chaos · Causal interpretation · Stochastic dynamics · Bohm-Hiley · Nelson · H-Theorem · Relaxation process · Born rule

7.1 Introduction

In the pilot wave formulations of quantum mechanics, the particle's position is always defined. In order to reproduce exactly the predictions of standard quantum mechanics, the ensemble of positions has to be distributed according to the Born rule $|\Psi(\boldsymbol{x}, t)|^2$. This is the so-called quantum equilibrium hypothesis (deBroglie 1927; Bohm 1952a,b). This condition, however, does not have the status of a

M. Hatifi
Quantum Machines Unit, Okinawa Institute of Science and Technology Graduate University, Onna, Okinawa, Japan

R. Willox
Graduate School of Mathematical Sciences, the University of Tokyo, Meguro-ku, Tokyo, Japan

T. Durt (✉)
Aix Marseille Université, CNRS, Centrale Marseille, Institut Fresnel UMR 7249, Marseille, France
e-mail: thomas.durt@centrale-marseille.fr

P. Castro et al. (eds.), *Advances in Pilot Wave Theory*, Boston Studies in the Philosophy and History of Science 344,
https://doi.org/10.1007/978-3-031-49861-9_7

postulate (contrary to the standard interpretation and to de Broglie-Bohm's seminal contribution (Norsen 2018)). In fact, it has been shown (Bohm and Vigier 1954; Nelson 1967; Bohm and Hiley 1989; Kyprianidis 1992; Valentini 1991a; Petroni 1995; Petroni and Guerra 1995; Guerra 1995; Efthymiopoulos et al. 2017, 2009; Valentini and Westman 2005; Norsen 2018) that the quantum equilibrium can possibly emerge as a consequence of a dynamical process.

In this paper, we study the process of convergence to quantum equilibrium for the de Broglie-Bohm dynamics and its stochastic extension à la Bohm-Hilley-Nelson (BHN) (Bohm and Vigier 1954; Bohm and Hiley 1989). The present work is a continuation of a previous paper Hatifi et al. (2018) where we studied the ergodicity of pilot wave dynamics and its possible interrelation with (bouncing oil) droplets dynamics (Hatifi 2019; Couder et al. 2010, 2005; Perrard et al. 2014). In the present paper, we highlight the role played by microstructures in the position distributions in differentiating the onset of equilibrium in both situations (deterministic/dBB and stochastic/BHN). In the case of dBB dynamics, our analysis confirms that, in accordance with previous studies (Bohm and Vigier 1954; Nelson 1967; Bohm and Hiley 1989; Kyprianidis 1992; Valentini 1991a; Petroni 1995; Petroni and Guerra 1995; Guerra 1995; Efthymiopoulos et al. 2017, 2009; Valentini and Westman 2005; Norsen 2018; Hatifi 2019), the onset of equilibrium can be understood in terms of the chaotic features (Lyapunov exponent) of the dynamics. In the case of BHN dynamics, we show that this picture is not relevant. Common to both approaches, however, the process of amplification of initial conditions provides a qualitative tool for characterizing the onset of equilibrium.

The paper is structured as follows. In Sect. 7.2, we describe the (deterministic) dBB and the (stochastic) BHN dynamics, and we discuss the status of time reversibility and its consequences regarding the onset of quantum equilibrium in both cases (Sect. 7.2.5).

In Sect. 7.3, we study the mechanism of homogenization and mixing of two ensembles initially separated spatially in the context of a 2D quantum harmonic oscillator.

In Sect. 7.4, we show how the presence of microstructures in the initial distribution of positions affects the process of convergence towards the quantum equilibrium. We show how the dispersion of microstructures relates to the largest Lyapunov exponent and, therefore, to the degree of chaos present in the dBB dynamics (Sect. 7.4.1); we also show how the presence of noise dramatically alters the dynamics: it accelerates the convergence to equilibrium and, at the same time, significantly and drastically reduces the interval of time for which the Lyapunov, exponential, process of amplification of initial conditions is valid (Sect. 7.4.2). We finally introduce a new measure (inspired by Bernoulli processes) to measure the degree of mixing of two ensembles initially separated in space (Sect. 7.5).

7.2 Two Pilot Wave Formulations of Quantum Mechanics and the Onset of Quantum Equilibrium

7.2.1 *The (Deterministic) de Broglie-Bohm Dynamics*

The dB-B theory (deBroglie 1927; Bohm 1952a,b) is a dynamical and deterministic formulation of quantum mechanics in which it is assumed that the positions of the particle exist at all times, i.e. independently of the observer. We will consider, in what follows, a single spinless and non-relativistic particle for which a quantum wave function $\Psi(\mathbf{x}, t)$ solves the Schrödinger equation:

$$i\hbar \frac{\partial \Psi(\mathbf{x}, t)}{\partial t} = -\frac{\hbar^2}{2m}\Delta\Psi(\mathbf{x}, t) + V(\mathbf{x}, t)\Psi(\mathbf{x}, t), \tag{7.1}$$

where $V(\mathbf{x}, t)$ is an external potential. In the standard formulation of quantum mechanics, the probability distribution of all particle positions $\mathcal{P}(\mathbf{x}, t)$ obeys the Born rule $\mathcal{P}(\mathbf{x}, t) = |\Psi(\mathbf{x}, t)|^2$. For convenience, let us express the wave function in polar form:

$$\Psi(\mathbf{x}, t) = R(\mathbf{x}, t)e^{i\, S(\mathbf{x},t)/\hbar}, \tag{7.2}$$

where $R(\mathbf{x}, t)$ and $S(\mathbf{x}, t)$ are two real functions. The probability distribution is then given by $\mathcal{P}(\mathbf{x}, t) = R(\mathbf{x}, t)^2$ and is conserved through the continuity equation:

$$\frac{\partial R(\mathbf{x}, t)^2}{\partial t} + \nabla \cdot \left(R(\mathbf{x}, t)^2 \frac{\nabla S}{m} \right) = 0, \tag{7.3}$$

By analogy with classical hydrodynamics, the phase-function $S(\mathbf{x}, t)$ is associated to a velocity field $\mathbf{v}(\mathbf{x}, t)$ given by:

$$\frac{d\mathbf{x}(t)}{dt} = \mathbf{v}(\mathbf{x}, t) = \frac{1}{m}\nabla S(\mathbf{x}, t)\bigg|_{\mathbf{x}=\mathbf{x}(t)}. \tag{7.4}$$

After integration of (7.4), the deterministic dB-B trajectory $\mathbf{x}(t)$ is obtained.

7.2.2 *The Bohm-Hiley-Nelson Dynamics (Non-deterministic)*

There also exists a stochastic extension of de Broglie-Bohm pilot wave theory (de la Peña et al. 2015; Bacciagaluppi 2005, 1999). In this approach, the particle is subject to a fluctuating background field, which induces a random motion in the dynamics. This idea was first imagined by Bohm and Vigier (1954) and then independently developed by Nelson in 1967 in a different context when he tried to derive quantum mechanics from classical random motions. More recently, we used this model in Hatifi et al. (2018) in order to quantify the degree of analogy that exists between pilot wave models and the dynamics of bouncing oil droplets. Briefly summarized, in the stochastic approach (de la Peña et al. 2015; Bacciagaluppi 2005, 1999), the particle evolves with a local velocity given by (7.4) supplemented by random fluctuations. In the Bohm-Hiley version of Nelson's dynamics (here denoted BHM in reference to Bohm-Hilley-Nelson) (Bohm and Hiley 1989), the evolution is then described by the following Ito differential equation

$$d\mathbf{x}(t) = \left[\frac{1}{m} \nabla S + \frac{\alpha}{2} \frac{\nabla |\Psi|^2}{|\Psi|^2} \right] \Bigg|_{x=x(t)} dt + \sqrt{\alpha} d\mathbf{W}(t), \tag{7.5}$$

where α is a free parameter related to the strength of the noise and $d\mathbf{W}(t)$ represents a Wiener process with the following properties

$$< dW_i(t) >= 0 \text{ and } < dW_i(t) dW_j(t') >= \frac{1}{2} \delta_{ij} \, \delta(t - t'). \tag{7.6}$$

The term $\gamma = \frac{\alpha}{2} \frac{\nabla |\Psi|^2}{|\Psi|^2}$ in (7.5) is called the *osmotic velocity* and can be interpreted as a drift term induced by the Brownian motion.

7.2.3 Quantum Non-equilibrium and dBB Dynamics

In the context of the dBB dynamics, it is possible to consider initial distributions of particle positions that initially, at $t = t_0$, differ from the Born rule. That is, for which $\mathcal{P}(\mathbf{x}, t_0) \neq |\Psi(\mathbf{x}, t_0)|^2$. This is the so-called quantum *non-equilibrium* condition. It has been shown (Valentini 1992, 1991a,b) that an out-of-equilibrium ensemble can tend to quantum equilibrium in a relaxation process, as we shall now briefly explain.

Let us consider a non-equilibrium distribution $\mathcal{P}(\mathbf{x}, t)$ which (because of (7.4)) evolves according to the continuity equation:

$$\frac{\partial \mathcal{P}(\mathbf{x}, t)}{\partial t} + \nabla \cdot \left(\mathcal{P}(\mathbf{x}, t) \frac{\nabla S}{m} \right) = 0. \tag{7.7}$$

It is worth noting that if the equilibrium condition $\mathcal{P}(\mathbf{x}, t) = |\Psi(\mathbf{x}, t)|^2$ is fulfilled at time t_0, the Born rule, and therefore standard quantum mechanics, are recovered at all later times (this is the so-called 'equivariance theorem' (deBroglie 1927; Bohm 1952a,b)).

Now, following Valentini and Westman (2005), if one defines $\mathcal{F} = \mathcal{P}/|\Psi|^2$, one deduces from (7.3) and (7.7) that

$$\frac{d\mathcal{F}(\mathbf{x}, t)}{dt} \equiv \frac{\partial \mathcal{F}(\mathbf{x}, t)}{\partial t} + \dot{\mathbf{x}} \cdot \nabla \mathcal{F}(\mathbf{x}, t) = 0. \tag{7.8}$$

Consequently,

$$\mathcal{F}(\mathbf{x}, t) = \mathcal{F}(\mathbf{x}_0, t_0), \tag{7.9}$$

and thus

$$\mathcal{P}(\mathbf{x}, t) = \frac{\mathcal{P}(\mathbf{x}_0, t_0)}{|\Psi(\mathbf{x}_0, t_0)|^2} |\Psi(\mathbf{x}, t)|^2, \tag{7.10}$$

where $\mathcal{P}(\mathbf{x}_0, t_0)$ is the distribution of the particle positions $\mathbf{x}_0$ at the initial time t_0. Thus, if initially

$$\frac{\mathcal{P}(\mathbf{x}_0, t_0)}{|\Psi(\mathbf{x}_0, t_0)|^2} \neq 1,$$

relaxation to the Born rule $\mathcal{P}(\mathbf{x}, t) = |\Psi(\mathbf{x}, t)|^2$ is ruled out because of the conservation of $\mathcal{F}(\mathbf{x}, t)$ along the dBB trajectories.[1]

However, Valentini (1992, 1991a,b) has found that a relaxation process is possible when the dynamics is complex enough (Valentini and Westman 2005; Colin and Struyve 2010; Towler et al. 2011; Colin 2012; Contopoulos et al. 2012; Efthymiopoulos et al. 2009; Abraham et al. 2014) and provided that coarse-grained approximations (denoted hereafter by a bar over the variable) are used instead of the exact, locally defined functions. Simply put, in this coarse-graining, space is divided into small cells (called coarse-grained (CG) cells) on which the quantity of interest X is averaged to give the coarse-grained version $\overline{X}$ of this quantity.

On the other hand, the initial distribution of the ensemble of positions should be taken smoothly enough, i.e. without any form of fine-grained microstructure on it. Because of (7.10), the process of relaxation is impossible at the microscopic level, and it is, therefore, necessary to choose a sufficiently large size for the CG cells (the relaxation process for various sizes of the CG cells is studied in Towler et al. (2011)). Moreover, if microscopic structures are indeed present in the initial distribution of positions, the convergence at the CG level is possible only if the dynamics is sufficiently chaotic, as we will see in the next Sect. 7.4.1, where we show that chaos is indeed necessary in order to disperse the microstructures at the level of the CG cells.

Following Valentini (1991a,b, 1992), the relaxation to quantum equilibrium can be quantified at the coarse-grained level by introducing an entropy-like function (also known in the literature as the Kullback-Leibler divergence (Jüngel 2016)):

$$\overline{H}(t) = \int d^3x \, \overline{\mathcal{P}} \ln \left(\overline{\mathcal{P}} / \overline{|\Psi|^2} \right). \tag{7.11}$$

[1] If on the contrary $\mathcal{F}(\mathbf{x}_0, t_0) = 1$ everywhere, the Born rule is satisfied everywhere and at all times which is nothing else than the equivariance theorem.

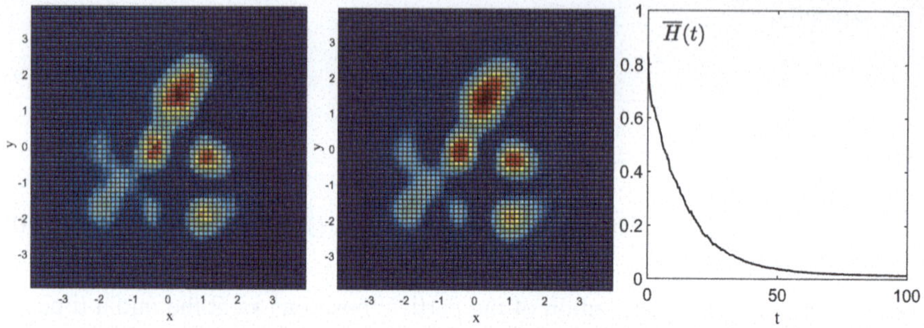

Fig. 7.1 This plot shows the convergence of $\overline{\mathcal{P}}$ to the quantum probability $\overline{|\Psi|^2}$, at the coarse-grained level, for a superposition of $M = 4^2 = 16$ modes in (7.12) . We started with an ensemble of $N = 300{,}000$ initial positions regularly distributed on $[-2, 2] \times [-2, 2]$. The coarse-graining domain, $[-4, 4]\times[-4, 4]$, was divided into 60×60 coarse-grained cells. We plot: (Left) the coarse-grained distribution $\overline{\mathcal{P}}$ at time $t = 100$. (Middle) The quantum probability $\overline{|\Psi|^2}$ at time $t = 100$. (Right) The evolution of the associated entropy function $\overline{H}(t)$ (7.11)

where $\mathcal{P}$ and $|\Psi|^2$ are averaged over CG cells to give $\overline{\mathcal{P}}$ and $\overline{|\Psi|^2}$. It is worth noting that the $\overline{H}$ function (7.11) is sensitive to the size of the coarse-grained cells (Towler et al. 2011) and also to the number of initial positions considered in the numerical simulations.

In Fig. 7.1, we plot the distribution $\overline{\mathcal{P}}$ obtained at equilibrium for a 2D quantum harmonic oscillator for the following state consisting of a superposition of $M = 4^2 = 16$ eigenstates ψ_n, equally weighted and with random individual phases θ_{n_x,n_y}:

$$\Psi(x, y, t) = \frac{1}{\sqrt{M}} \sum_{n_x=0}^{\sqrt{M}-1} \sum_{n_y=0}^{\sqrt{M}-1} e^{i\,\theta_{n_x,n_y} - i\omega(n_x+n_y+1)\,t}\, \psi_{n_x}(x)\, \psi_{n_y}(y)\,. \qquad (7.12)$$

Throughout the paper, we will use this particular case of the 2D quantum harmonic oscillator with 16 modes as the model of choice for testing the different approaches to equilibrium we wish to study (this model, conceived by S.Colin, was already implemented in Refs. Colin and Valentini (2013, 2014), Hatifi (2019)). It is worth emphasizing that since the potential well has a characteristic size $L_{ch} = 2\cdot\left(\frac{\hbar}{m\,\omega}\right)^{1/2}$, which in dimensionless units becomes $L_{ch} = 2$, the initial distribution of positions underlying the result in Fig. 7.1 was generated on the square $[-2, 2] \times [-2, 2]$. Coarse-graining was performed in the domain $[-4, 4] \times [-4, 4]$, which, as confirmed by numerical simulations, is such that all the trajectories remain in this domain.

7.2.4 Quantum Non-equilibrium in the Stochastic Formulation

In view of the non-deterministic character of the evolution, quantum *non-equilibrium* in the stochastic framework is fundamentally different from that for the deterministic case. As a matter of fact, the non-equilibrium distribution of particle positions $P(\mathbf{x}, t)$ evolves in a probabilistic space Ω according to a Fokker-Planck equation which is associated to the Ito equation (7.5),

$$\frac{\partial P}{\partial t} = \frac{\alpha}{2} \Delta P - \nabla \cdot \left(\frac{P}{m} \nabla S + \frac{\alpha}{2} \frac{P}{|\Psi|^2} \nabla |\Psi|^2 \right), \tag{7.13}$$

where $\Psi(\mathbf{x}, t)$ satisfies the Schrödinger equation (7.1). At the quantum equilibrium, $P(\mathbf{x}, t) = |\Psi(\mathbf{x}, t)|^2$, the diffusion velocity is compensated by the osmotic velocity, and the Fokker-Planck equation becomes the quantum continuity equation (7.3). In Hatifi et al. (2018), we derived a strong H-theorem showing that for any initial distribution, the ensemble will nearly always converge to the quantum probability:

$$\frac{dH_V}{dt} = -\frac{\alpha}{2} \int_\Omega d^3x \, (\nabla \mathcal{F})^2 \, \frac{|\Psi|^2}{\mathcal{F}} \quad \text{where} \quad H_V(t) = \int_\Omega d^3x \, P \ln \left(\frac{P}{|\Psi|^2} \right), \tag{7.14}$$

This property is also true at the microscopic level, so that the coarse-graining hypothesis is not necessary. As H_V, as well as $\mathcal{F}$, are positive-definite functions, we expect thus, when $\alpha \neq 0$, that the (fine-grained) Kullbak-Leibler divergence H_V decreases with time and after a while will converge to 0 (Hatifi et al. 2018), provided Ψ does not vanish everywhere in space (which is the case here because we only consider localized wave functions throughout the paper).

However, it should be noted that some level of coarse-graining is, of course, inevitable when we make numerical estimates of the various entropies that are of interest when describing the onset of equilibrium. The same is true when estimating the degree of mixing reached by two ensembles that are initially separated in space (as we shall do in the next section) as we must start from a finite set of initial positions which imposes a natural coarse-graining, even in the case of the stochastic dynamics.

7.2.5 Time Reversibility and Fine-Graining

There is a direct connection between fine-graining and time-reversal (Valentini and Westman 2005; Norsen 2018; Hatifi et al. 2018). The dBB theory is time-reversal invariant provided we replace t by $-t$ and $\Psi(\mathbf{x}, t)$ by $\Psi^*(\mathbf{x}, -t)$, in which case the velocity field $\mathbf{v}(\mathbf{x}, t)$ in Eq. (7.4) also gets replaced by $-\mathbf{v}(\mathbf{x}, -t)$, as should be expected. Now, if an initial non-equilibrium distribution relaxes to equilibrium at the coarse-grained level, then, under the dynamics generated by $\Psi^*(\mathbf{x}, -t)$, we would have a distribution that moves away from quantum equilibrium. This is possible only if the final distribution acquires a fine-grained microstructure. To some extent, time reversal is similar to spin echo experiments in nuclear magnetic resonance (NMR), which reveal the reversible part of the decoherence process.

When $\alpha \neq 0$, time-reversibility is broken, however, as well as determinism, and this affects the evolution of fine-grained structures in the BHN dynamics. In analogy with NMR, the dBB dynamics would be associated with the parameter T_1, which is accessible experimentally through spin echo experiments, while the intrinsically irreversible part of the evolution is expressed by the parameter T_2. The full decoherence time, resulting from both contributions, is expressed by the parameter T, with $1/T = 1/T_1 + 1/T_2$. These ideas are made explicit at the level of Figs. 7.2, 7.3, and 7.4. The plots in Fig. 7.4 were obtained by letting 36 clusters of 20 points evolve during a time t=4.66 with the direct dynamics and then during the same time with the time-reversed dynamics. The dBB dynamics being reversible, we obtain at the end the initial configuration (in terms of spin echo, this would mean that $1/T_2 = 0$). This is not so in the case of the BHN dynamics, which is not reversible in time. As we shall show in the rest of the paper, microstructures make

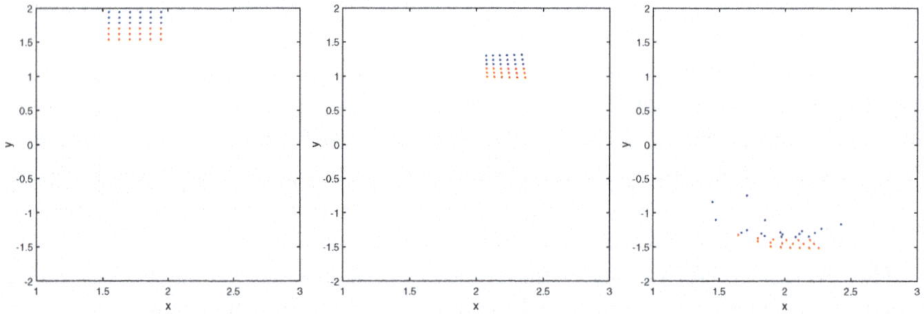

Fig. 7.2 We represent the positions in time of 360 red (blue) points obeying dBB dynamics. These points are initially concentrated in 18 very dense clusters of 20 points. We plot their positions at times $t = 0$, 1, and 4.66

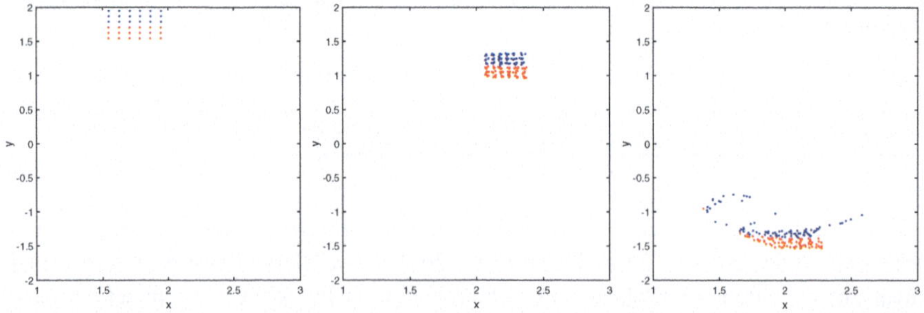

Fig. 7.3 Here, we plot the position of 720 points obeying BHN dynamics, at times $t = 0$, 1 and 4.66 with $\alpha = 10^{-4}$

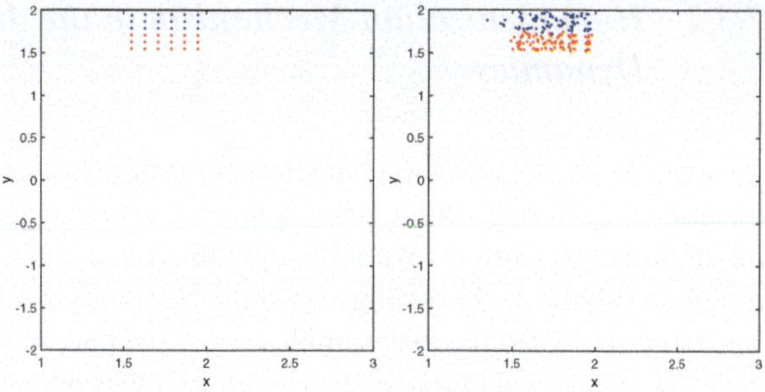

Fig. 7.4 At the left (right), we plot the position of the 720 points of Fig. 7.2 (7.3) subject to time-reversed dBB (BHN) dynamics between time t = 4.66 and time t' = 9.32

it possible to conceive operational tools aimed at differentiating the reversible and irreversible components of the dynamics.[2]

7.3 Homogenization and Mixing

In this section, we shall investigate the homogenization and mixing processes that arise in both the de Broglie–Bohm and Bohm–Hiley–Nelson models for the state (7.12) for the 2D quantum harmonic oscillator. The picture that we have in mind is the following: let us assume that at time t_0 we divide space into two distinct regions of $|\Psi|^2$ weight 1/2 each, say a red one and a blue one and that we consider an initial non-equilibrium distribution at time t_0 with a $\mathcal{P}$ weight and a constant value of $\mathcal{F}$ both equal to 2/3 in the red region and a $\mathcal{P}$ weight and a constant value of $\mathcal{F}$ both equal to 1/3 in the blue region. After the onset of equilibrium, at time t, we expect that in each cell the coarse-grained distribution $\mathcal{P}$ will be close to $|\Psi|^2$ ($\mathcal{F} = \mathcal{P}/|\Psi|^2 \approx 1$). This means that in the cell, one-half of the space will be "occupied by the red positions" and the other half "occupied by the blue positions", with a red density equal to twice the blue density. The fact that in each cell, we will find trajectories originating from the red and blue regions also reflects the **mixing** property of the dynamics. The fact that their densities conspire in order that $\overline{\mathcal{F}} = 1$ in each cell reflects what we call **homogenization**.

[2] It is not clear, for instance, whether the dynamics of droplets is reversible (like deterministic chaos) or irreversible (due to the presence of some Brownian motion). In a previous paper, Hatifi et al. (2018), we derived a method aimed at estimating the value of α, which quantifies the intensity of the Brownian contribution to the dynamics in the case of a 2D harmonic oscillator. This method could be applied when a ferromagnetic bouncing droplet is prepared in the presence of an external effective harmonic 2D potential (Hatifi 2019; Couder et al. 2010, 2005; Perrard et al. 2014).

7.3.1 Homogenization Mechanism in the de Broglie-Bohm Dynamics

In order to show how the characteristic time associated with the convergence to quantum equilibrium in the dBB dynamics also depends on the detail of the initial distribution of positions, we will study the homogenization mechanism for various initial distributions. In analogy with the thermodynamics of mixing (Fermi 1956), we start initially from two ensembles (say a red one, $\mathcal{R}$, and a blue one, $\mathcal{B}$) of particle positions, separated in space; the two ensembles contain a different number of initial points, N_R and N_B respectively, that are uniformly distributed in each set separately (see Fig. 7.5). We then track the elements of these ensembles over time and compute $\overline{H}(t)$ for each ensemble separately, as well as for the whole (see Fig. 7.6). In Fig. 7.6 we can see that the relaxation time is qualitatively the same for the red ($\mathcal{R}$), the blue ($\mathcal{B}$), and for the total ensemble ($\mathcal{R}+\mathcal{B}$): from an exponential fit of the entropy functions $\overline{H}(t)$ (as suggested in the reference Valentini and Westman (2005)) we find that these characteristic times are approximately $\tau_{\mathcal{R}} \sim 15.65$, $\tau_{\mathcal{B}} \sim 15.28$ and $\tau_{\mathcal{R}+\mathcal{B}} \sim 15.39$. Hence, the characteristic time of the relaxation process does not seem to be a useful measure when studying the role played by the initial position distribution in the dBB dynamics.

In order to understand how the (initial) non-equilibrium distribution homogenizes over time, we define the following *fraction numbers* on each CG cell for the red and the blue mixtures:

$$n_r^{\text{cell}(j)} = \frac{N_R^{\text{cell}(j)}}{N_B^{\text{cell}(j)} + N_R^{\text{cell}(j)}}, \quad n_b^{\text{cell}(j)} = \frac{N_B^{\text{cell}(j)}}{N_B^{\text{cell}(j)} + N_R^{\text{cell}(j)}}. \tag{7.15}$$

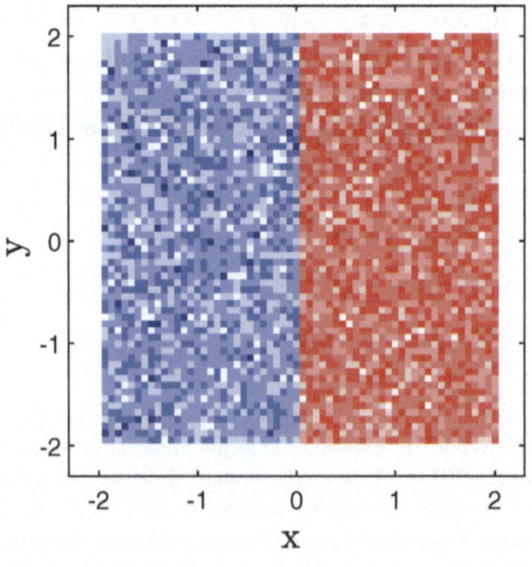

Fig. 7.5 This figure shows the initial distribution $\overline{P}$ consisting of two spatially separated sub-ensembles of particle positions: on the left, the ensemble $\mathcal{B}$ with $N_B = 10^5$ points and on the right the ensemble $\mathcal{R}$ with $N_R = 2 \cdot 10^5$ points. The x (y) coordinate of each point was chosen at random in the appropriate interval, which is $[-2, +2]$ along Y, $[-2, 0]$ along X for the blue ensemble and $[0, +2]$ along X for the red one

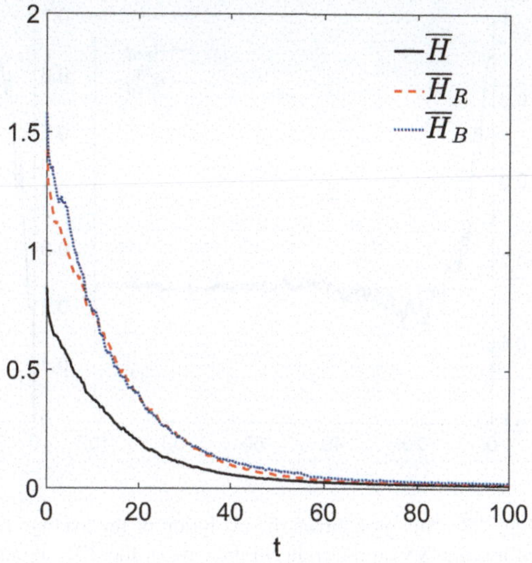

Fig. 7.6 This figure depicts the evolution of the entropy function $\overline{H}(t)$ (Eq. (7.11)) for the dBB dynamics, starting from the ensembles defined in Fig. 7.5. The dotted curve represents the ensemble $\mathcal{B}$, the dashed line corresponds to $\mathcal{R}$, and the solid line represents the total ensemble $\mathcal{B}+\mathcal{R}$

We also define the average fraction of the two mixtures:

$$\overline{n}_r = \frac{1}{N^{cell}} \sum_{j=1}^{N^{cell}} n_r^{\text{cell}(j)}, \quad \overline{n}_b = \frac{1}{N^{cell}} \sum_{j=1}^{N^{cell}} n_b^{\text{cell}(j)}. \tag{7.16}$$

Continuing the analogy with the thermodynamics of mixing, we define an entropy of mixing as:

$$\Delta S = -\overline{n}_r \ln(\overline{n}_r) - (1 - \overline{n}_r) \ln(1 - \overline{n}_r). \tag{7.17}$$

In Fig. 7.7, we plot the average fraction number of the red mixture $\overline{n}_r$ and the associated entropy of mixing ΔS. While the ensemble is converging to the quantum equilibrium, the mixture $\overline{n}_r$ converges to the fraction $\bar{n}_r^{th} = N_R/(N_R + N_B)$. This means that, on average, the same fraction of red or blue trajectories will be present in each cell. Moreover, the entropy becomes maximal after a time of the order of the characteristic time of convergence $t \sim \tau_{\mathcal{R}+\mathcal{B}} \sim 15.39$.

7.3.2 Homogenization Mechanism in the Bohm–Hiley–Nelson Dynamics

Let us now study the mechanism of homogenization in the framework of the stochastic dynamics. In particular, we focus on small fluctuations of the background field in (7.5) for which $\alpha \ll 1$. It is quite intuitive that if α is large, homogenization

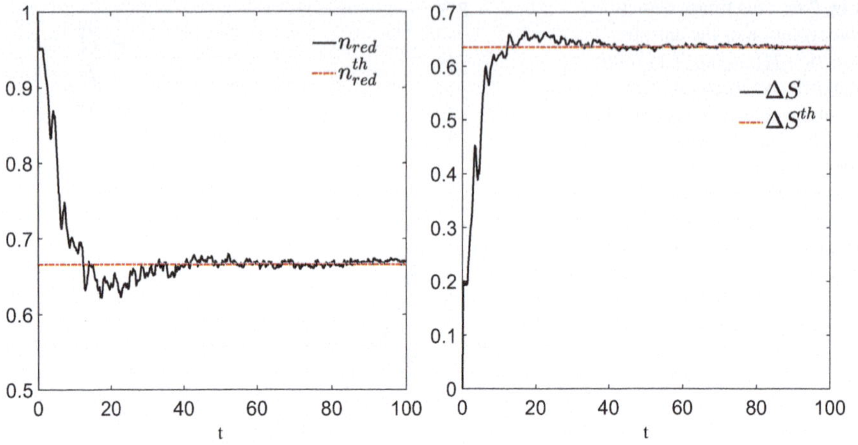

Fig. 7.7 This plot shows the evolution of the average red mixture $\overline{n}_r$ on the left and the entropy of mixing ΔS on the right, in the case of the dBB dynamics. We chose here: $N_R = 200{,}000$ and $N_B = 100{,}000$. The function $\overline{n}_r$ is converging to $\overline{n}_r^{th} = 2/3$

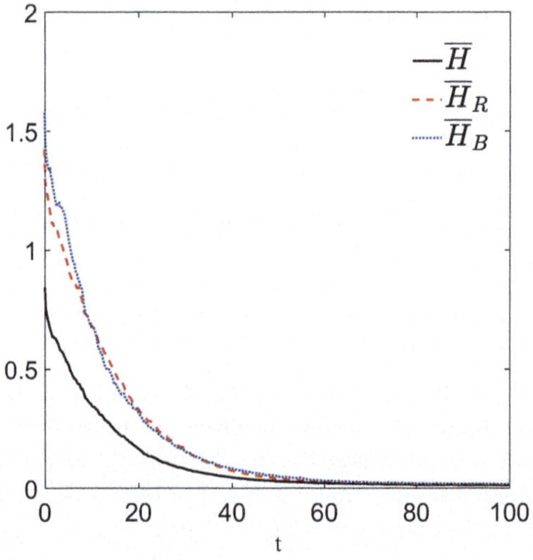

Fig. 7.8 This figure shows the evolution of the entropy functions $\overline{H}(t)$ (7.11) in the case of the stochastic dynamics, for $\alpha = 10^{-4}$, and starting from the ensembles defined in Fig. 7.5. The blue dotted curve is associated with the ensemble $\mathcal{B}$, the dashed line to $\mathcal{R}$ and the full line curve to the total ensemble $\mathcal{B} + \mathcal{R}$

will occur quickly because of the Brownian character of the evolution. As we shall see, however, compared to the dBB dynamics, the onset of equilibrium is also accelerated even for small values of α (small compared to unity, α being a dimensionless parameter). Starting from the distribution given in Fig. 7.5, we let each ensemble ($\mathcal{B}$ and $\mathcal{R}$) evolve according to the dynamics defined by equation (7.5). In Fig. 7.8, we

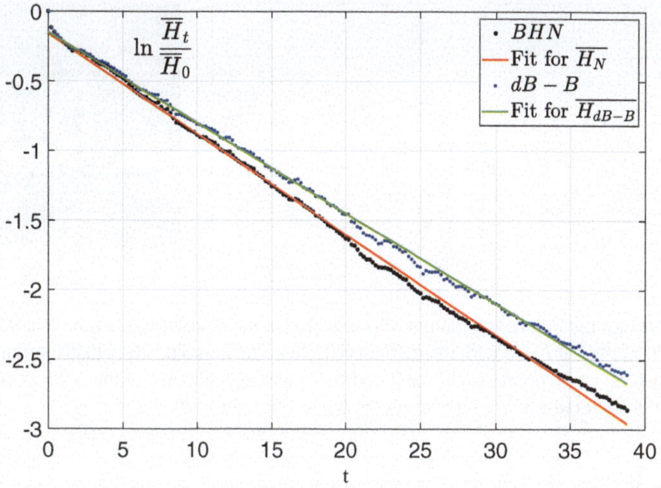

Fig. 7.9 In this plot, we compare the H-functions obtained for the de Broglie-Bohm dynamics and for the Bohm-Hiley-Nelson dynamics for $\alpha = 10^{-4}$. We plot here the logarithmic ratio $\ln(\overline{H}(t)/\overline{H}(t_0))$. For both dynamics, we considered the same initial ensemble of initial positions: $N = 300{,}000$ initial points regularly distributed on the domain $[-2, 2] \times [-2, 2]$. Using a linear fit, we estimated the characteristic time of convergence: approximately $\tau_{\mathcal{R}+\mathcal{B}} \sim 15.39$ for the dBB dynamics and $\tau_{\mathcal{R}+\mathcal{B}} \sim 13.85$ for the stochastic dynamics

plot the coarse-grained H functions corresponding to each ensemble: $\overline{H}$ represents the entropy associated with the full (red plus blue) ensemble, $\overline{H}_{red}$ and $\overline{H}_{blue}$ are respectively associated with the red and blue ensembles. As in the case of the dBB dynamics, the relaxation time is globally the same for the red, the blue, and the total ensemble. In Fig. 7.9, we compare the $\overline{H}$ functions obtained for the dBB dynamics and that obtained in the stochastic dynamics for $\alpha = 10^{-4}$. Even in the case where $\alpha \ll 1$ we observe an acceleration in the convergence for the stochastic dynamics, compared to its deterministic counterpart: it is almost 10% faster according to the values of the linear fit in Fig. 7.9. As we shall see in the next section, this effect is very sensitive to the initial distribution and gets amplified in the presence of microstructures.

7.4 Mixing and Homogenization in the Presence of Microstructures

7.4.1 Microstructures in the de Broglie–Bohm Dynamics

In the previous section, the initial positions were randomly distributed in each ensemble. Here, we study what happens to the relaxation process when the two ensembles exhibit organized microstructures. In Fig. 7.10 (Left), we illustrate a new

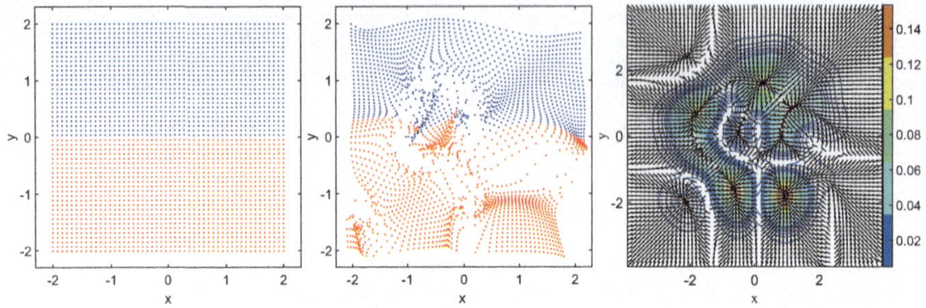

Fig. 7.10 We plot the initial distribution of points (Left), the distribution at time $t = 0.2$ (Middle), and the function $|\Psi|^2$ at $t = 0$ with its gradients (Right). The initial position distribution consists of 50×50 "clusters" of 20 points each, separated by an average distance of the order of $|\mathbf{d_0}| \sim 10^{-8}$ among them in each cluster. We choose $N_R = N_B = 50 \times 25 \times 20$

initial distribution for the red and the blue mixtures, in which the lower half of the domain $[-2, 2] \times [-2, 2]$ consists only of red points and the upper half only of blue ones. However, each such 'point' is, in reality, a cluster of points, with a microstructure made of 20 points, taken at random in a circle of radius $|\mathbf{d_0}| \sim 10^{-8}$ (the average distance between two points in the microstructure being then equal to $(4/\pi)d_0$ as can be shown by a straightforward calculation). In Fig. 7.10 (Middle), the clusters move along the dBB trajectories to areas where the density $|\Psi|^2$ is different from zero (Fig. 7.10 (Right)).

7.4.1.1 Remark: About Chaos, Homogenization, and Mixing in the dBB Dynamics

Intuitively, it is clear that if the dynamics are not sufficiently chaotic, these "clusters" will survive over time. This is illustrated in Fig. 7.11 where it is shown that when only $M = 2$ modes are present in (7.12), the H-function associated with the dBB dynamics does not decrease with time. It should be noted, however, that the number of modes in itself does not allow to decide whether or not equilibrium will be reached by the system. Indeed, a simple counter-example is provided by the coherent states of the $1D$ harmonic oscillator:

$$\Psi(x,t) = \left(\frac{1}{\pi}\right)^{\frac{1}{4}} e^{-\frac{1}{2}(x-\bar{x}_t)^2 + i\frac{\bar{p}_t x}{\hbar} + i\varphi(t)}, \tag{7.18}$$

where the variable φ is a global phase in which the energy is included and $\bar{x}_t$ (resp. $\bar{p}_t$) is the mean position (resp. momentum) of a classical oscillator at time t:

$$\bar{x}_t = \bar{x}_0 \, cos\,(\omega t) \quad \text{and} \quad \bar{p}_t = -m\omega\bar{x}_0 \, sin\,(\omega t)\,. \tag{7.19}$$

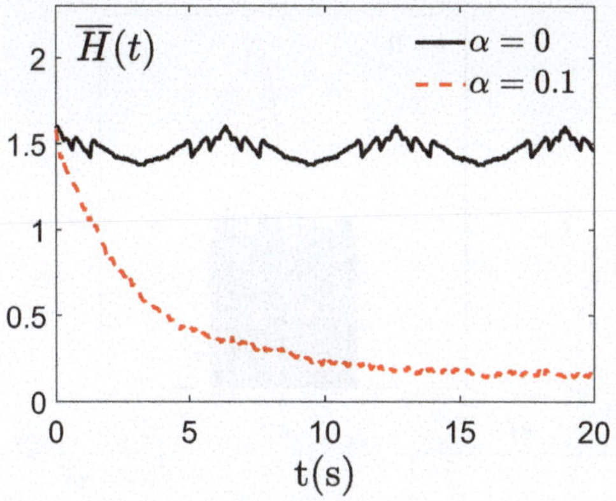

Fig. 7.11 In this plot we compare the H-functions obtained in the case of a superposition of just the two first modes of the harmonic oscillator (i.e. $\psi_{0,0}$ and $\psi_{0,1}$) for the de Broglie-Bohm dynamics ($\alpha = 0$) and for the Bohm-Hiley-Nelson dynamics (with $\alpha = 10^{-1}$). We considered for both dynamics the same initial ensemble of initial positions. We chose $N = 10{,}000$ initial points uniformly distributed in the domain $[-2, 2] \times [-2, 2]$

The velocity field can be shown to be equal to $v_{dBB} = -\omega\bar{x}_0 \, sin\,(\omega t)$, meaning that the dBB velocity is everywhere exactly the same as the velocity of the center of mass of the coherent state. Consequently, an arbitrary initial distribution of positions will oscillate over time around the origin $x = 0$ without deformation (see Fig. 7.12). Clearly, this system will never reach quantum equilibrium even for coherent states of high energy which count an arbitrarily high number of modes (energy modes are indeed Poisson distributed with a mean squared root deviation of the average number of photons in each mode equal to its mean value $< N >$, and an average energy equal to $(< N > +1/2)h\nu$).

Note, however, that the presence of noise changes the behavior of both systems. I.e. in the case of the BHN dynamics for the 2D harmonic oscillator (7.12) with only a few modes, as well as that for the coherent states (7.18), quantum equilibrium is reached in finite time (Hatifi et al. 2018).

For the dBB-dynamics, it is only in the case of *incoherent* superpositions of modes that an increase in the number of modes in the superposition of the pilot wave enhances the complexity of the dynamics (essentially by making the variation of the phase nontrivial which, in turn, increases the chaos as has been shown in Efthymiopoulos et al. (2017, 2009), Contopoulos et al. (2012), Abraham et al. (2014), Cesa et al. (2016).

7.4.1.2 Microstructures, Homogenization and Mixing in the dBB Dynamics

If we zoom in on an elementary volume of space, it will be seen that the clusters are always present up to a certain instant in time, after which they start to disappear.

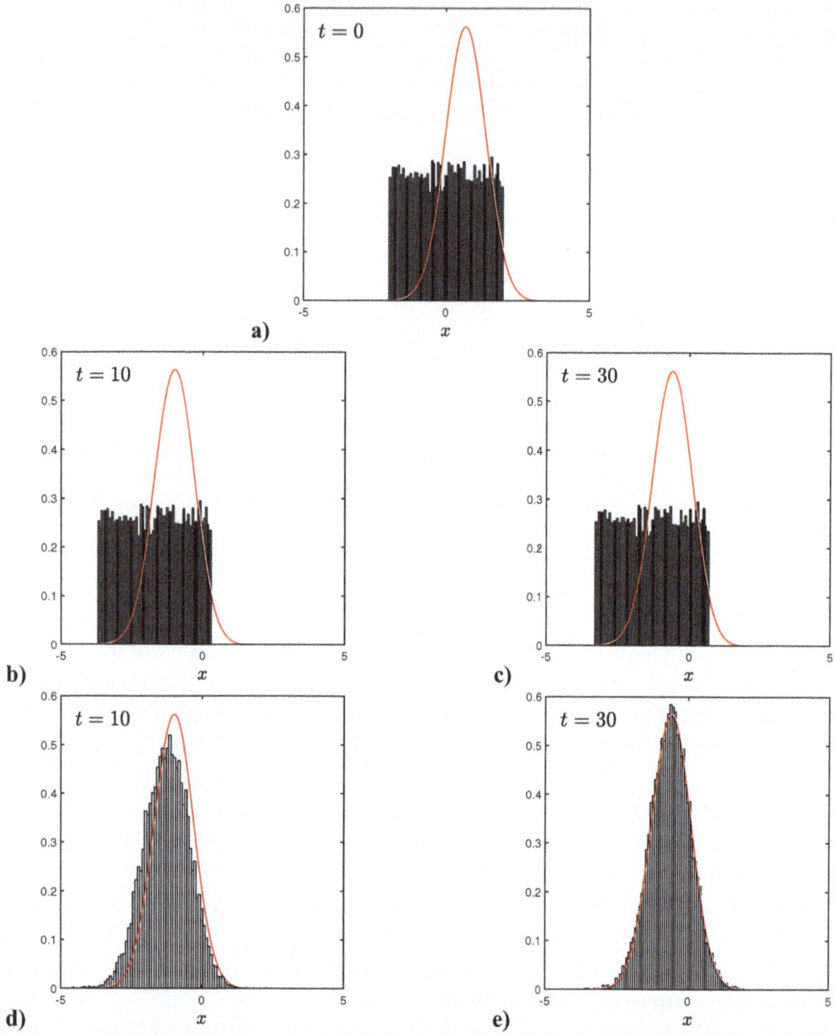

Fig. 7.12 Histograms of the positions of a non-equilibrium ensemble subject to the dBB dynamics (**b** and **c**) and to the BHN dynamics with $\alpha = 0.1$ (**d** and **e**) for the quantum coherent states (7.18). The full (red) curve corresponds to the quantum probability $|\Psi|^2$. Here we considered 10,000 initial conditions uniformly distributed in $[-4, 4]$ (see **a**))

Fig. 7.13a shows the evolution of the average distance $|\bar{\mathbf{d}}(t)|$ between the points of the microstructures inside each "cluster", for the case with $M = 16$ modes for the (incoherent) state (7.12). In the time interval $[0, \tau_{\mathcal{R}+\mathcal{B}}]$ (with $\tau_{\mathcal{R}+\mathcal{B}}$ as obtained in Fig. 7.9), $\ln\left(|\bar{\mathbf{d}}(t)|\right)$ is linear and we can estimate the time at which the clusters start to disappear (see Fig. 7.13b). This time is inversely proportional to the largest Lyapunov exponent, Λ, which characterizes the degree of chaos present in the

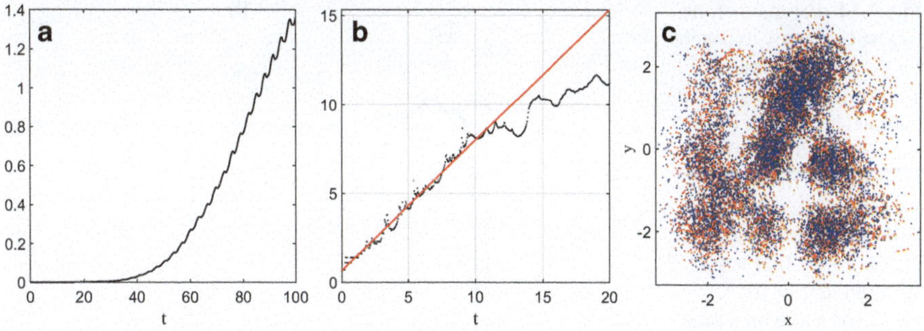

Fig. 7.13 This plot shows the evolution of the distances inside each "cluster"/microstructure in the case of the dBB dynamics. In (**a**) the averaged distance $|\overline{\mathbf{d}}(t)|$ between the points of the microstructures. (**b**) We plot here $\ln\left(|\overline{\mathbf{d}}(t)|/|\overline{\mathbf{d}_0}|\right)$, and we estimate the Lyapunov exponent as the slope of the full-line curve. (**c**) The distribution at time $t = 100$. We chose here 50×50 "clusters" of 20 points separated by an average distance $|\mathbf{d_0}| = 10^{-8}$ among them in each cluster. $N_R = N_B = 50 \times 25 \times 20$

dynamics: the largest Lyapunov exponent is related to the averaged "intracluster" distance $|\overline{\mathbf{d}}(t)|$ as

$$|\overline{\mathbf{d}}(t)| \sim |\overline{\mathbf{d}_0}|\, e^{\Lambda t}, \quad \text{so that} \quad \frac{\dot{d}(t)}{d(t)} \sim \Lambda. \tag{7.20}$$

If we estimate the Lyapunov time $\tau_\Lambda = 1/\Lambda$ with a linear fit in Fig. 7.13, we find an approximate value of $\tau_\Lambda \sim 1.4$. This estimate is compatible with the value $\Lambda = 0.6721$ computed using a standard algorithm (Benettin et al. 1976, 1980; Hatifi 2019) (see Fig. 7.14).

It is worth noting that this 'Lyapunov regime' during which the global dynamics in the system is correctly predicted by the value of the Lyapunov exponent, is limited in time: an exponential amplification of discrepancies in initial conditions is, de facto, limited in our case due to the confinement imposed by the harmonic potential. Therefore, the exponential growth saturates after a while, which can be observed at the level of the picture in the middle of Fig. 7.13. For the same reason, $|\overline{\mathbf{d}}(t)|$ will never reach values larger than more or less 2.5, as can be seen in Fig. 7.17.

7.4.2 Microstructures in the Bohm–Hiley–Nelson Dynamics

In Sect. 7.3.2, we have found that in the stochastic evolution, even a small diffusion coefficient induces a speed-up in the mechanism of homogenization. In this section, we identify the origin of this difference by looking at the evolution occurring at the microscopic level. Similar to the procedure in Sect. 7.4.1, we focus on the effect of

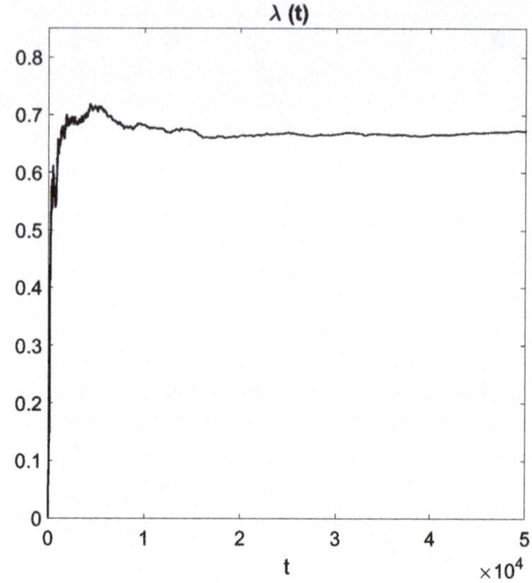

Fig. 7.14 Estimate of the largest Lyapunov exponent for the dBB dynamics according to the algorithm outlined in appendix 4.4 of Ref. Hatifi (2019). This estimate was made after averaging 150 orbits uniformly distributed $(x_0, y_0) \in [-2, 2] \times [-2, 2]$. We used $|\mathbf{d_0}| = 10^{-8}$, $t = 5.10^4$ and with a time step $\Delta t = 10^{-3}$

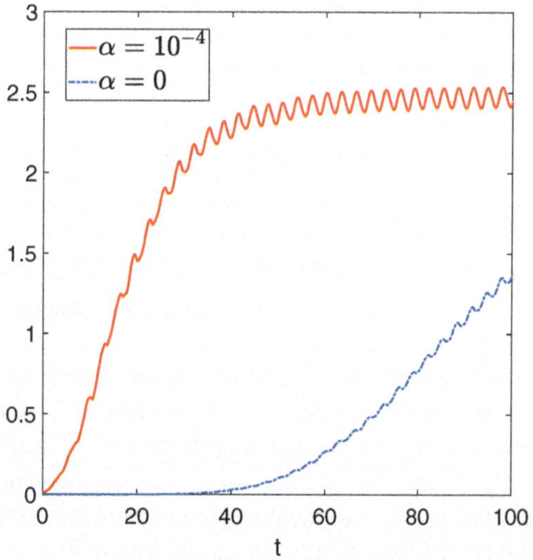

Fig. 7.15 This plot shows the evolution of the interdistance $|\overline{\mathbf{d}}(t)|$ of the microstructures inside each "cluster". The red full line curve corresponds to the stochastic dynamics, and the blue dotted curve corresponds to the dBB dynamics. We chose here 50×50 "clusters" of 20 points separated by a distance $|\mathbf{d_0}| \sim 10^{-8}$ among them in each cluster

microstructures on the relaxation process, but here in the framework of the (BHN) stochastic dynamics.

In Fig. 7.15 we plot the averaged distance $|\overline{\mathbf{d}}(t)|$ between the points in each microstructure under a stochastic BHN evolution. The distance $|\overline{\mathbf{d}}(t)|$ grows quickly compared to the dBB dynamics. The mechanism of homogenization is thus more efficient in the case of the stochastic dynamics, as pointed out in Sect. 7.3.2: even with a small diffusion coefficient ($\alpha = 10^{-4}$) the Brownian character of the evolution implies a quick homogenization of the microstructures.

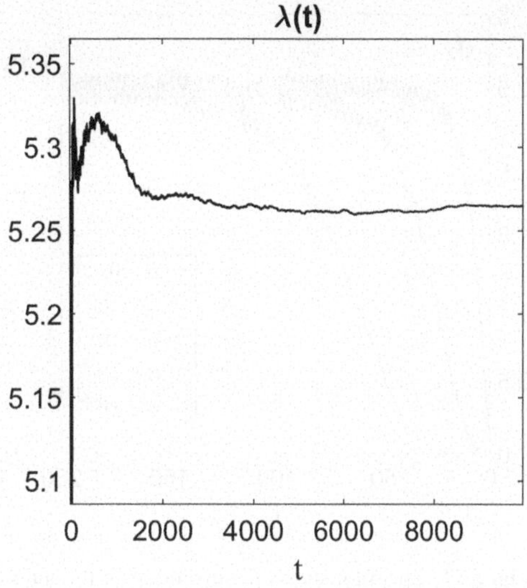

Fig. 7.16 Estimate of the largest Lyapunov exponent for the stochastic dynamics. This estimate was made after averaging 150 orbits uniformly distributed. We used $|\mathbf{d_0}| = 10^{-8}$, $t = 10^4$ and with a time step $\Delta t = 10^{-3}$

Visually, from Fig. 7.15, we can estimate the convergence time $\tau_{\mathcal{R}+\mathcal{B}}$ which is necessary for the clusters to grow to a size comparable to the size of the domain where the particle is confined.

To estimate the chaos in the BHN dynamics, we make use of the procedure developed in the case of the dBB dynamics (Sect. 7.4.1), which appeared to be well adapted in that case. The main result is encapsulated in Fig. 7.16 in which we plot the largest Lyapunov exponent computed following Benettin et al. (1976, 1980). The numerical value for the Lyapunov exponent converges to $\Lambda \sim 5.26$. The Lyapunov time τ_Λ is then given by $\tau_\Lambda \sim 1/\Lambda \sim 0.19$, which is almost ten times smaller than that for the dBB dynamics.

A direct estimate based on Fig. 7.15 shows, however, that the Lyapunov exponent is ill-suited for characterizing the dynamics in the presence of the BHN noise. Actually if we try to fit the red full line curve of Fig. 7.15 with an exponential curve (as we would for a Lyapunov regime as we found for the de Broglie-Bohm case) or with a curve that is quadratic in time (purely Brownian motion) we fail.[3] The dynamics in the BHN case is thus clearly different. Although the precise details of this dynamics remain to be investigated, it is apparent that the spread of neighboring trajectories is much faster, at least initially, than one would expect if the underlying mechanism were purely chaotic as it is in the dBB case. One tentative explanation might be that the combination of Brownian motion and chaos, which appears during the passage close to a zero of the wave function, actually accelerates the behavior, even for a very small amount of noise ($\alpha = 10^{-4}$ in

[3] Purely Brownian motion would lead to a spread at time t of the order of $\sqrt{\alpha \cdot t}$. Taking $\alpha = 10^{-4}$ and $t = 40$, we should find a spread of the order of $6{\cdot}10^{-2}$, which is quite smaller than what we can see at the level of Fig. 7.15.

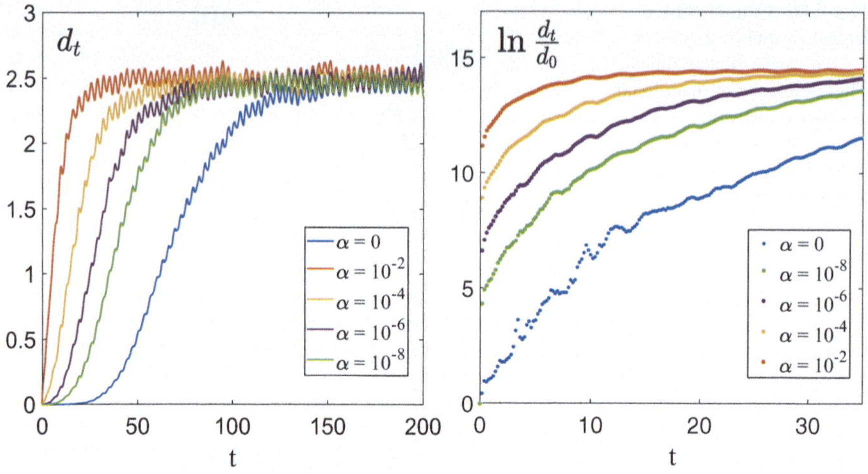

Fig. 7.17 This plot shows the evolution of the inter distance $|\bar{\mathbf{d}}(t)|$ of the microstructures inside each "cluster" of two points. Here the initial distance between the two points is $\mathbf{d}_0| = 10^{-6}$. α=0 corresponds to the dBB dynamics

the present case). Our observation is confirmed by the plots in Fig. 7.17, which clearly show that exponential growth in time (*à la* Lyapunov) only occurs and then, again, only initially, in the absence of noise ($\alpha = 0$). A kind of inflationary process (non-exponential but very sudden and violent) appears to occur for very short times whenever $\alpha \neq 0$.

7.5 Convergence to the Bernoulli Statistics: A New Measure of Mixing

In Sect. 7.3, we looked at the statistics at the coarse-grained level. Here, we pay particular attention to the statistics at the level of the microstructures, with the aim of introducing a measure of mixing and homogenization which does not depend on any coarse-graining. To do so, we recorded all the trajectories ($N = N_B + N_R$) corresponding to each ensemble $\mathcal{R}$ and $\mathcal{B}$ of Fig. 7.5, for ensembles with and without microstructures. We then projected all the trajectories along one axis (for example, the y-axis) and computed the probability of having a particular sequence of blue (n_B) and red (n_R) points in a sub-interval of $n = n_R + n_B$ points (see Fig. 7.18).

We illustrate in Fig. 7.19a, b the *occurrence probability* of having $\{n_B = 1, n_R = 1\}$ in a sub-interval $n = 2$, without and with clusters of 20 microstructures (with $d_0 = 10^{-8}$), as defined and used in the previous sections. The plot on the r.h.s. of this figure shows the fundamental difference between the deterministic and the stochastic dynamics in the presence of the microstructures. In fact, the dBB

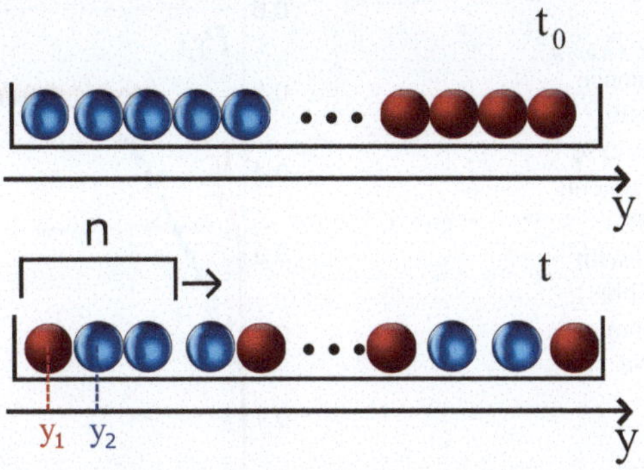

Fig. 7.18 At $t = t_0$ two ensembles of initial positions are placed on both sides of a box. The ensembles evolve in time according to the dBB (or stochastic) dynamics. During the evolution, we look at the sequence of blue/red points in sub-intervals of n points

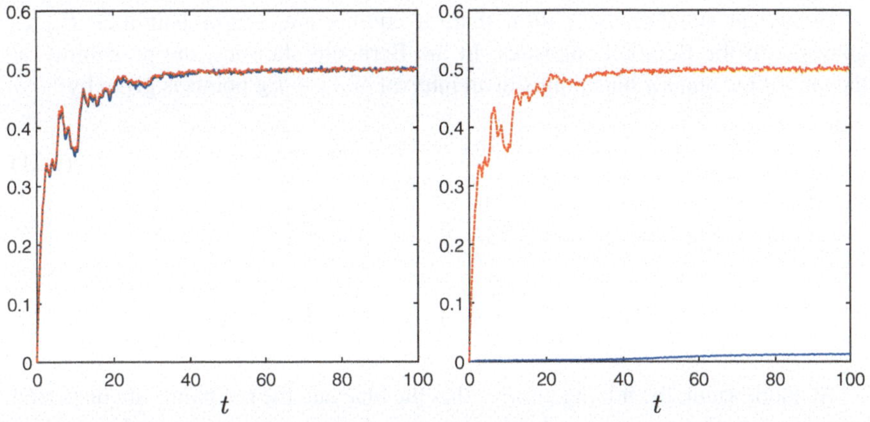

Fig. 7.19 Plot of the occurrence probability P_{n_B,n_R} of having a sequence of $\{n_B = 1, n_R = 1\}$: (Left) without microstructures and (Right) with clusters of 20 microstructures and with a spacing $|\mathbf{d_0}| = 10^{-8}$. The dBB dynamics is associated with the full line curve (blue) and the stochastic dynamics to the dashed line (red). We chose here $\alpha = 10^{-4}$

evolution, contrary to the stochastic dynamics, is strongly affected by the presence of the microstructures because of the conservation law (7.8). When the spacing inside the microstructures d_0 increases, the mixing occurs more quickly, as can be seen in Fig. 7.20 where $P_{1,1}$ is plotted (in this case, we chose pairs of points with a spacing $d_0 = 10^{-6}$). This figure has to be compared with Fig. 7.17 in which the initial spacing d_0 is the same. It indicates that the probability $P_{1,1}$ saturates after a time that is the same as that required to saturate the distance due to the confinement imposed by the potential well (roughly at $t = 50$ for the BHN dynamics and at $t = 100$ for the dBB dynamics).

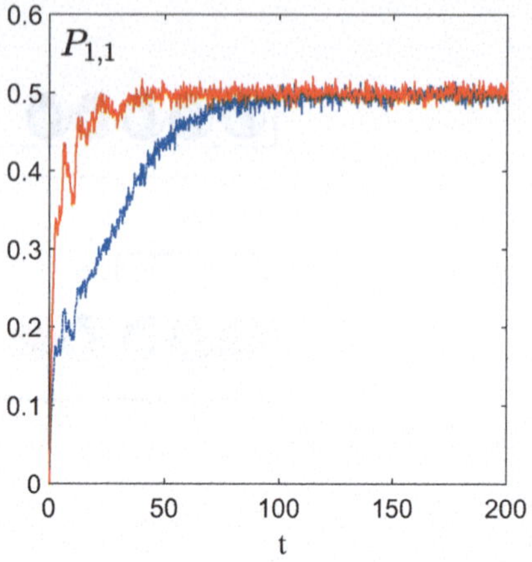

Fig. 7.20 Plot of the occurrence probability P_{n_B,n_R} (7.21) of having a sequence of $\{n_B = 1, n_R = 1\}$ with microstructures (in this case we chose pairs with a spacing $|\mathbf{d_0}| = 10^{-6}$). The dBB dynamics is associated with the dashed line curve (blue) and the stochastic dynamics to the full line (red). We chose here $\alpha = 10^{-4}$

Numerical simulations confirm that, at equilibrium, the probabilities P_{n_B,n_R} converge to the Bernoulli statistics. In the Bernoulli statistics, the probability of having n_R red and n_B blue points in an interval of $n_R + n_B$ points is given by

$$P_{n_B,n_R} = \frac{(n_B + n_R)!}{n_B!n_R!} P_R^{n_R} P_B^{n_B} \tag{7.21}$$

or in terms of a parameter λ defined by $N_B = \lambda N_R$:

$$P_{n_B,n_R} = \frac{(n_B + n_R)!}{n_B!n_R!} \left(\frac{1}{1+\lambda}\right)^{n_R} \left(\frac{\lambda}{1+\lambda}\right)^{n_B} \tag{7.22}$$

At equilibrium, the mixing ensures that the blue and the red points are dispersed in such a way that the most probable sequence of points in a sub-interval $n = n_R + n_B$ respects the ratio $n_B/n_R = \lambda$, on average.

To take the analysis one step further, we plot in Fig. 7.21 the histograms of P_{n_B,n_R} at the equilibrium (at $t = 100$) for both the dBB and the BHN dynamics for different combinations $\{n_B, n_R\}$.

Obviously, the plots 23 a, b, and d are indistinguishable from the Bernoulli statistics plotted in Fig. 7.22.

Comparing (quantitatively) the difference between the probability distributions of Fig. 7.21 with the Bernoulli statistics, we find

$$\sum_{n_R=0}^{5}\sum_{n_B=0}^{5} |P_{n_B,n_R} - P^{dBB}_{n_B,n_R}| = 0.07 \quad \text{and} \quad \sum_{n_R=0}^{5}\sum_{n_B=0}^{5} |P_{n_B,n_R} - P^{BHN}_{n_B,n_R}| = 0.06, \tag{7.23}$$

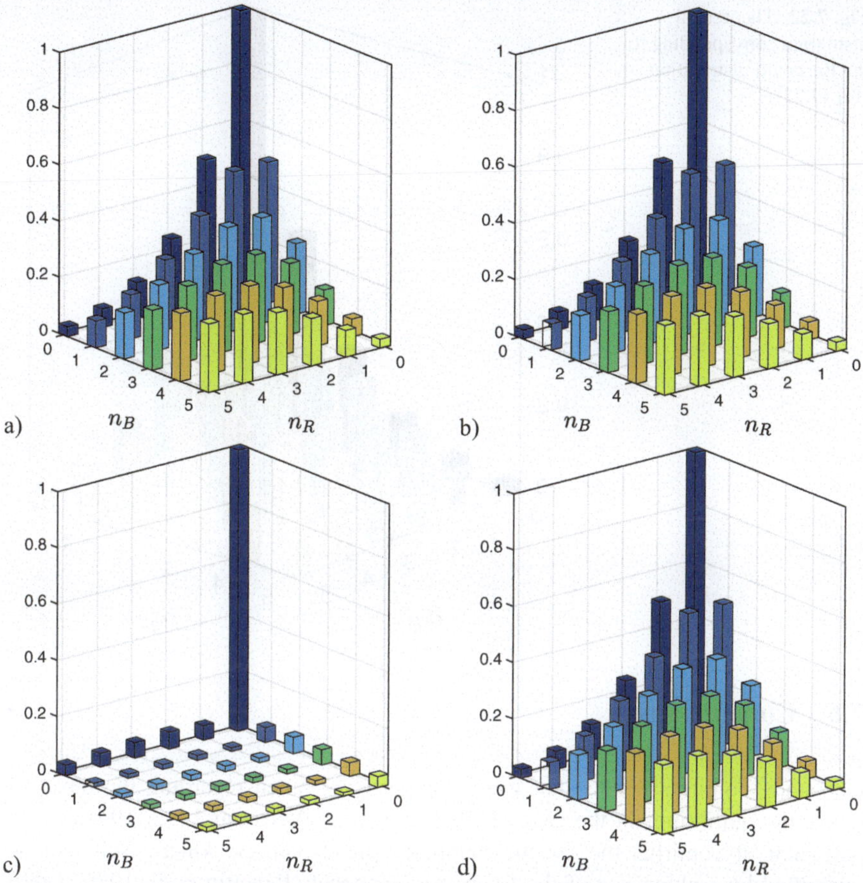

Fig. 7.21 Probability diagrams P_{n_B,n_R} at $t = 100$ and $|\mathbf{d_0}| = 10^{-8}$ for: (**a**) dBB (without microstructures), (**b**) Stochastic (without microstructures), (**c**) dBB (with microstructures), (**d**) Stochastic (with microstructures)

without microstructures and

$$\sum_{n_R=0}^{5}\sum_{n_B=0}^{5}|P_{n_B,n_R} - P^{dBB}_{n_B,n_R}| = 7.40 \quad \text{and} \quad \sum_{n_R=0}^{5}\sum_{n_B=0}^{5}|P_{n_B,n_R} - P^{BHN}_{n_B,n_R}| = 0.04, \tag{7.24}$$

with microstructures (in the case of an initial spacing $d_0 = 10^{-8}$ as in Fig. 7.20).

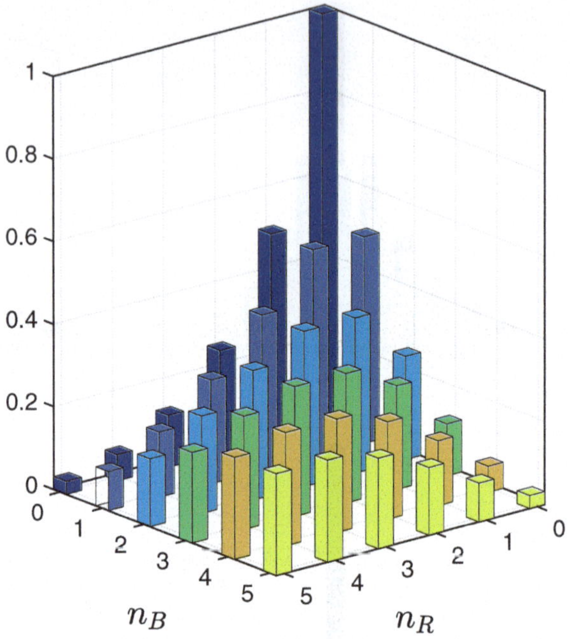

Fig. 7.22 Theoretical histogram corresponding to the Bernoulli distribution (Eq. (7.22))

7.6 Conclusions

Our study confirms results that were previously obtained regarding the onset of quantum equilibrium in the case of dBB dynamics (Valentini 1992, 1991a,b). In particular, it confirms the chaotic nature of the dynamics, which itself can be quantified by making use of the Lyapunov exponent (Benettin et al. 1976, 1980) (Sect. 7.4.1).

In the case of the dBB dynamics, it confirms that, resulting from the reversibility of the dynamics, the fine-graining of the initial distributions plays a crucial role regarding the onset of equilibrium, as we emphasized in Sect. 7.2.5. In particular, our analysis highlights the role played by **microstructures** in the initial distribution regarding the onset of quantum equilibrium. In Sect. 7.4.1, we explained this role in relation to the Lyapunov exponent, which is known to play a prior role in the onset of quantum equilibrium in dBB dynamics (Efthymiopoulos et al. 2017; Contopoulos et al. 2012; Efthymiopoulos et al. 2009). We confirmed our analysis by the introduction of a new measure based on Bernoulli statistics (Sect. 7.5). In the presence of noise, however (BHN dynamics), our results show that the Lyapunov analysis is largely irrelevant when microstructures are present: the Lyapunov analysis does not allow us to quantify the emergence of the quantum equilibrium (Sect. 7.4.2). In this case (i.e. for stochastic dynamics), our results rather show that the convergence to equilibrium becomes insensitive to the initial distribution after some time due to the sudden and violent blurring of initial microstructures, which get, in a sense, dislocated by the BHN noise through a non-exponential but very intense kind of inflationary process.

The two main lessons of our paper are the extreme sensitivity of the time necessary for the onset of equilibrium

- to the initial distribution in the deterministic case (Sect. 7.4.1) and
- to the presence of noise in the stochastic case (Sect. 7.4.2) as confirmed by the use of a measure based on Bernoulli statistics (Sect. 7.5).

It is worth insisting on the fact that it is only at the level of microstructures that a clear distinction appears between the dBB and the BHN dynamics. When initial distributions are coarse-grained, the effects of those dynamics are nearly indistinguishable as we can see from Fig. 7.23 where we compare the averaged fraction number of the red mixture (7.16) and the corresponding entropy of mixing (7.17) for both dynamics (see also Fig. 7.19, Left). The mixture $\overline{n}_r$ converges slightly faster to the ratio $N_R/(N_R + N_B)$ in the case of the stochastic dynamics compared to its deterministic counterpart but this effect is qualitatively weak (in accordance with plots a and b of Fig. 7.21), contrary to what happens when microstructures are present (as can be seen for instance from the comparison of the plots c and d of Fig. 7.21).

The appearance of an inflationary-like burst of initial inhomogeneities of the function $\mathcal{F}$ in the presence of a stochastic noise opens interesting and promising perspectives regarding the onset of quantum equilibrium just after the Big Bang, which has been scrutinized by several authors in the past (Colin and Valentini 2013; Valentini 2010; Underwood and Valentini 2015; Pinto-Neto and Struyve 2019), in the framework of dBB dynamics. BHN dynamics provides a consistent

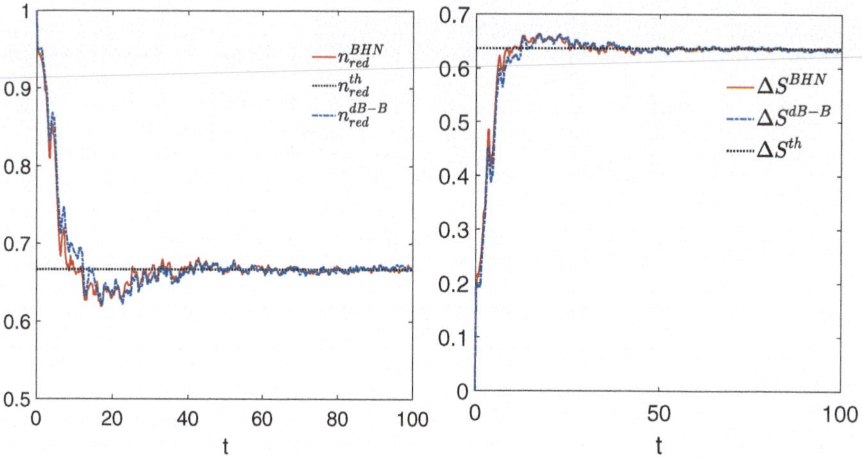

Fig. 7.23 This plot shows the evolution of the average red mixture $\overline{n}_r$ (Left) and the entropy of mixing ΔS for the stochastic evolution (7.5) (Right). As before: $N_R = 200{,}000$ and $N_B = 100{,}000$. No microstructure is present here. The function $\overline{n}_r$ is converging to $\overline{n}_r^{th} = 2/3$

model for introducing noise at the level of dBB dynamics, which drastically affects, as we have shown here, the onset of quantum equilibrium in presence of fine-graining/microstructures.

Acknowledgments The authors gratefully acknowledge funding and support from the John Templeton Foundation (grant 60230, Non-Linearity and Quantum Mechanics: Limits of the No-Signaling Condition, 2016–2019) and a FQXi Physics of what happens grant (Quantum Rogue Waves as Emergent Quantum Events, 2015–2017). RW would also like to thank FQXi for support through a mini-grant (Grant number FQXi-MGA-1819), which enabled him to elucidate crucial aspects of the relaxation to quantum equilibrium.

We are grateful to Samuel Colin, MH in particular, who spent two months in Rio de Janeiro in order to learn new simulation techniques under his supervision in the framework of the aforementioned project Non-Linearity and Quantum Mechanics: Limits of the No-Signaling Condition. The Bohmian trajectories as well as the measures of the Kullback-Leibler divergence were, for instance, all simulated with a code originally conceived by S. Colin (see also Refs. Colin and Valentini (2013), Colin and Valentini (2014), Hatifi (2019), Hatifi et al. (2018)).

References

Abraham, E., S. Colin, and A. Valentini. 2014. *Journal of Physics A: Mathematical and Theoretical* 47: 395306.
Bacciagaluppi, G. 1999. *Foundations of Physics Letters* 12: 1. ISSN 0894-9875.
Bacciagaluppi, G. 2005. In *Endophysics, Time, Quantum and the Subjective*, ed. R. Buccheri, M. Saniga, and A. Elitzur, 367–388. Singapore: World Scientific.
Benettin, G., L. Galgani, and J.-M. Strelcyn. 1976. *Physical Review A* 14: 2338.
Benettin, G., L. Galgani, A. Giorgilli, and J.-M. Strelcyn. 1980. *Meccanica* 15: 9.
Bohm, D. 1952a. *Physical Review* 85: 166.
Bohm, D. 1952b. *Physical Review* 85: 180.
Bohm, D., and B. Hiley. 1989. *Physics Reports* 172: 93.
Bohm, D., and J.-P. Vigier. 1954. *Physical Review* 96: 208.
Cesa, A., J. Martin, and W. Struyve. 2016. *Journal of Physics A: Mathematical and Theoretical* 49: 395301.
Couder, Y., S. Protière, E. Fort, and A. Boudaoud. 2005. *Nature* 437: 208.
Couder, Y., A. Boudaoud, S. Protière, and E. Fort. 2010. *Europhysics News* 41: 14.
Colin, S. 2012. *Proceedings of the Royal Society A* 468: 1116.
Colin, S., and W. Struyve. 2010. *New Journal of Physics* 12: 043008.
Colin, S., and A. Valentini. 2013. *Physical Review D* 88: 103515.
Colin, S., and A. Valentini. 2014. *Proceedings of the Royal Society* A: 20140288.
Contopoulos, G., N. Delis, and C. Efthymiopoulos. 2012. *Journal of Physics A: Mathematical and Theoretical* 45: 165301.
deBroglie, L. 1927. *Journal de Physique et Le Radium* 8: 225.
de la Peña, L., A.M. Cetto, and A. Valdés Hernández. 2015. *The Emerging Quantum*. Berlin: Springer.
Efthymiopoulos, C., C. Kalapotharakos, and G. Contopoulos. 2009. *Physical Review E* 79: 036203.
Efthymiopoulos, C., G. Contopoulos, and A.C. Tzemos. 2017. *Annales de la Fondation Louis de Broglie* 42: 133.
Fermi, E. 1956. *Thermodynamics*. Mineola: Dover Publications.
Guerra, F. 1995. *The Foundations of Quantum Mechanics*, 339–355. Dordrecht: Kluwer Academic Publishers.

Hatifi, M. 2019. *Beyond Pilot Wave Dynamics: Non-linearity and Non-equilibrium in Quantum Mechanics*. http://www.theses.fr/2019ECDM0006.
Hatifi, M., R. Willox, S. Colin, and T. Durt. 2018. *Entropy* 20: 780.
Jüngel, A. 2016. *Entropy Methods for Partial Differential Equations*. Springer Briefs for Mathematics. Berlin: Springer.
Kyprianidis, P. 1992. *Foundations of Physics* 22: 1449.
Nelson, E. 1967. *Mathematical Notes* 131: 2381.
Norsen, T. 2018. *Entropy* 20: 422.
Perrard, S., M. Labousse, M. Miskin, E. Fort, and Y. Couder. 2014. *Nature Communications* 5: 3219.
Petroni, N.C. 1995. *Quantum Communications and Measurement*, 43–52. Berlin: Springer.
Petroni, N.C., and F. Guerra. 1995. *Foundations of Physics* 25: 297.
Pinto-Neto, N., and W. Struyve. 2019. In *Applied Bohmian Mechanics: From Nanoscale Systems to Cosmology*, ed. X. Oriols and J. Mompart, 2nd ed., 607–656.
Towler, M., N.J. Russell, and A. Valentini. 2011. *Proceedings of the Royal Society A* 468: 990.
Underwood, N.G., and A. Valentini. 2015. *Physical Review D* 92: 063531.
Valentini, A. 1991a. *Physics Letters A* 156: 5.
Valentini, A. 1991b. *Physics Letters A* 158: 1.
Valentini, A. 1992. Ph.D. thesis, SISSA.
Valentini, A. 2010. *Physical Review* D 82: 063513.
Valentini, A., and H. Westman. 2005. *Proceedings of the Royal Society A* 461: 253.

Chapter 8
Non-quantum Behaviors of Configuration-Space Density Formulations of Quantum Mechanics

Philipp Roser and Matthew T. Scoggins

Abstract The trajectories of the pilot-wave formulation of quantum mechanics and hence its empirical predictions may be recovered via the dynamics of a density function on the configuration space of a system, without reference to a physical wave function. We label such formulations '*CSD* frameworks.' But this result only holds if a particular, apparently *ad hoc* condition, broadly speaking equivalent to the single-valuedness of the wave function in standard quantum mechanics, is imposed. Here we relax this condition. We describe the types of scenarios in which this would lead to deviations from quantum mechanics. Using computational models we ask how the degree of 'non-quantumness' of a state, suitably defined, changes with time. We find that it remains constant in time even under non-trivial dynamics, and argue that this implies that a *dynamical* justification of the Wallstrom condition is unlikely to be successful. However, we also make certain observations about stationary states in CSD frameworks, which may offer a way forward in justifying the Wallstrom condition.

Keywords Configuration-space density formulations · CSD frameworks · Hamilton-Jacobi Many Worlds · Hydrodynamical approach · Wallstrom condition

8.1 Introduction

Quantum mechanics is well established as a framework of extraordinary empirical success. Its conceptual foundations however remain a matter of debate. The challenge to provide a physical interpretation of the mathematical structure of quantum theory has engendered a great variety of "interpretations", "models" and

P. Roser (✉) · M. T. Scoggins
Department of Physics and Astronomy, Western Washington University, Bellingham, WA, USA
e-mail: roserp@wwu.edu; scoggim@wwu.edu

P. Castro et al. (eds.), *Advances in Pilot Wave Theory*, Boston Studies in the Philosophy and History of Science 344,
https://doi.org/10.1007/978-3-031-49861-9_8

"approaches" over the last one hundred or so years. The literature on these questions is vast.[1]

In at least one such family of approaches (Tipler 2010; Sebens 2014; Hall et al. 2014; Schmelzer 2011a; Smolin 2015; Boström 2015; Roser 2015) the wave function is not considered a physically fundamental entity. Instead, a continuous[2] ensemble of interacting "classical-like" systems of varying density in configuration space forms the basic ontological foundation, together with a field of configuration-space velocities. From these the wave function can be reconstructed mathematically (up to one single degree of freedom, the global phase). This ensures compliance with the PBR theorem (Pusey et al. 2012), for example. Different authors have referred to these approaches variously by "Hamilton-Jacobi Many-World Theory" (Tipler 2010), "Many Interacting Worlds" (Hall et al. 2014), and other names, although confusion with the conceptually very different many-worlds theory of Everett (1956) needs to be avoided. For this reason we will label such approaches simply as configuration-space density (CSD) frameworks. The origins of this type of foundation can be traced back to (incomplete) ideas of Madelung's "hydrodynamical" approach (Madelung 1927). A more detailed summary of the similarities and differences between frameworks within this family is found elsewhere (Roser 2015).

CSD approaches all share one feature: In order to recover the phenomenology of quantum mechanics, a quantization condition must be added "by hand", namely the condition that the line integral of the velocity field in configuration space around any closed loop can only take discrete values. This condition is key for the quantized nature of angular-momentum values, for example. In other frameworks this condition is guaranteed by the single-valuedness of the wave function, which is absent in CSD approaches. This necessary *ad hoc* addition detracts from the conceptual appeal of CSD theories.

In this paper we consider the consequences of relaxing this condition in the context of a CSD approach to non-relativistic quantum mechanics. We ask if the "non-quantumness" of this generalized framework actually poses an empirical problem, or if it in some way disappears dynamically. We find that at least in the context of fairly simple systems the problem does *not* disappear dynamically and the challenge posed to CSD-type models remains.

In Sect. 8.2 we summarize the CSD framework and make some general observations. In Sect. 8.3 we explain the problem of quantized values in situations with periodic boundary conditions. We also identify in what sort of systems the problem arises and highlight some more subtle aspect of the issue. In Sect. 8.4 we describe

[1] No reasonably sized list of references would do this justice. For an overview of traditional approaches, see e.g. Wallace (2008). A reader looking for a broad exposure to different approaches may also find Schlosshauer (2011) useful.

[2] Some authors (Sebens 2014; Hall et al. 2014) however argue that the uncountably infinite number of systems in configuration space is only an approximation to a dense but countable number of systems for reasons we will not critique here. The resulting theory would be experimentally distinguishable from standard quantum theory, at least in principle.

how one may approach an investigation of the wider dynamics that CSD allows, and in Sect. 8.5 we present the specific case study of the 2D harmonic oscillator. Section 8.6 constitutes a discussion of stationary states in CSD frameworks, which may yet provide insights with regards to the condition. We conclude in Sect. 8.7.

8.2 Quantum Mechanics as a Quantum-Hamilton-Jacobi Theory of an Ensemble

A framework that reproduces the empirical results of (non-relativistic) quantum mechanics may be formulated as follows. Consider a classical n-dimensional configuration space that represents the degrees of freedom of the system. For concreteness, consider a system of N particles of mass m_i, $i \in \{1, ..., N\}$ in three-dimensional space, $n = 3N$, although much of our discussion generalizes freely to other systems. The Schrödinger equation for the wave function ψ of this system may be split into its real and imaginary parts, giving the quantum Hamilton-Jacobi equation and a continuity equation for the system:

$$-\frac{\partial S}{\partial t} = \sum_{I=1}^{N} \frac{(\nabla_I S)^2}{2m_I} + V(x) + Q[|\psi|] \tag{8.1}$$

$$\frac{\partial |\psi|^2}{\partial t} + \nabla \cdot (|\psi|^2 \nabla_I S/m_I) = 0 \tag{8.2}$$

where $Q = -\sum_{I=1}^{N} \frac{\hbar^2}{2m_I}(\nabla_I^2|\psi|)/|\psi|$ is the quantum potential. Here V is the potential function describing this system and S is the phase of ψ. This constitutes the entry point to Bohm's second-order formulation of pilot-wave theory (Bohm 1952; Holland 1993). If a large number of identical systems described by these equations form a density ρ in configuration space, one can readily show that *if* at some initial time t_0 it is the case that $\rho = |\psi|^2$ and that the configuration-space velocity of the system is $\vec{v}_i = \nabla_i S/m_i$, *then* the equality $\rho = |\psi|^2$ will continue to hold at all times, and the empirical predictions of quantum mechanics can be recovered. There are, of course, various wider conceptual issues concerning the interpretation of probability that may be relevant to a more thorough discussion here, but we leave these aside in the present paper and instead refer the reader to the extensive literature on the subject.

The question arises how to explain the equality $\rho = |\psi|^2$ for some initial time. Some authors have proposed typicality arguments (Dürr et al. 1992). However, it has also been shown theoretically (Valentini 1991) and through extensive simulation (Valentini and Westman 2004) that the condition $\rho = |\psi|^2$ constitutes a thermodynamic-like equilibrium: If $\rho \neq |\psi|^2$ at some early time, then under suitably general conditions ρ *approaches* $|\psi|^2$. It may therefore be suggested that if ρ (for some type of system, or the universe) was in this quantum *non*-equilibrium at some point in the distant past, ρ today would be in practice experimentally indistinguishable from exact equality with $|\psi|^2$. It should be noted that the velocity condition $\vec{v}_I = \nabla_I S/m_i$ on the other hand is unstable and deviations from it at some initial time will lead to chaotic behaviour (Colin and Valentini 2014).

Here we consider a different proposal. We can eliminate any reference to ψ entirely by redefining the quantum potential term as $Q = -\sum_{i=I}^{N} \frac{\hbar^2}{2m_I}(\nabla^2\sqrt{\rho})/\sqrt{\rho}$ and re-interpreting S as a Hamilton principal function analogue for this system. This constitutes the dynamics in conjunction with the continuity equation for ρ. Up to a subtlety whose discussion will constitute much of the present paper, the evolution of the system will remain unchanged from the (equilibrium) pilot-wave prediction. The fundamental dynamical quantities in this framework are the configuration-space density ρ and the velocity field $\vec{v}_i$ (also defined on configuration space and itself a $3N$-dimensional vector field). A quantity $\psi \equiv \sqrt{\rho}e^{iS}$ can be reconstructed mathematically up to a single global phase factor from knowledge of ρ and the spatial gradient of S in the form of the velocity field. The global phase factor is physically irrelevant even in quantum mechanics, so its disappearance poses no issue.

Formulations along these lines with varying details have been proposed by a number of authors (Tipler 2010; Sebens 2014; Hall et al. 2014; Schmelzer 2011a; Boström 2015; Roser 2015; Poirier 2010), and have been employed for their convenience in numerical methods (see Wyatt (2005) for a comprehensive overview). The conceptual foundation can arguably be traced back to Madelung's (incomplete, three-space based) "hydrodynamic" approach.

The absence of the wave function does not imply that this theory is "classical," except in the most misleading sense. The dependence of Q on the configuration-space ensemble density ρ has no analogue in classical mechanics. The local configuration-space ensemble density affects motion of a single system in this ensemble. This implies that the other elements of the ensemble must themselves be real constituents of nature, not merely mathematical artifacts. Each point in the configuration space is a "universe" or "world" that is equal in ontological status to ours. As a result, some (Hall et al. 2014) have dubbed this formulation "many interacting worlds." Care should be taken to distinguish it from the conceptually very different Everettian many-worlds approach (Everett 1956). We will use the term *actual world* in order to distinguish the configuration-space position and trajectory we "inhabit."

Aside: "Ensemble" in this context does not mean the same as the term in thermodynamics, for example, where an "ensemble" refers to a large number of identical systems *in the actual world*, which are not interacting (except possibly through a classical potential) and are described by mathematically identical configuration spaces, which allows their collection to be represented as a density in a single configuration space. The density itself is however irrelevant for the calculation of a single trajectory, in stark contrast with the dynamics of the formulation of quantum mechanics discussed here.

The existence of the quantum potential Q (as a function of configuration-space density rather than of wave function amplitude) can be motivated starting from classical mechanics by imposing the condition that no two configuration-space trajectories intersect, or similarly that the Hamilton Principal Function for the system is differentiable. This condition itself may be motivated by more fundamental conditions, such as determinism (Tipler 2010). We will not critique this discussion here.

One further remark: Rather than beginning with modification of the Hamilton-Jacobi equation, one can introduce the "force" $-\nabla_i Q$ at the level of Newton's Law of Motion. The crucial difference is that by doing this we do not assume that the velocity field in configuration space at some initial time t_0 is equal to the gradient of some function S. That is, we do not assume the existence of a single Hamiltonian Principal Function for all trajectories. This therefore allows for a much wider range of initial conditions for the set of configuration-space trajectories. The situation here also differs from classical mechanics where non-actual trajectories are not presumed physically real, so S is constructed to determine the evolution of our actual system only. The mathematical existence of other trajectories evolving according to the same function S in classical mechanics is, in a sense, an afterthought. Furthermore, in contrast with the present framework, the shape of the other trajectories in classical mechanics has no consequence for the configuration-space motion of the actual system.

8.3 Wallstrom's Objection and Responses, and the Relevance of Nodes

In the standard quantum formalism, as well as in the Pilot-Wave formulation, discrete sets of values ("quantum numbers") arise for a large range of systems. Textbook examples include the discrete energy levels of a potential well and the possible angular momentum values of a hydrogen atom. In the first of these two cases the discreteness arises because of boundary conditions imposed on the wave function at the edges of the well, while in the second example, crucially, the *single-valuedness of the wave function* forces the phase to undergo a change that is an integer multiple of 2π around any closed path (e.g. Holland (1993), although the fact was first noted by Dirac (1931)),

$$\oint_{\text{any loop}} \nabla S \cdot d\vec{\ell} = n \cdot 2\pi\hbar, \qquad n \in \mathbb{Z}. \tag{8.3}$$

Note that the weaker condition that ∇S be single-valued does not imply that n be an integer. Therefore, if the configuration-space density ρ together with the velocity field $\vec{v}$ are taken to be the fundamental variables, imposing the above condition is unmotivated, except *a posteriori* in order to reproduce (albeit not explain) experimental observation.

That this requirement needs to be added "by hand" in the configuration-space density based approach was prominently pointed out by Wallstrom (1994) as a criticism of Madelung's hydrodynamic formulation. We will therefore refer to condition (8.3) in this context as the *Wallstrom condition*. However, the point had been raised significantly earlier by Takabayasi (1952).

Starting in Sect. 8.4 we will consider the consequences of *not* imposing the Wallstrom condition. For the remainder of the present section we will briefly review how some authors have addressed the issue and then make some more detailed observations about the condition.

Some works proposing a formulation based on configuration-space density fail to address the issue entirely. Others add the Wallstrom condition by hand, acknowledging its *ad hoc* nature (e.g. Sebens (2014)), perhaps suggesting that the current theory is only an effective theory and that a more fundamental one may exist from which the Wallstrom condition would follow naturally.

Among explicit attempts to address the issue are the following. Schmelzer (2011a,b) proposes a regularity condition for the Laplacian of ρ and shows that this is mathematically equivalent to demanding that the wave function only have simple zeros, so that the loop integral around any single node is $2\pi\hbar$ exactly. This deviates from the predictions of standard quantum mechanics, which is taken to be an approximation, raising a number of other questions.

Smolin's proposal (Smolin 2015) based on a "principle of maximum variety" differs significantly from the others in that it concerns the dynamics of similar systems in our *actual world* and would be experimentally distinguishable from standard quantum mechanics, at least in principle, especially for systems with a small number of identical copies in the universe. We will not discuss its merits in the present paper, but only note that it too encounters the need to explain the Wallstrom condition. The work proposes the existence beables with phase-like properties that take the role of momenta in order to do so.

Elsewhere Roser (2015) considered an action integral in which the quantum potential arises as a curvature term in a Weyl geometry (on configuration space), based on earlier insights by Santamato (1984a,b). Here the function S is introduced by adding the most general possible total time derivative to the Lagrangian. Doing so does not change the equations of motion (Goldstein et al. 2000). The time derivative that is added must be $\mathbb{R}^1$. However, $\mathbb{R}^1$ is the tangent space of *two* possible spaces, $\mathbb{R}^1$ itself and the closed space $\mathbb{S}^1$. The most general total time derivative therefore naturally includes a phase-like function (defined on $\mathbb{S}^1$), giving the function S the properties required to satisfy the Wallstrom condition. Once again we omit further discussion of the details here.

The problem also arises in Nelson's stochastic foundation of quantum mechanics. In this context Derakhshani (2015, 2016) has proposed a solution based on *Zitterbewegung* at the fundamental level.

It has been tacitly assumed here and in the various works mentioned above that without the Wallstrom condition such a wave function-free formulation cannot account for the empirical data obtained from experiments involving microscopic systems. This assumption may ultimately prove accurate. However, it warrants investigation. Below we will investigate the consequences of relaxing the Wallstrom condition. In preparation to do so, we will first make some formal observations.

If a function S is defined on the configuration space $\mathcal{C}$, that is, $S : \mathcal{C} \to \mathbb{R}$, and $m\vec{v} = \nabla S$, then for *any* loop $\mathcal{L}$ in $\mathcal{C}$ it is the case that

$$\oint_{\mathcal{L}} m\vec{v} \cdot d\vec{\ell} = 0. \tag{8.4}$$

Therefore, S cannot be defined globally on $\mathcal{C}$ if the integral is to match the corresponding integral obtained with the gradient of the phase of the wave function. This result is not spectacular but it does imply that S cannot be thought of as a classical Hamilton Principal function.

However, S may be defined *locally* in $\mathcal{C}$, that is, for some open, simply connected neighbourhood $\mathcal{D} \subset \mathcal{C}$, provided $\mathcal{D}$ does not contain a point where the density ρ vanishes. Indeed, in the standard quantum description the phase of the wave function ψ is continuous everywhere except at nodes (points where $\psi = 0$), at which it is undefined. In quantum mechanics, analytic extension of the phase through the node is only possible if the integral along a loop that encircles the node vanishes. In the absence of a wave function we will use the term "node" to refer to a point where $\rho = 0$.

The integral is invariant under continuous deformations of $\mathcal{L}$ provided the deformation does not move $\mathcal{L}$ across a node. Any loop that does not encircle a node could therefore be shrunk to a point and the corresponding integral must therefore vanish. "Encircling" does, of course, only make sense for nodes that are $(n-2)$-dimensional (points in two-dimensional space, lines in three-dimensional space, etc.)—nodes of lower dimension would allow for continuous deformation of $\mathcal{L}$ to a point (such as point node in three-dimensional space). However, $n-2$-dimensional nodes occur generically in quantum mechanics since $\psi = 0$ constitutes two real conditions. In density-based approaches, ρ vanishes and is a minimum (assuming differentiability of ρ), which also constitutes two conditions. Hence an $(n-2)$-dimensional nodes occur generically here also. Spherical harmonics (e.g. in the context of hydrogen-like atoms) present a textbook example of this: States where the quantum number usually denoted by m is non-zero have a one-dimensional node along one axis, while states with $m = 0$ do not, and the loop integral of ∇S is proportional to $2\pi\hbar \cdot m$. Edge and screw dislocations (see Holland (1993), sec. 4.11) present another.

In the numerical investigations below we therefore exclusively focus on scenarios with nodes.

One further remark: In quantum mechanics, in any closed system the value of the loop integral Eq. (8.3) of $\vec{v}$ around a node cannot change over time. We refer to n in Eq. (8.3) as the *vorticity* of the node. Change is impossible because kinematically possible values are discrete but evolution of ψ is continuous. It is therefore impossible for a system to move from one vorticity to another. Some useful quantum models allow discrete jumps between states of different vorticity but these constitute effective models of some underlying theory, not a fundamental description (e.g. a Hamiltonian used to calculate transition probabilities between orbitals). This result no longer holds in the CSD formulation. There is no *kinematical* reason the vorticity of a system cannot change. Whether or not it *does* change becomes a matter of the *dynamics*. Further, in ψ-based formulations nodes cannot be created or destroyed except via pair creation or annihilation of nodes of opposite vorticity (Underwood 2018).

8.4 Analytical and Numerical Investigation of Non-quantum States

Relaxing the Wallstrom condition opens up a broader physics. Broadly speaking, there are two paths forward in order to investigate the dynamics of this physics. First, one can directly consider the density version of the quantum Hamilton-Jacobi equation (or alternatively Newton's Law with the added quantum potential) in conjunction with the continuity equation for ρ. This system of equations determining ρ and the vector field $\vec{v}$ on configuration space is non-linear and general properties are difficult to determine analytically. This approach does lend itself to numerical investigation however. Second, one can re-introduce a "wave function-like" function that is multi-valued on the configuration space. Nodes are analogues of branch points (or higher-dimensional equivalents) of this function. This line of inquiry will be explored further in future investigations.

For this paper we report on the numerical investigation into the Wallstrom-relaxed CSD formulation based on the first approach. Our numerical investigation involved choosing an initial state and fixed grid size, and then numerically calculating initial values for Q, $\rho = |\psi|^2$ and $v = \frac{\nabla S}{m}$ for each location on the grid. Evolution follows equations (1) and (2), and trajectory tracking is done with a 4th order Runge-Kutta method (see Appendix).

One approach we pursue here is to begin with a quantum state and add a small non-quantum perturbation. We then track how the size of this perturbation evolves over time. In order to track the degree of "non-quantumness", that is, the deviation from integer vorticity, we introduce a parameter ν as

$$\nu = \frac{1}{2\pi\hbar}\left(\oint_0^{2\pi} m\vec{v}\cdot d\vec{\ell}\right) - n, \tag{8.5}$$

where n is the (integer) vorticity corresponding to the value of the loop integral of the initial quantum state relative to which the non-quantumness is measured. For example, in two-dimensional scenarios we begin with initial fields ρ and $\vec{v}$ that correspond to a quantum state—field values that could have been derived from a single-valued wave function—and then add a small perturbation to the velocity field of the form

$$\vec{v}_P = \hbar \nu_0 \frac{\hat{\theta}}{r} = \frac{\hbar \nu_0}{x^2 + y^2} \begin{pmatrix} -y \\ x \end{pmatrix}, \tag{8.6}$$

where the coordinates are assumed centered on the node. Note that this velocity-field perturbation may be expressed as the gradient of a function in any open simply-connected neighborhood that does not encompass the origin (the node), namely as

$$\vec{v}_P = \nabla S_P, \qquad \text{where } S_P = \hbar \nu_0 \arctan\left(\frac{y}{x}\right). \tag{8.7}$$

As expected, the function S_P would be multi-valued if analytically continued over the entire configuration space minus the node. The parameter ν_0 is the initial degree of non-quantumness, $\nu_0 = \nu(0)$. Given that any velocity perturbation we consider is curl-free except at the node, this velocity-field perturbation is generic: Any other perturbation could be expressed as the gradient of the sum of S_P and a function that is analytic on the entire configuration space and therefore leaves the vorticity unchanged.

8.5 Case Studies: The 2D Harmonic Oscillator

The harmonic oscillator presents a paradigm example when studying both classical and quantum systems. Its relevance reaches from small oscillations of mechanical systems to quantum field theory and cosmological perturbations. In the context of the present investigation the one-dimensional harmonic oscillator is of little interest since in one dimension nodes cannot be encircled by loops. However, the two-dimensional harmonic oscillator is sufficient to display non-quantum behavior. We study it in this section. The algorithmic details are briefly described in Appendix.

In standard quantum mechanics, the one-dimensional oscillator has eigenstates

$$\psi_j(x) = \left(\frac{m\omega}{\pi\hbar}\right)^{\frac{1}{4}} \frac{1}{\sqrt{2^j j!}} H_j\left(\sqrt{\frac{m\omega}{\hbar}} x\right) e^{-\frac{m\omega}{2\hbar}x^2}$$

for $j = 0, 1, 2, \ldots$. A two-dimensional basis for the space of all allowed states is formed by

$$\psi_{jk} = \psi_j(x)\psi_k(y). \tag{8.8}$$

We are interested in states with nodes, since such states allow generalization to the non-quantum realm. A node here is an $(n-2) = 0$-dimensional locus, a point. In particular, we require nodes of non-zero vorticity. Such states may be constructed as the sum of two or more basis states with relative phase factors. For example,

$$\psi = \psi_{01} + i\psi_{10}. \tag{8.9}$$

It is straightforward to verify by explicit calculation that the vorticity n of such a state is a non-zero integer (here $n = 1$). Computed trajectories orbit the node. Speed falls off with distance from the node. Testing a variety of such states that fall within configurations accessible by standard quantum mechanics served to corroborate reliability of the numerical computation. Explicit computation in CSD (via ρ and $\vec{v}$ as fundamental quantities rather than ψ) shows n remains constant, as one would expect given that the initial state of the system has a quantum-mechanical counterpart. That n is constant is however a result of the *dynamics* in CSD, unlike in standard quantum mechanics, where non-integer n is kinematically impossible.

To these states we added a perturbation of the form of equation (8.6). The stability of the vorticity remained unchanged. Note though that a product state of the general form (8.9) does not remain a stationary state if the velocity field is perturbed while the density function is not.[3] In particular, an increase in the initial speed must lead to an initial outwards motion of trajectories from the node (since the instantaneous acceleration remains unchanged upon this perturbation) and hence a change in the radial dependence of $\rho(r)$.

The stability of the vorticity extends to situations of time-dependent potentials. We investigated several choices of $V(x, t)$ to simulate the influence of an external interaction, such as the temporary proximity of another particle. In all cases the vorticity remained unaffected. We also added non-quantum perturbations to other scenarios in which nodes occur, such as the 'dislocated wave' (see Holland (1993), 4.11) and an infinite circular well, and found similar results.

The result of the simulation is the demonstration of the *dynamical* stability of the degree of non-quantumness of this simple system, including under various forms of perturbations that can be taken to simulate interactions.

8.6 Stationary States in CSD Frameworks

Stationary states in quantum mechanics correspond to the energy eigenfunctions of the system, the set of which is discrete in scenarios such as this one. An analysis of a textbook derivation of the 2D harmonic oscillator, the circular infinite well, or the hydrogen atom, for example, shows that the discreteness arises exactly as a result of imposing the Wallstrom condition, or, in other words (in standard quantum terminology), as a result of the single-valuedness of the wave function. Every quantum stationary state is also stationary in CSD (in the sense that $\rho(x)$ and $\vec{v}(x)$ are

[3] See the discussion on stationary states in Sect. 8.6.

time independent), but the absence of the Wallstrom condition creates a continuous rather than discrete infinity of such basis states, characterized by continuous values of vorticity n. There is no wave function in the quantum description of particular system that corresponds to stationary states of non-integer n.

It is conceivable however to develop a pseudo-quantum description of such states in terms of multi-valued wave functions (specifically, multi-valued phases), or states defined on configuration spaces with a branch cut, with the node as its branch point. "Point" here is to be understood as defined on a suitably chosen two-dimensional subspace. This we leave to future work.

The analysis of non-integer vorticity states presents a mathematical challenge since the radial functions are, depending on the particular details of the scenario, various special functions of non-integer order. In the harmonic oscillator studied in Sect. 8.5 these functions are the Hermite polynomials. In other cases the properties of non-integer-order Bessel functions of the first kind (infinite well) or non-integer-order generalized Laguerre polynomials (hydrogen atom) need to be studied in more depth.

In most of the literature on special functions non-integer orders are delegated to a footnote or very brief discussion only.[4] Series or integral representations for such functions exist that are well-defined for non-integer cases, often in terms of the Gamma function (e.g. Arfken 1985). For Bessel functions of the first kind, for example (relevant to the circular well),

$$J_n(x) = \sum_{s=0}^{\infty} \frac{(-1)^s}{s!\Gamma(n+s+1)} \left(\frac{x}{2}\right)^{n+2s} \tag{8.10}$$

holds for non-integer n. However, it is interesting to note that there are *qualitative* differences between non-integer and integer-order special functions. For the functions (8.10) only integer-order ones are entire, while non-integer ones are multi-valued with a singularity at zero when defined over $\mathbb{C}$.

We posit here, speculatively, the possibility that this line of investigation may yet reveal insights into what makes integer vorticity special, and perhaps ultimately explain the empirical absence of non-integer vorticity even in CSD frameworks. The thought here is that integer vorticity arises because when expanded around the node the *radial* functions $\rho(r)$, which are linked via the dynamics to the velocity field, have special properties in exactly those cases in which the integral (8.3) has integer n. A more detailed analysis of the stationary states in CSD frameworks is required. This constitutes future work.

[4] Arfken (1985), for example, only dedicates two paragraphs to the subject on non-integer Bessel functions over the course of sixty pages on the subject as a whole.

8.7 Conclusion

In all cases we studied the degree of non-quantumness is conserved. This is akin to angular momentum conservation around the node across the ensemble. The conservation might therefore not come as a surprise, although it must not be presupposed given that the theory studied here is notably distinct from quantum mechanics. We noted that the *kinematic* conservation of vorticity is not a feature of CSD theories. Here angular momentum *across the ensemble* is dynamically conserved. From the CSD perspective, it can be argued that this result indicates that a kinematical constraint of the vorticity is unnecessary structure. However, this argument is undermined by the fact that the absence of non-integer vorticity in experimental data is yet to explained. Barring the attempts outlined in Sect. 8.3 or Sect. 8.6 such a justification remains outstanding.

In standard quantum mechanics the experimental evidence for quantized vorticity are the outcomes of what one would term "angular momentum measurements"—really quantized eigenvalues of a mathematical operator one calls the "angular momentum operator." This interpretation of mathematical operators as physical measurements makes (even) less sense in CSD frameworks, but the interpretation and terminology is irrelevant for what is observed and what standard quantum mechanics does, and what CSD frameworks as they stand right now do not predict.

Of course, more complex and higher-dimensional systems are yet to be investigated, so the generalization of the results of the numerical investigation is conjecture only. However, the insights so far stifle hope that the Wallstrom challenge might have been resolved dynamically with the quantized vorticities of quantum mechanics (and pilot-wave theory) as some form of stable equilibrium towards which systems tend generically under their own evolution. Any proposal for a CSD-type theory must therefore explicitly address the Wallstrom condition.

Investigating CSD theories in the non-quantum regime is, of course, not limited to merely tracking the degree of non-quantumness. They may have interesting dynamics in their own right with qualitative differences to dynamics of the quantum regime. A study of non-quantum stationary states as outlined in Sect. 8.6 may provide a systematic entry point towards further study of this nature.

Appendix: Computational Methods

The data and code used in this paper are available at https://github.com/mscoggs/no_wave_qm. Simulation is performed on a uniform square grid, where initial density and velocity field are specified, then Q, ρ, v and $V(x_i)$ are updated at each time step. In order to ensure computational reliability of the result each simulation is repeated with smaller time step sizes until a further decrease in the time step has negligible effect on the evolution. The algorithm also allows tracking of individual trajectories based on user-selected initial positions. Trajectory evolution is calculated using a fourth-order Runge-Kutta method.

References

Arfken, G. 1985. *Mathematical Methods for Physicists*, 3rd ed. San Diego: Academic Press.

Bohm, D. 1952. A suggested interpretation of the quantum theory in terms of 'hidden' variables. I and II. *Physical Review* 85: 166–194.

Boström, K.J. 2015. Quantum mechanics as a deterministic theory of a continuum of worlds. *Quantum Studies: Mathematics and Foundations* 2. quant-ph: 1410.5653.

Colin, S., and A. Valentini. 2014. Instability of quantum equilibrium in Bohm's dynamics. *Proceedings of the Royal Society A* 470: 20140288. quant-ph: 1306.1576.

Derakhshani, M. 2016. A suggested answer to Wallstrom's criticism: Zitterbewegung stochastic mechanics I + II. quant-ph: 1510.06391, 1607.08838, 2015.

Dirac, P.A.M. 1931. Quantized singularities in the electromagnetic field. *Proceedings of the Royal Society A* A133: 60.

Dürr, D., S. Goldstein, and N. Zanghí. 1992. Quantum equilibrium and the origin of absolute uncertainty. *Journal of Statistical Physics* 67: 843–907.

Everett, H. 1956. The theory of the universal wavefunction. PhD thesis, Princeton University.

Goldstein, H., C. Poole, and J. Safko. 2000. *Classical Mechanics*, 3rd ed. Boston: Addison-Wesley.

Hall, M., D.-A. Deckert, and H. Wiseman. 2014. Quantum phenomena modelled by interactions between many classical worlds. *Physical Review X* 4: 041013. quant-ph: 1402.6144.

Holland, P.R. 1993. *The Quantum Theory of Motion: An Account of the de Broglie-Bohm Causal Interpretation of Quantum Mechanics*. Cambridge: Cambridge University Press.

Madelung, E. 1927. Quantentheorie in hydrodynamischer Form. *Zeitschrift für Physik* 40: 322–326.

Poirier, B. 2010. Bohmian mechanics without pilot waves. *Chemical Physics* 370: 4.

Pusey, M., J. Barrett, and T. Rudolph. 2012. On the reality of the quantum state. *Nature Physics* 8: 475. quant-ph: 1111.3328.

Roser, P. 2015. Quantum mechanics as the dynamical geometry of trajectories. quant-ph: 1507.08975. https://arxiv.org/abs/1507.08975

Santamato, E. 1984a. Geometric derivation of the schrödinger equation from classical mechanics in curved Weyl-spaces. *Physical Review D* 29 (2): 216.

Santamato, E. 1984b. Statistical interpretation of the Klein-Gordon equation in terms of the spacetime Weyl curvature. *Journal of Mathematical Physics* 25: 2477.

Schlosshauer, M. 2011. *Elegance and Enigma*. Berlin: Springer.

Schmelzer, I. 2011a. The paleoclassical interpretation of quantum theory. quant-ph: 1103.3506.

Schmelzer, I. 2011b. A solution for the Wallstrom problem of Nelsonian stochastics. quant-ph: 1101.5774.

Sebens, C. 2014. Quantum mechanics as classical physics. quant-ph.

Smolin, L. 2015. Quantum mechanics and the principle of maximal variety. quant-ph: 1506.02938.

Takabayasi, T. 1952. On the formulation of quantum mechanics associated with classical pictures. *Progress of Theoretical Physics* 8: 143–182.

Tipler, F. 2010. Hamilton-Jacobi many-worlds theory and the Heisenberg uncertainty principle. quant-ph: 1007.4566.

Underwood, N. 2018. Extreme quantum nonequilibrium, nodes, vorticity, drift, and relaxation retarding states. *Journal of Physics A* 51: 055301. quant-ph: 1705.06757.

Valentini, A. 1991. Signal locality, uncertainty, and the subquantum H-theorem. I. *Physics Letters A* 156: 5.

Valentini, A., and H. Westman. 2004. Dynamical origin of quantum probabilities. *Proceedings of the Royal Society A* 461: 253–272.
Wallace, D. 2008. Philosophy of quantum mechanics. In *The Ashgate Companion to Contemporary Philosophy of Physics*. London: Routledge.
Wallstrom, T.C. 1994. Inequivalence between the Schrödinger equation and the Madelung hydrodynamic equations. *Physical Review A* 49: 1613.
Wyatt, R. 2005. *Quantum Dynamics with Trajectories*, Springer

Chapter 9
The Doubochinski Pendulum – A Paradigm for Quantization Through Nonlinear Interactions

Jonathan Tennenbaum

Abstract The Doubochinski pendulum is a simply-constructed nonlinear oscillator having a discrete series of stable amplitudes, which can be seen to mimic the quantum behavior of microphysical objects in several respects. Apart from this system's intrinsic interest, the specific form of phase-modulated interaction (so-called argumental interaction) which gives rise to amplitude "quantization" in the Doubochinski pendulum, might prove relevant to developing a more realistic and more intelligible form of quantum physics in the future. In this paper I describe the experimentally-observed behavior of the Doubochinski pendulum in some detail, and sketch some elements of the theory of the pendulum, focusing on insights into the mechanism of amplitude quantization and related phenomena in this system. I shall also briefly touch on a related model system, consisting of a low-frequency mechanical oscillator interacting with a high-frequency wave, which likewise displays a discrete set of stable amplitudes. Finally, I shall discuss some results of A. I. Shumaev and Z. A. Maizelis concerning the statistical distribution of quantized amplitudes of the Doubochinski pendulum as a function of initial conditions, which bears some analogy to probability densities in quantum mechanics.

Keywords Argumental interactions · Macroscopic quantum effect · Phase modulation · Multistability · Distribution function

9.1 Introduction

In the early days of quantum mechanics Louis DeBroglie suggested that quantum phenomena might be the result of nonlinear interactions of particles with what he called a "hidden subquantum medium".

J. Tennenbaum (✉)
International Collaborator, Centre for Philosophy of Sciences of the University of Lisbon (CFCUL), University of Lisbon, Lisbon, Portugal
e-mail: tennenbaum@landbridge.de

P. Castro et al. (eds.), *Advances in Pilot Wave Theory*, Boston Studies in the Philosophy and History of Science 344,
https://doi.org/10.1007/978-3-031-49861-9_9

Here I want to describe an elementary example of how quantization can arise from nonlinear interactions in a "perfectly classical" one-dimensional macroscopic system, which might provide a kind of paradigm for efforts to develop a nonlinear quantum physics in the direction of DeBroglie's original idea.

At first glance the system I am going to describe might seem quite unrelated to the hydrodynamic models which are the main topic of this conference, but there may be some hidden similarities. Besides that, the system I am going to describe is extremely interesting and worthy of attention in and of itself.

Starting in a very simple-minded way from classical physics, we ask: where can you get a macroscopic system that you can easily investigate – that you can literally get your hands on – which displays something analogous to the quantization of energy levels in an atom? The obvious place to look, nowadays, is in the domain of dynamic systems possessing the property of multistability, i.e. having a discrete set of attractors, each considered as analogous to a quantum state, such that for a large range of initial conditions the system will inevitably be "captured" by one or the other of these attractors. A sufficiently large perturbation might cause the system to "jump" into the basin of another attractor. For nonlinear systems the end-state can be highly sensitive to the initial conditions, giving rise to apparently indeterministic, often chaotic behavior with statistical distributions of outcomes, typical of the microscopic world.

At the time when quantum physics was born, knowledge concerning nonlinear dynamic systems – and specifically those possessing multistability – was very limited. Today the situation is very different; we know a great variety of such systems, we have a great number of theoretical and empirical investigations of them, powerful numerical methods and above all computational capabilities beyond the wildest dreams of our colleagues back in the 1920s.

There are many sorts of macroscopic systems possessing a discrete array of stable or quasi-stable modes. Now, if we want to take multistability as a starting-point for thinking about quantization, then we would want to narrow down the range of examples and look for something we could call fundamental or universal. This is where the Doubochinski pendulum comes in, as a kind of paradigm: a one-dimensional system which is easy to build and study, which displays a discrete set of stable energy states, analogous to the quantized energy levels of an atom. Moreover, I think the specific type of nonlinear interaction which gives rise to the quantized modes of this system, has some right to be regarded as fundamental and may play a much bigger role in the physical world, than might appear at first glance.

The Douboschinski pendulum was originally invented in 1968–1969 by Danil Doubochinski and his brother Yakov while students at Moscow University, originally as a model for studying nonlinear mechanisms of acceleration of particles in oscillating fields. The device consists of a low-friction gravitational pendulum with a proper frequency (frequency for small displacements) in the range 0.5–1 Hz and a small permanent magnet affixed to its end. A narrow solenoid coil is installed under the pendulum's equilibrium position, and supplied by alternating current of some fixed higher frequency, typically between 30 and 1000 Hz. The apparatus is designed in such a way, that the oscillating magnetic field generated by the solenoid,

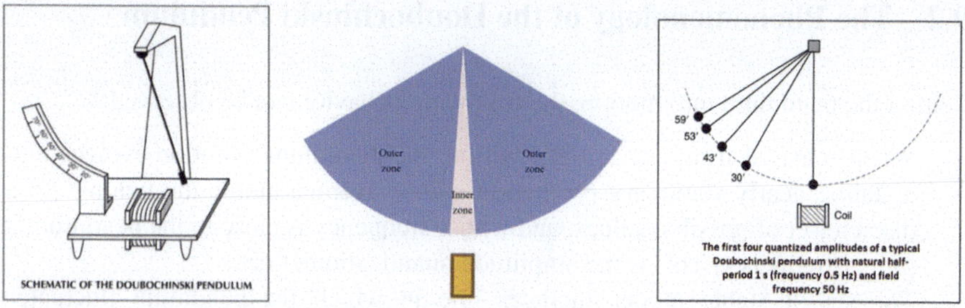

Fig. 9.1 Left: Design of the Doubochinski pendulum. Middle: A significant exchange of energy with the external source occurs only over a limited section of the pendulum's trajectory – the "interaction zone" around the electromagnet –, outside of which the pendulum swings freely. Right: The values of the stable amplitudes depend on the frequency of the electromagnet and the pendulum's natural period (for free small oscillations); but are practically independent of the field strength and the width of the interaction zone

exerts a significant influence on the pendulum only within a narrow *"interaction zone"* above the solenoid, outside of which the force on the pendulum drops rapidly to zero. As we shall see, this spatial inhomogeneity permits the pendulum to self-regulate its exchange of energy with the solenoid via shifts in its phase relative to that of the oscillating field (Fig. 9.1).

After describing the remarkable phenomena displayed by the Doubochinski pendulum I shall devote some space to *heuristic* considerations concerning the existence and values of its quantized amplitudes. As it turns out, the detailed behavior of the pendulum and the mechanism by which it maintains its stable oscillatory modes, are extremely complicated, and a half-way rigorous treatment would go beyond the limits of this overview.

I shall argue that Doubochinski's pendulum provides much more than a suggestive analogy with quantum phenomena. The pendulum is representative of a whole class of nonlinear oscillatory systems based on so-called "argumental interactions", in which the *phase* (argument) of the force acting between two or more oscillators depends on their relative positions as well as the time. What the latter signifies is best illustrated by the differential equation of Doubochinski's pendulum, which I shall give below. Argumental interactions provide a mechanism for coupling and exchange of energy between oscillators whose frequencies can differ by orders of magnitude. I suspect that studying interactions of this type may provide a key to understanding the subquantum level of reality.

At the end of this paper I shall briefly discuss some newer results concerning the behavior of the Doubochinski pendulum, which strengthen the analogy with quantum physics.

9.2 The Phenomenology of the Doubochinski Pendulum

Setting the pendulum into motion, the following behavior can be observed:[1]

1. When released from any initial position, the pendulum's motion evolves into a stable, nearly stationary oscillation whose average amplitude belongs to a discrete set of possible values, and whose frequency is close to the pendulum's proper frequency. This is the amplitude quantization effect.
2. The set of stable modes includes one in which the pendulum fluctuates around the vertical position, executing small-amplitude forced vibrations at the frequency of the solenoid field. This mode we can call the "ground state", while the other modes, with non-zero average amplitudes, constitute the "excited states" of the system.
3. In each of these modes, the pendulum's frictional losses over a given period of oscillation are compensated in a self-regulated fashion by a transfer of energy from the field of the solenoid, thereby sustaining the pendulum's oscillation in a quasi-stationary regime near to its own proper frequency.
4. The pendulum's motion in a given regime is never strictly periodic, but constantly fluctuates in phase, frequency and amplitude around one of the discrete set of values. (This fluctuational "play" is essential to the mechanism of stability of the quantized oscillatory regimes.)
5. The values of the quantized amplitudes vary as a function of the *frequency* of the alternating current supplied to the electromagnet, but are virtually independent of the *strength* of the current over a wide range. For very strong fields the pendulum's behavior becomes chaotic.
6. Provided that the pendulum's resistance is small, the values of the quantized amplitudes are essentially independent of the coefficient of dissipation.
7. Within certain limits the values of the quantized amplitudes are independent of the dimensions of the solenoid.
8. The specific stable mode into which the pendulum motion evolves – it's "choice" of final state – is extremely sensitive to the initial conditions, i.e. the height at which the pendulum is released and the phase of the alternating current at the moment of release. This dependence has a *stochastic character*.
9. In particular, even after the pendulum motion has evolved into one of one of the modes, there is a nonzero probability of abrupt decay into a lower-amplitude state, including decay into the least-energy "ground state". In this sense we could speak more precisely of quasi-stability rather than 100% stability. In experimental observation, collapses to lower-amplitude states nearly always occur in the "dangerous" initial period, when the pendulum motion is evolving into a given mode (Fig. 9.2).

[1] A short video demonstration of the Doubochinski pendulum can be seen here: https://www.youtube.com/watch?v=lWx4eKYd9C8&ab_channel=DanilDoubochinski

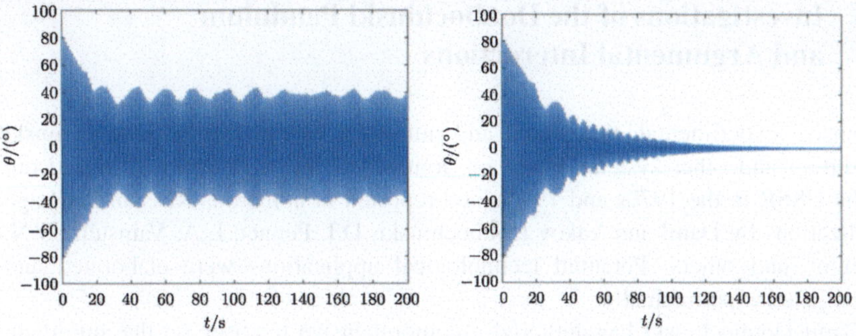

(a) $\theta_0 = 85.1°$, $\varphi_0 = 0.14°$

(b) $\theta_0 = 84.5°$, $\varphi_0 = 0.14°$

Lan Ziheng, Chen Jiahao, Gao Yantai, et al. The principle of electromagnetic "quantum" pendulum, and its experimental study. *Mechanics in Engineering*, 2016, 38(4): 471-475 (in Chinese)

Fig. 9.2 Nearby initial conditions can lead to qualitatively different evolutions. Here a Doubochinski pendulum with proper frequency 0.9 Hz and a field frequency of 50 Hz, is released at identical phases of the field from initial angles of 85.1° and 85.5°. Left: Post-release transition to a stable amplitude following release at 85.1°. Note the constant amplitude fluctuations. Right: Decay to the "ground state" after release at 85.5°. From measurements made on a Doubochinski pendulum at the School of Physics and Electronics, University of Electronic Science and Technology of China, Chengdu, China. (Lan Ziheng, Chen Jiahao, Gao Yantai, et al. The principle of electromagnetic "quantum" pendulum, and its experimental study, *Mechanics in Engineering*, 2016, 38(4): 471–475 (in Chinese))

10. Finally, it is important to emphasize that the alternating field frequency can be orders of magnitude larger than that of the pendulum (1000× or more in suitably-designed experiments). The higher the frequency, the larger the density of the quantized amplitudes.
11. In an impressive variant of the pendulum experiment, a whole array of pendula of different lengths and periods of oscillation is maintained in motion, each near its own proper frequency and each having its own discrete set of quasi-stable amplitudes, by interaction with the field of a single electromagnet fed with fixed-frequency alternating current.

This case is particularly suggestive, when we consider the hypothesis, that the quantum behavior of microscopic systems might be due to interactions with an oscillatory subquantum medium or field. Apart from such speculations, the Doubochinski pendulum exemplifies the ability of argumental interactions to transfer energy between oscillating systems whose frequencies can differ by orders of magnitude.

9.3 Investigations of the Doubochinski Pendulum and Argumental Interactions

Extensive experimental, theoretical and numerical studies of the Doubochinski pendulum and other systems based on argumental interactions were carried out in the USSR in the 1970s and 1980s, and reported in numerous Russian-language publications by Danil and Yakov Doubochinski, D.I. Penner, L.A. Vainstein, V.N. Damgov and others. Potential technological applications were elaborated and investigated experimentally.[2]

Danil Doubochinski has authored a comprehensive treatise[3] on the amplitude quantization effect and argumental interactions, which includes the results of the investigations made in the Soviet Union as well as more recent work by Doubochinski and others.

Initially, the work done in the Soviet period seems to have attracted little attention in the West, and was nearly forgotten. Over the last 20 years the situation has changed, however. Today the Doubochinski pendulum has become fairly widely known, even finding its way into some school and university classrooms. Significant new scientific publications have appeared.[4] In the meantime Danil Doubochinki and I jointly elaborated some fundamental ideas, inspired by the study of amplitude quantization and argumental interactions, concerning the general nature of physical objects.[5]

[2] Danil Doubochinski has continued this effort, leading to a series of patented inventions. See http://danil-doubochinski.com. Of particular interest and potential economic impact, I think, is a new method for vibratory atomization of liquids, claimed to produce far smaller droplets, with far less energy, than standard ultrasonic methods. This invention could greatly enhance the effectiveness of evaporative cooling.

[3] The full text of the treatise, Данил Дубошинский, АРГУМЕНТНЫЕ КОЛЕБАНИЯ – МАКРОФИЗИЧЕСКИЙ КВАНТОВЫЙ ЭФФЕКТ is can be downloaded at http://www.sciteclibrary.ru/texsts/rus/stat/st6832.pdfA shorter version in English, Doubochinski's Macrophysical Quantum Effect (DMQE): Experimental and Theoretical Argumentation, can be downloaded at http://www.sciteclibrary.ru/eng/catalog/pages/12065.html. Unfortunately, the detailed mathematical theory of argumental interactions is included only in the Russian version.

[4] See the bibliography at the end of this paper.

[5] Danil Doubochinski and Jonathan Tennenbaum, On the General Nature of Physical Objects and their Interactions, as Suggested by the Properties of Argumentally-Coupled Oscillating Systems, *arXiv: General Physics*, published 8 August 2008.

9.4 Source of Quantized Behavior of the Doubochinski Pendulum – Heuristic Considerations

Evidently the Doubochinski pendulum is a highly nonlinear system. At this point it is impossible to give an exact mathematical analysis of its behavior, i.e. one which does not require essential assumptions and simplifications, including the use of averaging methods. Nevertheless, a theoretic framework exists, going back to the original work of the Doubochinski brothers and their collaborators, which accounts for the existence of the quantized amplitudes and various other aspects of the system's behavior in a fairly satisfactory way, and permits one to calculate the values of the amplitudes in good agreement with experiment.

In the following discussion I take a very simple-minded heuristic standpoint, avoiding the complexities of a rigorous treatment.

Evidently the amplitude quantization and other novel characteristics of the Doubochinski pendulum, originate in the specific character of the interaction between the pendulum and the electromagnet, and especially the fact, that the effect of the alternating magnetic field on the pendulum depends sensitively both on the time and on its position in relation to the interaction zone. For simplicity I shall assume that the field is spatially homogeneous within the interaction zone, and drops off immediately to zero outside the zone.

Within that zone, the oscillating magnetic field, acting on the permanent magnet fixed to the end of the pendulum, causes the pendulum to experience an alternating series of accelerating and decelerating impulses, depending on the momentary polarity of the current. That train of impulses is interrupted when the pendulum leaves the zone. The net result will depend on the *transit time* – the time taken by the pendulum to traverse the interaction zone – and the phase of the current at the moment of entry. If the transit time corresponds to an *even number of half-periods* of the electromagnet, then the effects of the accelerating and decelerating half-cycles will essentially cancel out, and the pendulum will exit the interaction zone with the same energy as when it entered. That is the exceptional case, however. Nearly always, a non-zero net acceleration or deceleration will occur, depending on the phase of the field at the moment of entry (Fig. 9.3).

This suggests that the pendulum, by varying its phase relative to that of the field, might be able to self-regulate its interaction with the electromagnet, in such a way that the accelerating phases predominate just enough to compensate the frictional losses on average.

The simplest hypothetical case would be a *stationary mode*, with a constant phase relation between the field and the pendulum, and in which the energy gained in any given passage through the zone, is exactly equal to the energy loss through dissipation over the time from its exit until its next entry into the zone. In that case the pendulum enters the zone with the same velocity each time, and receives the same packet of energy each time.

In this context we should keep in mind, that each time, after reaching its maximum height, the pendulum *reverses* its direction and enters the zone again

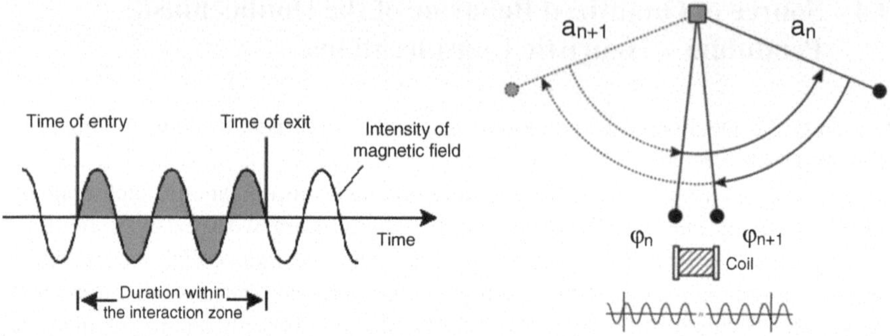

Fig. 9.3 The effect of a single passage of the pendulum through the "interaction zone" in the diagram) depends on the relationship of the phases of the magnetic field at the moments of entry and departure from the zone. In the example shown at left, the pendulum has left the zone after 5 half-cycles of the field, with 2 decelerating and 3 accelerating half-cycles, thereby resulting in a net acceleration. A 90° shift in phase at entry produces the opposite result. Right: the pendulum's motion can be tracked as an iterative process P: $(a_n, \phi_n) \rightarrow (a_{n+1}, \phi_{n+1})$, where a_n = amplitude of the nth half-period of pendulum and ϕ_n = the phase of the force acting on pendulum at the moment of nth entry into interaction zone

in the *opposite* direction. Thus, in order to receive a constant amount of energy, the phase of the field at the moment of reentry must be exactly opposite to that of the previous entry, i.e. displaced by a half-period. In other words, the time between successive entries into the zone – equal to a half-period **T/2** of the pendulum – must be an *odd-number multiple* of half-period **τ/2** of the oscillating field.

Thus, $T/2 = (2n + 1)\tau/2$, $T/\tau = 2n + 1$ is a *necessary* condition for a stationary mode of the system. Assuming they exist, the stationary modes form a discrete, "quantized" series.

The possibility of fulfilling the condition $T/\tau = 2n + 1$ lies in the well-known anisochronicity of the simple gravitational pendulum: the fact, that for large displacements the period **T** grows as a function of the amplitude, in a manner approximated by Bernoulli's formula $T \approx T_0(1 + \frac{\theta^2}{16})$, where θ is the angular displacement (in radians) at the highest point of the pendulum's swing and T_0 is the natural (or proper) period, i.e. the period for infinitesimal displacements. Using Bernoulli's formula – which is fairly good for amplitudes up to about 60° – we get the following approximate expression for the amplitudes of stationary modes:

$$\theta_n \approx 4((2n + 1)(\tau/T_o) - 1)^{1/2}$$

The lowest excited mode being determined by the least integer n for which $2n + 1 > T_o/\tau$.

To take a concrete example: in a typical demonstration device the pendulum has a natural period $T_o = 2$ s ($f = 0.5$ Hz) and the field frequency is $F = 50$ Hz ($\tau = 0.02$ s). In this case $T_o/\tau = 100$, and the first four excited modes will be for n = 50, 51, 52, 53 and T/τ = 101, 103, 105, 107. The corresponding angular

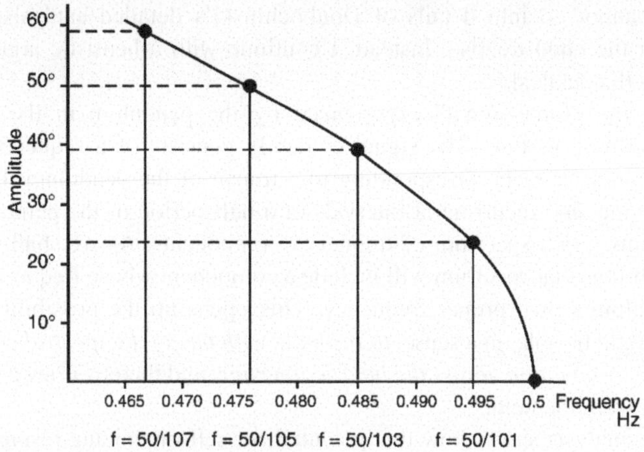

Fig. 9.4 Approximate determination of the first four stable amplitudes of the Doubochinski pendulum using Bernoulli's formula for the anisochronicity of a free pendulum

amplitudes, calculated from the formula above and converted into degrees, are roughly 23°, 39°, 50° and 60° (Fig. 9.4).

As it turns out, these calculated values are quite close to the amplitudes of the first four stable modes actually observed in experiments with Doubochinski's pendulum. Using a more precise, calibrated formula for the anisochronicity, this approach can easily be extended to higher-order modes, which can be observed with a well-constructed, low-friction device.

Note that the values of the dissipation, field strength and width of the interaction zone play no role in this calculation. In fact, as I mentioned above, the observed amplitudes are virtually independent of these parameters as long as the dissipation is low, the field strength is not too high, and the interaction zone is narrow.

9.5 Self-Modulation

It is important to emphasize that the above approach tells us nothing about whether strictly stationary modes are actually possible. We have also learned nothing about how the pendulum self-regulates its exchange of energy with the field in order to compensate for dissipation and maintain an amplitude in a manner that is stable for small perturbations.

Observation shows that stable modes are never strictly stationary, but involve constant phase and amplitude fluctuations. According to Doubochinski, these fluctuations are necessary features of the quantized modes.

In fact, although the existence of the stable modes of the Doubochinski pendulum is an experimental fact, that can be observed in suitable numerical simulations as well as physical systems, it is very difficult to prove *priori* that stable modes should exist.

Here, I cannot go into details of Doubochinski's detailed analysis, which is presented in the cited treatise. Instead, I continue with a heuristic argument that originates in that analysis.

Consider the "force-signal" experienced by the pendulum in the course of its nearly-periodic motion. The signal evidently consists of a sequence of short oscillatory "bursts", each corresponding to a transit of the pendulum through the interaction zone, and recurring at intervals of a half-period of the pendulum. This signal contains a rich spectrum of frequency components. As we shall see, under certain conditions that spectrum will include a component whose frequency is close to the pendulum's own proper frequency. This opens up the possibility, that the pendulum might be able, in a sense, *to resonate with one of the spectral components evoked by its own motion across the interaction zone*, and thereby draw a net energy contribution from the field.

A simple analysis shows how this potential for self-organizing resonant energy transfer arises.

The Doubochinski pendulum satisfies a differential equation of the form

$$\ddot{x} + 2\beta\dot{x} + \omega_0^2 \sin(x) = A\,\varepsilon(x)\sin(\Omega t) \quad \varepsilon(x) = \begin{matrix} 1 & -\delta \le x \le \delta \\ 0 & |x| > \delta \end{matrix} \tag{9.1}$$

where x denotes the pendulum's angle relative to the equilibrium position, ω_0 is the natural (proper) frequency of the pendulum, and Ω denotes the frequency of the magnetic field. The factor $\varepsilon(x)$ in the force term on the right side of the equation corresponds to the assumption, that the field drops off instantly to zero outside the interaction zone. It is easy to see, however, the following analysis does not depend essentially on the specific form of the function ε.

Needless to say, the equation of the Doubochinski pendulum cannot be solved in closed form. Assessing the behavior of the solutions of such nonlinear differential equations can be as much an art as a science!

Continuing in a heuristic fashion, assume that the pendulum's angular position x(t) is periodic with a frequency ω close to the pendulum's proper frequency ω_0. The function $\varepsilon(x) = \varepsilon(x(t))$ can then be expanded in a Fourier-series:

$$\varepsilon(x) = c_0 + \sum\nolimits_{n=1}^{\infty} (c_n \cos n\omega t + d_n \sin n\omega t)\,.$$

The force term in the equation of the pendulum will then take the form:

$$\begin{aligned} A\,\varepsilon(x)\sin\Omega t &= Ac_0 \sin\Omega t + A\sum_{n=1}^{\infty} (c_n \sin\Omega t \cos n\omega t + d_n \sin\Omega t \sin n\omega t) \\ &= Ac_0 \sin\Omega t \\ &\quad + \frac{A}{2}\sum_{n=1}^{\infty} c_n (\sin(\Omega - n\omega t) + \sin(\Omega + n\omega t)) \\ &\quad + d_n (\cos(\Omega - n\omega t) - \cos(\Omega + n\omega t)) \end{aligned}$$

From this it follows, that if the frequency of the pendulum is equal to an integral multiple of the field frequency, $\Omega = N\omega$, then the force function A $\varepsilon(x)$ $\sin\Omega t$ will contain a spectral component of frequency ω, namely:

$$F(t) = \frac{A}{2}\,(c_{N-1} - c_{N+1})\sin\omega t + \frac{A}{2}\,(d_{N-1} + d_{N-1})\cos\omega t$$

whose frequency coincides with that of the pendulum. If we could leave out all the other terms in the series then the equation (9.1) would turn into the differential equation of forced oscillations of a damped pendulum near to resonance. In that case, assuming the amplitude of the forcing function F(t) is not too large, the result is sustained oscillations. (For large amplitudes there is a transition to chaos.)

Now I should caution that this argument is highly non-rigorous and even misleading in the sense that the global behavior of Doubochinski's pendulum differs radically from that of a forced damped pendulum. But I think it provides some plausibility for the ability of Doubochinski's pendulum to sustain itself in an approximately constant amplitude through a mechanism of resonant exchange of energy with self-induced frequency components of the field. As before, the condition $\omega = \Omega/N$ leads to a discrete set of possible modes, whose amplitudes are determined by the period-amplitude dependency of the pendulum. (For reasons given earlier, N must be odd number.)

Apart from ignoring the higher-frequency terms of the Fourier series, the heuristic argument still fails to identify the mechanism of regulation of energy exchange through fluctuations in the amplitude and the phase relationship between the pendulum and the field. A comprehensive analysis begins by setting

$$x = a(t)\ \sin(\omega t + p(t))\,,$$

where the amplitude a(t) and phase shift f(t) are arbitrary, slowly-varying functions of the time. This leads to a more complicated set of differential equations, whose behavior can be analyzed using the averaging method of Krylov-Bogoliubov. Details can be found in the cited treatise and other publications by Doubochinski.

9.6 Argumental Interaction of an Oscillator with a Wave

In his famous investigation of the law of blackbody radiation, leading to the hypothesis of the quantum of action, Max Planck initially considered the exchanges of energy between a field of electromagnetic radiation and a collection of microscopic Hertzian oscillators (or "resonators") located in a cavity with reflecting walls. Adopting a deliberately simple theoretical model, Planck seems not to have considered the possible role of *spatial motion or elastic deformation* of the tiny resonators, which would inevitably experience vibratory mechanical forces through their interaction with the field. Spatial displacement would modulate the phase of

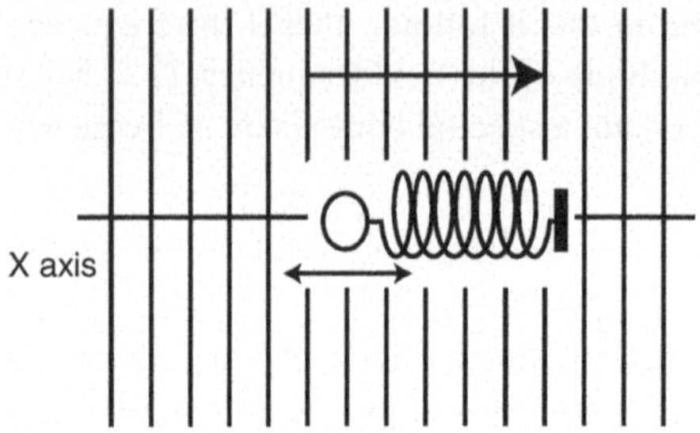

Fig. 9.5 A harmonic oscillator interacting with a wave

the field experienced by a resonator (and its constituent parts), leading to a nonlinear interaction that might produce a different result, than if the resonators were assumed to be rigid and fixed in space.

It would no doubt be extremely difficult to carry out an exact analysis of this problem with the methods of classical physics. In the context of their systematic study of argumentally-coupled oscillating systems, Danil and Yakov Doubochinski studied a much simpler, one-dimensional theoretical model of the argumental interaction of an oscillator with a wave, which leads to quantized amplitudes in a similar way to the Doubochinski pendulum.[6] Although their results has no direct relevance to the problem just mentioned, they do suggest how a certain kind of quantization might naturally arise, in a slightly analogous situation, within the domain of classical physics.

The theoretical model consists of an oscillator, constituted by a particle with mass $m = 1$ attached to a nonlinear spring and free to move in the x-direction, together with a longitudinal wave, propagating along the x-axis and exerting a force $A\sin(\Omega t - kx)$ on the particle (Fig. 9.5).

The corresponding differential equation has the form:

$$\ddot{x} + 2\beta\dot{x} + \omega_0^2\, x + u\,(x) = A\sin\,(\Omega t - kx) \tag{9.2}$$

Using theoretical and numerical methods, the Doubochinskis showed that under certain conditions, when the wave frequency Ω is large compared to the proper frequency ω_0 of the oscillator, the interaction will give rise to a discrete array of stable modes of the oscillator. For the particular case where $\Omega = N\omega_0$ the stable amplitudes are given approximately by extrema of the Bessel function J_N. Interestingly, these stable amplitudes persist even when the coefficient of dissipation is zero.[7]

[6] D.B. Doubochinski, Ya.B. Duboshinski, Wave excitation of an oscillator having a discrete series of stable amplitudes, *Dokl. Akad. Nauk SSSR*, 265, N°3, 605 (1982) [*Sov. Phys. Doklady* 27, 564 (1982)].

[7] Generally speaking I think it would be misleading to characterize the stable modes of systems based on argumental interactions as "dissipative structures" in the sense of Prigogine.

As far as I know, no attempt has been made so far to discover or create real physical systems whose behavior is described by equation (9.2). Certainly Doubochinski's model cannot be applied in any obvious, direct way to the scenario investigated by Max Planck. Nevertheless, one might speculate that the development of quantum mechanics might have taken a different path, if the results, described here, had been known at that time.

9.7 Distribution Functions

In closing I want to mention a very interesting paper by Vyacheslav V. Shumaev and Zakhar A. Maizelis, published in 2017.[8] These authors derive an exact equation for the Doubochinski pendulum under realistic conditions and present a method for theoretical prediction of its behavior using so-called envelope functions. They compare the theoretical analysis with the results of numerical simulations for a very large set of initial conditions, expressed in terms of the initial energy e_0 (proportional to the height at which the pendulum is released) and the phase of the field at the moment of entry into the interaction zone. The results give a glimpse of the phase portrait of the system, which agrees broadly with the results of the theoretical analysis.

Not surprisingly, the stable amplitudes appear as attractors. The basins of these attractors are shaped in a very complicated way, however. The data and the theoretical analysis both confirm that the pendulum, when released from a given height may "chose" any one of the stable amplitudes lower than its initial one, including the "ground state"; or occasionally even a higher amplitude. In general the "choice" depends very sensitively on the phase of the field at the moment of release. Regarding this phase as a random parameter, one can estimate the probability that the pendulum, released from a given height, will evolve into a given stable amplitude. We thereby obtain a probability distribution for the quantized states. The authors point out the analogy with the probability density functions in quantum mechanics, which are derived from the wave functions of the relevant systems.

Does something analogous to a quantum mechanical wave function exist in the case of the Doubochinski pendulum? The authors do not directly address that question; but they demonstrate that the observed probability distribution of the stable amplitudes can be predicted, approximately, from what they call the *envelop function*. This envelop function is the solution of a set of differential equations obtained by a smoothing procedure similar to that used in the original theory of the pendulum, referred to above. For details I refer the reader to their excellent paper.

[8] A. I. Shumaev, Z. A. Maizelis, Distribution functions of argumental oscillations of the Duboshinskiy pendulum, Journal of Applied Physics 121, (2017), doi: 10.1063/1.4979800.

9.8 Conclusion

I hope my brief introduction to the Doubochinski pendulum and related aspects will help to inspire further work in this fascinating area, especially concerning its possible relevance for developing a coherent theory of the subquantum level of reality, which lies at the basis of quantum phenomena and most probably also to the existence of particles of matter in our Universe.

References

Данил Дубошинский, Аргументные колебания – макрофизический квантовый эффект. http://www.sciteclibrary.ru/texsts/rus/stat/st6832.pdf

Cintra, Daniel, and Pierre Argoul. 2016a. Nonlinear argumental oscillators: Stability, criterion and attractor's capture probability, 2016 *arXiv*:1606.08993v1 [nlin.CD]. https://doi.org/10.48550/arXiv.1606.08993

———. 2016b. Nonlinear argumental oscillators: A few examples of modulation via spatial position. *Journal of Vibration and Control* 23 (18). https://doi.org/10.1177/1077546315623888.

Doubochinski, Danil. *Doubochinski's Macrophysical Quantum Effect (DMQE): Experimental and theoretical argumentation*. http://www.sciteclibrary.ru/texsts/eng/stat/st4624eng.pdf

Doubochinski, Danil, and Jonathan Tennenbaum. 2008. On the general nature of physical objects and their interactions, as suggested by the properties of argumentally-coupled oscillating systems. *arXiv: General Physics,* published 8 August 2008.

Luo, Yao. 2020. Subharmonic frequency response in a magnetic pendulum. *American Journal of Physics* 88: 115. https://doi.org/10.1119/10.0000038.

Luo, Yao, Wenkai Fan, et al. 2009. Multi-stability in Doubochinski's Pendulum. *arXiv:1904.02908v1 [physics.class-ph],* published on 5 Apr 2019.

Pedersen, Henrik, et al. 2021. Investigation of argumental oscillations of a physical pendulum. *European Journal of Physics* 42 (2): 27. https://doi.org/10.1088/1361-6404/abcee4.

Shumaev, A.I., and Z.A. Maizelis. 2017. Distribution functions of argumental oscillations of the Duboshinskiy pendulum. *Journal of Applied Physics* 121. https://doi.org/10.1063/1.4979800.

The above treatise contains an extensive bibliography of the scientific literature on the Doubochinski pendulum and argumental interactions up to about 2010.

Ziheng, Lan, Chen Jiahao, Gao Yantai, et al. 2016. The principle of electromagnetic "quantum" pendulum, and its experimental study. *Mechanics in Engineering* 38 (4): 471–475. (in Chinese).

Chapter 10
Completing the Quantum Ontology with the Electromagnetic Zero-Point Field

Luis de la Peña and Ana María Cetto

Abstract This text begins with a series of critical considerations on the initial interpretation of quantum phenomena observed in atomic systems. The bewildering explanations advanced during the construction of quantum mechanics are shown to have distanced the new theory from the rest of scientific knowledge, by introducing indeterminism, acausality, nonlocality, and even subjectivism as part of its interpretative framework. The conclusion drawn from this unsatisfactory interpretative landscape is that quantum mechanics lacks a key ontological ingredient. Arguments are given in favour of the random zero-point radiation field (ZPF) as the element needed to complete the quantum ontology. The (wave-mediated) quantum stochastic process is shown to be essentially different from Brownian motion, and more amenable to an analogy with the hydrodynamic case. The new perspective provided by the introduction of the ZPF is used to explain some salient features of quantum systems, such as the stationary atomic states and the transitions between them, and the apparent nonlocality expressed in the entangled states. Notably, the permanent presence of the field drastically affects the dynamics of the (otherwise classical) particle, which eventually falls under the control of the field. This qualitative change is reflected in the transition from the initial classical description in space-time, to the final quantum one in the Hilbert space. The clarification of the mechanism of quantization leads us to consider the possibility that a similar phenomenon occurs in other physical systems of corpuscles subjected to an oscillating background, of which the walking-droplet system is a paradigmatic example.

Keywords Quantum ontology · Zero-point field · Quantum stochasticity · Wave-particle behavior · Classical-quantum transition

L. de la Peña · A. M. Cetto (✉)
Instituto de Física, Universidad Nacional Autónoma de México, Mexico City, Mexico
e-mail: ana@fisica.unam.mx

P. Castro et al. (eds.), *Advances in Pilot Wave Theory*, Boston Studies in the Philosophy and History of Science 344,
https://doi.org/10.1007/978-3-031-49861-9_10

10.1 Introduction

A careful study of quantum mechanics (QM) as it is systematically presented and understood in present-day publications, from the dominant Copenhagen perspective or any of the alternatives—which can be counted by the dozen—leads to the conclusion that it must be indeed an incomplete theory. Only an incomplete theory can be 'completed' with such a plentiful variety of alternatives, some even in clear contradiction amongst them. But rather than the incompleteness considered by Einstein —any statistical description is incomplete by nature—we refer here to an essential ingredient that is missing. This assertion, quite significant for a theory that is considered to be the basis of half (if not more) of modern physics, gets reaffirmed by a critical analysis of the origin of a few of its most representative and fundamental postulates—some implicit, others explicit.

One such postulate that calls for immediate attention is related to the early discovery by Heisenberg of the hazardous motions of the electrons. Being unaware of any known explanation, Heisenberg proposed to consider such motions as a trait of *nature* and *postulated* that the electron (as any other quantum particle) has an inherently indeterministic and acausal behaviour. This postulate was blindly accepted by the physics community of the time, and was never subjected to experimental corroboration, not even after the advent of quantum field theory, which inherited it. It is interesting to note that even the community of philosophers of science involved in quantum theory, otherwise so demanding, accepted Heisenberg's postulate as a significant part of reality. With this step, the phenomenological description—of the *save-the-phenomenon* type—provided by quantum mechanics was adopted and taken for a fundamental, first-principles construction.

On the other hand, the Schrödinger description and variants thereof contain, either tacitly or explicitly, a nonlocal element that is present in all instances except for the (rather unphysical) case of constant probability density in the entire space. Physicists became accustomed to use (even inadevertently) quantum nonlocality and accept nonlocal behaviour as a quantum trait; of recently this point has taken on increased importance. Being locality a property that is expected of any fundamental physical theory, the conceptual—and philosophical—difficulties with usual quantum theory continued to accumulate.

On top of this, we recall the description of the atom à la Heisenberg as an entity that lives in an abstract, mathematical space—a Hilbert space, a well defined mathematical structure—with no indication at all of what is taking place in real space-time. Such a description leaves us without a meaningful and transparent picture of what the atom is and what its electrons are doing. A further difficulty is related to the introduction of 'quantum jumps' between states, which were (and still are, to a large extent) taken as a capricious quantum trait, not amenable to further analysis. Strictly speaking, we rely on a powerful formal description of the atom, with no associated intelligible picture of it.

Despite the fact that the difficulties mentioned here (among others) strongly suggested the need to adopt a statistical perspective of the quantum phenomenon

(Ballentine, 1998), this possibility was dismissed in general (and adamantly opposed by the Copenhagen school in particular), opening the door to another infelicitous ingredient, the observer. The introduction of this *active* character in order to 'explain' the reduction of the characteristic quantum mixtures to the pure states observed, added a subjective ingredient to the already odd quantum scheme.

10.2 The Missing Ingredient

From a narrative as the one just presented, one concludes indeed that, for historical reasons, in the process of construction of quantum mechanics some fundamental ingredient was left aside. The absence of an appropriate ontological element turned the physical situation into a mystery. The missing ingredient being so fundamental, its disclosure should, as a minimum, explain the quantum motions that go under the indeterministic name of quantum fluctuations, as well as the origin of the apparent nonlocality, the existence of stationary states and the spontaneous transitions between them.

Since all atomic constituents, whether charged or not, have an electromagnetic structure, and electromagnetic interactions are ubiquitous in the atomic world, a natural candidate to fill the ontological gap is the random zero-point electromagnetic field (ZPF). This stochastic field was well known since its introduction into the quantum world by Max Planck in 1912 (Planck, 1912), yet it was largely ignored for decades. Taken as a real electromagnetic field that pervades the entire space, one may envisage that any quantum particle with electromagnetic properties is in permanent contact with it and therefore acquires an essentially stochastic motion (Marshall, 1963). In addition, by serving as a bridge that connects the individual particles of a system, this field is expected to induce correlations between their motions even when they do not interact directly, thus leading to an apparently nonlocal behaviour, as is manifested e. g. in quantum entanglement.

More broadly, consideration of the ZPF as a fundamental ontological constituent offers a range of possibilities to explain quantum phenomena. An important one is the fact that the electron extracts energy from the ZPF, which can make up for the loss of energy due to the radiation of the accelerated charge. As noted by Nernst as early as 1916 (Nernst, 1916), this may explain a most intriguing and persistent mystery, namely atomic stability: how is it that the electron is permanently radiating, but the atom is stable? What's more: a precise compensation of the mean energy radiated to the field by the mean energy absorbed from it can take place only for certain very specific orbital motions, offering in principle an *explanation* of atomic quantization. Simple calculations or estimates taking into account the known properties of the ZPF give support to these conjectures. The theory based on the assumption of the ZPF as an essential ingredient, called stochastic electrodynamics (SED), is not yet fully developed, but it has already produced a variety of positive and promising results (Boyer, 1975; de la Peña and Cetto, 1977, 1996; de la Peña et al., 2015 and references therein).

One important lesson of the theory developed so far is that, as a result of the permanent action of the background field on an otherwise classical particle, a *qualitative, irreversible change in the dynamics* takes place: the particle ceases to behave classically and acquires properties that are considered inherently quantum (Cetto et al., 2021; Cetto and de la Peña, 2022). The effect of the field is not a mere perturbation, which means that a perturbative approach to the SED problem is doomed to produce erroneous results in general. No perturbative calculation to any order will give rise to a qualitatively different behaviour; the *new* situation requires a *new* description. And this is just the quantum one. Briefly: QM ceases to be a mechanical theory to become an *electrodynamic* theory.

The introduction by Planck of the zero-point field and its associated energy $\mathcal{E} = \hbar\omega/2$ per mode, can be considered as significant and groundbreaking as his introduction of the quantization of the energy $\mathcal{E} = \hbar\omega$ interchanged between matter and field, which led him to his famous blackbody formula. On one hand, the zero-point term means a definitive departure from classical electromagnetism by establishing a nonzero energy for the ground state of the field. Further, it provides the basis for an understanding of the quantum phenomenon, as proposed by SED. The fact that the total energy, integrated over the entire range of frequencies from 0 to ∞, has an infinite value, has been used as an argument to deny the reality of this field. This problem, however, is not unique to SED; it is shared by quantum theory, which deals with it in the form of vacuum fluctuations. The infinite energy of the vacuum is in fact an open problem for cosmology, and different attempts to solve it can be found in the literature (see e. g. Santos, 2022).

In this regard it is worth mentioning the proof by Unruh (1995) that the black-hole evaporation process is insensitive to the high-frequency regime. Unruh ascribes this to the time scales involved in the process, which are inversely proportional to the mass of the black hole, hence relatively long compared with Planck or atomic scales, and concludes that 'if the state is the vacuum state at high frequencies, it remains the vacuum.' In SED (as in nonrelativistic QM) the particles interact predominantly with modes of low frequencies, whence the high-frequency modes have no effect on the dynamics; particles are essentially transparent to them. Moreover, at frequencies higher than (double) the Compton frequency, relativistic (high-energy) processes such as particle creation and annihilation take place. Therefore in the calculation of (nonrelativistic) radiative corrections, in which the entire spectrum intervenes in principle, it is legitimate to introduce a cutoff at the Compton frequency (as is usually done in nonrelativistic QED).

Coming back to the conceptual, philosophical considerations, we conclude that the mystery and magic that have accompanied the quantum world over decades, may be dissolved in principle by considering the presence of the ZPF as a real, physical field in permanent contact with matter. Its introduction as an inseparable ingredient of the ontology of any quantum system allows us to recover determinism, causality, locality and objectivity. The corollary is that the classical and the quantum worlds are not two distinct, separate worlds, each obeying its own rules; there is a single world in which they coexist.

10.3 On the Nature of Quantum Stochasticity

A phenomenological theory called stochastic mechanics (alternatively, stochastic quantum mechanics, SQM) was initiated by the mathematician Nelson (1967) with the purpose of describing quantum mechanics as a stochastic phenomenon without the need to specify the source of stochasticity. A more general formulation of SQM was developed later, which serves to describe the dynamics of two distinct types of stochastic process, in the Markov approximation: the classical, Brownian-motion type and the quantum one (de la Peña, 1969; de la Peña and Cetto, 1996).

An important feature of SQM is the appearance of two (statistical) velocities on an equal footing: the flux (or flow) velocity $\boldsymbol{v}$ and the diffusive (or stochastic) velocity $\boldsymbol{u}$. These basic kinematic elements for the description are obtained by averaging over the ensemble of particles in the neighborhood of $\boldsymbol{x}$ at times close to t. When the time interval Δt is taken small but different from zero, the two velocities are obtained, namely (see e. g. Nelson, 1967; de la Peña and Cetto, 1996)

$$\boldsymbol{v}(\boldsymbol{x}, t) = \overline{\frac{\boldsymbol{x}(t + \Delta t) - \boldsymbol{x}(t - \Delta t)}{2\Delta t}} = \hat{\mathcal{D}}_c \boldsymbol{x}, \tag{10.1}$$

with the systematic derivative operator given by

$$\hat{\mathcal{D}}_c = \frac{\partial}{\partial t} + \boldsymbol{v} \cdot \nabla, \tag{10.2}$$

and

$$\boldsymbol{u}(\boldsymbol{x}, t) = \overline{\frac{\boldsymbol{x}(t + \Delta t) + \boldsymbol{x}(t - \Delta t) - 2\boldsymbol{x}(t)}{2\Delta t}} = \hat{\mathcal{D}}_s \boldsymbol{x}, \tag{10.3}$$

with the stochastic derivative operator given by

$$\hat{\mathcal{D}}_s = \boldsymbol{u} \cdot \nabla + D\nabla^2, \tag{10.4}$$

and the diffusion coefficient

$$D = \frac{\overline{(\Delta x)^2}}{2\Delta t}, \tag{10.5}$$

assumed to be constant. The symbol $\overline{(\cdot)}$ denotes ensemble averaging.

The two time derivatives (10.2) and (10.4), applied to the velocities (10.1) and (10.3), give rise to four different accelerations, which are used to construct a couple of generic dynamical equations. In the absence of an external electromagnetic field these are the time-reversal invariant generalization of Newton's Second Law, and the time-reversal non-invariant equation leading to the continuity equation, respectively,

$$m\left(\hat{\mathcal{D}}_c \boldsymbol{v} - \lambda \hat{\mathcal{D}}_s \boldsymbol{u}\right) = \boldsymbol{f}, \tag{10.6}$$

$$m\left(\hat{\mathcal{D}}_c \boldsymbol{u} + \hat{\mathcal{D}}_s \boldsymbol{v}\right) = 0, \tag{10.7}$$

with λ a free, real parameter, and $\boldsymbol{f} = -\nabla V$ the external force acting on the particle. Since the magnitude of λ can be absorbed into the value of D, one takes $\lambda = \pm 1$. The specific dynamical properties of the system strongly depend on the sign of this parameter: $\lambda = -1$ implies an irreversible dynamics, of the Brownian-motion type. By contrast, $\lambda = 1$ implies a reversible stochastic dynamics and leads after some algebra to the Schrödinger-like equation

$$-2mD^2\nabla^2\psi(\boldsymbol{x}, t) + V(\boldsymbol{x})\psi(\boldsymbol{x}, t) = 2imD\frac{\partial\psi(\boldsymbol{x}, t)}{\partial t}, \tag{10.8}$$

and its complex conjugate, where $\psi(\boldsymbol{x}, t)$ is a complex function whose squared modulus is the density distribution

$$\rho(\boldsymbol{x}, t) = |\psi(\boldsymbol{x}, t)|^2, \tag{10.9}$$

and

$$\boldsymbol{v} = iD\left(\frac{\nabla\psi^*}{\psi^*} - \frac{\nabla\psi}{\psi}\right), \quad \boldsymbol{u} = D\left(\frac{\nabla\psi^*}{\psi^*} + \frac{\nabla\psi}{\psi}\right). \tag{10.10}$$

The Schrödinger equation proper is obtained from (10.8) by taking

$$D = \hbar/2m. \tag{10.11}$$

10.3.1 A Possible Connection with Hydrodynamics

By subtracting Eq. (10.7) from (10.6) and introducing the *access* (or backward) velocity, which is the linear combination of the velocities $\boldsymbol{v}$ and $\boldsymbol{u}$

$$\boldsymbol{v}_a = \boldsymbol{v} - \boldsymbol{u} = (\mathcal{D}_c - \mathcal{D}_s)\boldsymbol{x}, \tag{10.12}$$

one obtains after some rearrangements

$$\frac{\partial}{\partial t}\boldsymbol{v}_a + \boldsymbol{v}_a \cdot \nabla \boldsymbol{v}_a - D\nabla^2 \boldsymbol{v}_a = -\frac{1}{m}\nabla(V + 2V_Q), \tag{10.13}$$

where V_Q stands for the quantum potential,

$$V_Q = -\frac{1}{2}m\boldsymbol{u}^2 - mD\nabla \cdot \boldsymbol{u} = -2mD^2\frac{\nabla^2\sqrt{\rho}}{\sqrt{\rho}}. \tag{10.14}$$

Incidentally, this equation shows that the quantum potential V_Q is determined by the spatial density $\rho(x, t)$, a function that depends on what is happening in the entire space, and thus contains nonlocal information. Equation (10.13) corresponds in hydrodynamics to the Navier-Stokes equation for an incompressible, viscous fluid, if mD is taken for the kinematic viscosity ν and $(V + 2V_Q)/m$ is identified with the total pressure divided by the fluid density ρ_o. With this identification, Planck's constant corresponds to $\hbar \Longleftrightarrow 2\rho_o\nu$. Interestingly, however, this is not an equation for the flow velocity $\boldsymbol{v}$ but for the access velocity $\boldsymbol{v}_a$. This velocity represents the coarse time-scale local average of the displacement from time $t - \Delta t$ to time t, i. e.,

$$\boldsymbol{v}_a(\boldsymbol{x}) = \frac{\langle \boldsymbol{x}(t) - \boldsymbol{x}(t - \Delta t)\rangle}{\Delta t} = \frac{\langle \Delta_- \boldsymbol{x}\rangle}{\Delta t}, \tag{10.15}$$

where the average is taken over the ensemble of particles that cross the point $\boldsymbol{x}$ at time t, taking into account that at an earlier time $t - \Delta t$ those particles had a distribution of positions $\boldsymbol{x}' = \boldsymbol{x}(t - \Delta t)$. It should be borne in mind that Δt must be much smaller than the characteristic time of the systematic motion, but large enough as to embrace the most closely spaced, rapid changes in $\boldsymbol{x}$. If the system satisfies an ergodic principle, t may also represent the different times at which the same particle crosses the point $\boldsymbol{x}$ again and again.

A question that comes to mind is whether this analogy between the quantum and the hydrodynamic cases can be extended to include the effect of a walking droplet on the fluid and, ultimately, obtain the 'quantumlike' behaviour of the fluid-droplet system (Bush, 2015). In particular, are the statistics of the horizontal trajectories of walking droplets the equivalent of the quantum statistics represented by ρ, as is suggested in several papers on the subject?

For a system composed of a fluid layer acted on by a vibrating force and by the bouncing droplet, the total pressure appearing in the Navier-Stokes equation must include the external pressure P on the fluid surface due to the bouncing droplet, and the gravity term g must include the effective acceleration due to the vibrations, $g(t) = g + \gamma(t)$. Further, the resulting Navier-Stokes equation has to be complemented with the equation of motion for the droplet under the action of the fluid. Since with each bouncing, the droplet pressure on the fluid surface modifies the value of $h(\boldsymbol{x})$, and the dynamics of the droplet depends in its turn on the force exerted upon it by the modified fluid surface, one ends up with a coupled, nonlinear system of equations that is difficult to analyze with a view to establishing a (direct) comparison with SQM, viz. QM.

The accumulated memory effects on the wave field, and the dependence of the droplet-fluid interaction on the relative phase and varying shape of the fluid surface during contact, seem to be essential points to consider in this regard. The first of these suggests focusing on hydrodynamic analogs of the stationary states in quantum mechanics, when the wave field has become stabilized and a well-defined histogram of the droplet positions is obtained. The second point suggests introducing a random element in the description of the horizontal motions, which could serve to bring to the surface the counterpart of the diffusion coefficient of SQM—or equivalently, of the stochastic velocity $\boldsymbol{u}$. It seems to us of considerable importance that $\boldsymbol{u}$ may acquire values comparable to those of $\boldsymbol{v}$.

To give precision to the above considerations, we rewrite Eq. (10.13) in the form

$$\frac{D\boldsymbol{v}_a}{Dt} = -\frac{1}{m}\nabla(V - mD\nabla \cdot \boldsymbol{v}_a + 2V_Q). \tag{10.16}$$

This equation, which is just another form of the Schrödinger equation, offers a statistical description of the moving quantum *particles* of the problem under scrutiny, the analog of the droplets. We should then perhaps reconsider our previous 'hydrodynamic' point of view and take Eq. (10.16), and thus (10.13), as the quantum equivalent (or analog) of Newton's equation of motion for the walking droplets. The mean local force acting on the droplets contains then, in addition to the expected classical components, a term similar to the quantum potential with its associated nonlocality.

10.4 The Wave Element in SED

A usual starting point for the analysis of the particle dynamics in SED is the (nonrelativistic) equation of motion, known as Braffort-Marshall equation

$$m\ddot{\boldsymbol{x}} = \boldsymbol{f}(\boldsymbol{x}) + m\tau\dddot{\boldsymbol{x}} + e\boldsymbol{E}(t), \tag{10.17}$$

where $m\tau\dddot{\boldsymbol{x}}$ stands for the radiation reaction force, with $\tau = 2e^2/3mc^3$ (de la Peña and Cetto, 1996). For an electron, $\tau \approx 10^{-23}$ s. $\boldsymbol{E}(t)$ represents the electric component of the ZPF taken in the long-wavelength approximation, with time correlation given in the continuum limit by $(j, k = 1, 2, 3)$

$$\left\langle E_k(s)E_j(t)\right\rangle = \delta_{kj}\varphi(t-s), \tag{10.18a}$$

where the spectral function

$$\varphi(t-s) = \frac{2\hbar}{3\pi c^3}\int_0^\infty d\omega\,\omega^3\cos\omega(t-s) \tag{10.18b}$$

corresponds to an energy $\hbar\omega/2$ per mode. This establishes a crucial distinction between classical, Brownian-type processes, governed by a white noise, and the present stochastic process which is governed by a stationary, correlated (wave) field with a highly coloured spectrum.

A statistical treatment of the one-particle problem, starting from Eq. (10.17) and involving the construction of a generalized Fokker-Planck equation, has been shown to lead in the radiationless approximation to the Schrödinger equation (de la Peña et al., 2014, 2015); this is the approximation that corresponds to quantum mechanics. The stationary states are obtained as the solutions that satisfy the energy-balance condition, as predicted by Nernst over a century ago. When the radiative terms are not neglected, the theory reproduces in addition the corresponding corrections in coincidence with non-relativistic quantum electrodynamics (Cetto et al., 2013; de la Peña et al., 2015). In particular, the formulas for the radiative lifetimes are obtained for the excited states that in the Schrödinger approximation appear as stationary.

Interestingly, in the process leading to the Schrödinger equation via the Fokker-Planck equation, which focuses on the statistical properties of the dynamics, the wave element is not conspicuous; it remains as concealed as in SQM. Yet the ZPF eventually leaves its indelible mark through the appearance of Planck's constant in the Schrödinger 'wave' equation.

An alternative route followed in a subsequent development of SED, called linear stochastic electrodynamics (LSED), takes us to the quantum equations in its matrix formulation (de la Peña et al., 2015; see also Cetto et al., 2021 and Cetto and de la Peña, 2022 for more recent work). As is well known, the wave element is absent from the final (Heisenberg) equations, which have a purely mechanical aspect. However, in the SED process leading to these equations (see Sect. 10.5 below), the relevant field modes with which the particle interacts appear explicitly. This allows us to investigate the role played by such field modes when two (not directly interacting) identical particles are simultaneously connected to them. As a result, the entanglement of particles finds an explanation in the correlation between their motions established through the common field modes (de la Peña et al., 2010, 2015). In this regard, it is appropriate to mention the connection with the hydrodynamic work by Borghesi et al. (2014), who investigate experimentally the energy stored in the wave field for two coupled walking droplets and how it conveys an interaction between them. More generally, the observation that the background field acts as a bridge between particles has important consequences for the statistics of (identical) quantum particles.

10.4.1 Compton's Frequency Revindicated

In de Broglie's pioneering work on wave mechanics, the Compton frequency is well known to have played a key role as the frequency of the particle's internal clock. (Later it was understood by some that de Broglie's wavelength has actually a statistical meaning.) Nevertheless, the nature of the associated wave in de Broglie's theory, which was basic for the development of Schrödinger's theory, was left unidentified by de Broglie, and in fact it remains to date concealed at the core of the foundations of quantum mechanics.

A variety of recent works, both theoretical and experimental, point to a revival of the internal clock conjectured by de Broglie, which in turn may have an important impact on our understanding of the quantum phenomenon. Among such works we have on one hand a series of experiments in which high-energy electron beams are channelled through silicon crystals (Catillon et al., 2008; Osche, 2011), providing evidence of a resonance suggestive of de Broglie's clock. Ironically, these findings have received apparently almost no attention, except for an essay by Hestenes (2008) and a Monte-Carlo-based analysis by Bauer (2019) showing consistency with Dirac's description of the free particle.

On the other hand there is a growing series of both experimental and theoretical work carried out in the field of bouncing droplets referred to above (Bush, 2015), in which a high-frequency vibration—somehow parallel to de Broglie's clock—is shown to play a central role in producing phenomena that suggest a hydrodynamic quantum analogy. Also in other fields of physics, high-frequency vibrations imposed on a material system seem to have an important influence, inducing qualitative changes in the dynamics of the system. Worth mentioning are a couple of recent extensions to a generic elastic system with a bead (Borghesi, 2017) where also the internal clock of high frequency appears as a key ingredient.

De Broglie's frequency has had occasional appearances in SED. A first attempt to establish contact with it is the work by Surdin (1982) where the electromagnetic nature of de Broglie's wave is specified as follows from SED, and some experiments dealing with interference and diffraction of neutrons are considered. In de la Peña and Cetto (1996) and de la Peña et al. (2015) we present a preliminary view on the functioning of the modes of the ZPF of Compton's frequency, and their relation to de Broglie's clock, with a few simple but interesting results. In dealing with some quantum problems within the SED framework we have been induced to introduce, as an ancillary step, a revised version of de Broglie's idea, appropriately updated in accordance with the demands of SED, which means assigning an *electromagnetic* nature to the de Broglie wave. In the process leading to the appearance of de Brogle's wave, the assumption of a resonant interaction of the electron with the ZPF modes of Compton's frequency has been the starting point. Essentially this amounts to distinguishing the modes of the ZPF of Compton's frequency from the rest of the field.

More recently, we have put forward an estimate of the duration of a transition between atomic states—the infamous quantum jumps—based on the assumption that the transition is triggered precisely by a resonance of the atomic electron with modes of the zero-point radiation field of Compton's frequency (de la Peña et al., 2020). The theoretical result, given essentially by the expression $(\alpha\omega_C)^{-1}$, where $\alpha \sim 1/137$ is the fine structure constant and ω_C the Compton angular frequency for the electron, lies well within the range of the experimentally estimated values, which is of the order of attoseconds (10^{-18} s) (Ossiander et al., 2018; Minev et al., 2019). Incidentally, one can still come across articles negating quantum transitions, in adherence to Bohr's dictum—and any other kind of apparent discontinuities, for that matter (e.g. Jadszyk, 2014)—or taking them as a sudden increase of our knowledge of the system (e. g., Zeh, 1993; Plenio and Knight, 1998) rather than a physical phenomenon.

10.5 On the Classical-to-Quantum Transition

The transition from the classical to the quantum regime represents a most delicate point of the emerging-quantum theory. A detailed account of the dynamics leading to quantization, and of the conditions under which quantization is obtained, is still a pending task, although some important steps have been achieved.

Notably, by applying a Hamiltonian treatment to the composite (initially noninteracting) particle-field system, it is possible to show that in the classical-to-quantum transition, the field takes control of the response of the particle. As a result of the interplay of stochasticity and dissipation, the particle loses memory of its initial conditions and ends up responding linearly and resonantly to a well-defined set of field modes, which depends in each instance on the specific problem (i. e., on the external forces acting on the particle) (Cetto et al., 2021; Cetto and de la Peña, 2022). In the new situation the dynamics is no longer described by the phase-space mechanical variables, but by the coefficients of the response of the particle to those field modes. The response coefficients turn out to be nothing less than the matrix elements of the operators $\hat{x}$ and $\hat{p}$, which satisfy the basic quantum commutator $\left[\hat{x}, \hat{p}\right] = i\hbar$, just as was advanced by Heisenberg in his pioneering work on quantum mechanics (Heisenberg, 1925); this reflects the fact that the Hamiltonian evolution preserves the original symplectic structure.

The Hilbert-space formalism represents thus a compact and elegant way of describing the response of the particle to a set of (relevant) field modes once the quantum regime has been attained, at the cost of a space-time description of what is really happening inside the atom. Interestingly, the field has disappeared once more from the picture, this time leaving its indelible mark through the appearance of Planck's constant in the basic commutator.

It should be noted that prior to the onset of the quantum regime, the dynamics is irreversible and memory accumulates in the near background field, which is an indication of a highly non-Markovian process. In the quantum regime, by contrast, the stochastic process is Markovian, the Markovianity referring to the loss of memory between quantum-mechanical processes such as atomic transitions. This feature is related with the times involved in the description: we recall from Sect. 10.3 that the time interval Δt introduced in SQM to derive the QM equations entails a statistical coarse-grain description of the dynamics. The minimum times involved in (non-relativistic) QM are smaller than the orbital periods and the lifetimes of excited states, but certainly larger than the inverse of the Compton (or Zitterbewegung) frequency, or even the time involved in a 'quantum jump' (which as discussed above, is an estimated 100 times larger than the Compton time). For this reason, QM as we know it is unable to describe with a sufficiently high resolution in time the trajectories of particles, transitions between states, or other continuous processes.

10.6 Final Comments

The experience with SED, as illustrated by the results discussed in the preceding sections, is that the main features characteristic of quantum-mechanical systems emerge from the permanent interaction of the random zero-point radiation field with matter; for this reason it can be said that QM is an emergent theory. By adding the ZPF to the quantum ontology, a physically coherent, local and objective picture of the quantum phenomenon is obtained, free of ad-hoc postulates and mysterious elements.

It seems hard to rule out the possibility that analogous quantization phenomena take place in other realms of physics, in which a stationary, vibrating background acts permanently on one or more corpuscles. The (hydrodynamic) walking-droplet system and the (electrodynamic) atomic-matter system are perhaps two prominent examples of a more general phenomenon with interesting ramifications. It remains to be seen whether a common ground can be found that allows us to deepen our mutual learning, further identify commonalities, and establish the extent of the analogies, for the benefit of a more comprehensive, satisfactory understanding of the quantum phenomenon.

References

Bauer, M. 2019. De Broglie clock, electron channeling, and time in quantum mechanics. *Canadian Journal of Physics* 97(1): 37.

Borghesi, C. 2017. Equivalent quantum equations in a system inspired by bouncing droplets experiments. *Foundations of Physics 47*: 933.

Borghesi, C., J. Moukhtar, M. Labousse, A. Eddi, E. Fort, and Y. Couder. 2014. Interaction of two walkers: Wave-mediated energy and force. *Physical Review E 90*: 063017.

Boyer, T.H. 1975. Random electrodynamics: The theory of classical electrodynamics with classical electromagnetic zero-point radiation. *Physical Review D 11*: 790.

Bush, J.W.M. 2015. Pilot-wave hydrodynamics. *Annual Review of Fluid Mechanics 47*: 269.

Catillon, P., et al. 2008. A Search for the de Broglie particle internal clock by means of electron channeling. *Foundations of Physics 38*: 659.

Cetto, A.M., and L. de la Peña. (2022). Role of the electromagnetic vacuum in the transition from classical to quantum mechanics. *Foundations of Physics 52*: 84.

Cetto, A.M., L. de la Peña, and A. Valdés-Hernández. 2013. Atomic radiative corrections without QED: Role of the zero-point field. *Revista Mexicana de Física 59*: 433.

Cetto, A.M., L. de la Peña, and A. Valdés-Hernández. 2021. On the physical origin of the quantum operator formalism. *Quantum Studies: Mathematics and Foundations 8*: 229–236. https://doi.org/10.1007/s40509-020-00241-7

de la Peña, L. 1969. New formulation of stochastic theory and quantum mechanics. *Journal of Mathematical Physics 10*: 1620.

de la Peña, L., and A.M. Cetto. 1977. Derivation of quantum mechanics from stochastic electrodynamics. *Journal of Mathematical Physics 18*: 1612.

de la Peña, L., and A.M. Cetto. 1996. *The Quantum Dice*. Dordrecht: Kluwer Academic.

de la Peña, L., A. Valdés-Hernández, and A.M. Cetto. 2010. Entanglement of particles as a result of their coupling through the common background zero-point radiation field. *Physica E 42*: 308; Valdés-Hernández, A., L. de la Peña, and A.M. Cetto. 2011. Bipartite entanglement induced by a common background (zero-point) radiation field. *Foundations of Physics 41*: 843.

de la Peña, L., A.M. Cetto, and A. Valdés-Hernández. 2014. The zero-point field and the emergence of the quantum. *International Journal of Modern Physics E 23*(9): 1450049; Cetto, A.M., L. de la Peña, and A. Valdés-Hernández. 2015. Specificity of the Schrödinger equation. *Quantum Studies: Mathematics and Foundations 2*: 275.

de la Peña, L., A.M. Cetto, and A. Valdés-Hernández. 2015. *The Emerging Quantum*. Berlin: Springer.

de la Peña, L., A.M. Cetto, and A. Valdés-Hernández. 2020. How fast is a quantum jump? *Physics Letters A 384*: 126880.

Heisenberg, W. 1925. Über quantentheoretische Umdeutung kinematischer und mechanischer Beziehungen. *Zeitschrift für Physik 33*: 879.

Hestenes, D. 2008. *Reading the Electron Clock*. arXiv:0802.3227v1.

Jadszyk, A. 2014. Are there quantum jumps? In *Quantum Fractals. From Heisenberg's Uncertainty to Barnsley's Fractality*. Singapore: World Scientific.

Marshall, T.W. 1963. Random electrodynamics. *Proceedings of the Royal Society A 276*: 475.

Minev, Z.K., et al. 2019. To catch and reverse a quantum jump mid-flight. *Nature 570*: 7760, 200. https://doi.org/10.1038/s41586-019-1287-z

Nelson, E. 1967. *Dynamical Theories of Brownian Motion*. Princeton: Princeton University Press. *Quantum fluctuations*, Princeton University Press, Princeton.

Nernst, W. 1916. Über einen Versuch, von quantentheoretischen Betrachtungen zur Annahme stetiger Energieänderungen zurückzukehren. *Verhandlungen der Deutschen Physikalischen Gesellschaft 18*: 83.

Osche, G.R. 2011. Electron channeling resonance and de Broglie's internal clock. *Annales de la Fondation Louis de Broglie 36*: 61.

Ossiander, M., et al. 2018. Absolute timing of the photoelectric effect. *Letter to Nature 561*: 20.

Planck, M. 1912. Über die Begründung des Gesetzes der schwarzen Strahlung. *Annalen der Physik 37*: 642.

Plenio, M. B., and P.L. Knight. 1998. The quantum-jump approach to dissipative dynamics in quantum optics. *Reviews of Modern Physics 70*: 101.

Santos, E. 2022. *Realistic Interpretation of Quantum Mechanics*. Cambridge: Cambridge Scholars Publisher, Ch. 7.

Surdin, M. 1982. De Broglie's wave revisited. *Foundations of Physics 12*: 873.

The statistical description of quantum mechanics was proposed and defended by several authors, Einstein being the most prominent of them. For a more recent discussion see L. E. Ballentine 1998. *Quantum Mechanics: A Modern Development*. Singapore: World Scientific.

Unruh, W.G. 1995. Sonic analogue of black holes and the effects of high frequencies on black hole evaporation. *Physical Review 51*(6): 2827.

Zeh, H.D. 1993. There are no quantum jumps, nor are there particles! *Physics Letters A 172*: 189.

Chapter 11
Are Hidden-Variable Theories for Pilot-Wave Systems Possible?

Louis Vervoort

Abstract Recently it was shown that certain fluid-mechanical 'pilot-wave' systems can strikingly mimic a wide range of quantum properties, including single particle diffraction and interference, quantization of angular momentum etc. How far does this analogy go? The ultimate test of quantumness of such systems is a Bell-test. Here the premises of the Bell inequality are re-investigated for particles accompanied by a pilot-wave, or more generally by a resonant 'background' field. We find that two of these premises, namely outcome independence and measurement independence, may not be generally valid when such a background is present. Under this assumption the Bell inequality is possibly (but not necessarily) violated. A class of hydrodynamic Bell experiments is proposed that can test this claim. Such a Bell test on fluid systems can provide a wealth of new insights on the different loopholes for Bell's theorem. Finally, it is shown that certain properties of background-based theories can be illustrated in Ising spin-lattices.

Keywords Hidden variables · Bell's theorem · Nonlocality · Fluid mechanics · Bell's inequality · Background field · Spin lattices

11.1 Introduction

Since the birth of quantum mechanics, physicists have been intrigued by its counterintuitive features, such as the collapse of the wave function, the uncertainty relations, wave-particle duality, the probabilistic nature of the theory, etc. Many attempts have been made to restore a more classic character to quantum mechanics, notably by efforts aiming at deriving the theory from a more fundamental, deeper-lying theory – maybe even a deterministic one. Such a hidden-variable theory (HVT) would contain yet unknown variables which, once integrated-out, yield quantum

L. Vervoort (✉)
Higher School of Economics, Moscow, Russia
e-mail: lvervoort@hse.ru

P. Castro et al. (eds.), *Advances in Pilot Wave Theory*, Boston Studies in the Philosophy and History of Science 344,
https://doi.org/10.1007/978-3-031-49861-9_11

theory. This would after all be a familiar situation in physics: quantum mechanics would have the status of an 'effective' theory, as other theories. But this quest for sub-quantum theories, initiated by such giants as Einstein, de Broglie, Schrödinger etc., has been restrained, not only by the unprecedented precision and efficiency of quantum theory, but also by certain mathematical results that seem to prove the impossibility of constructing any reasonable HVT. These are the so-called 'no-go' theorems, among which Bell's theorem (Bell 1964, 1981, 2004) but also the Kochen-Specker theorem (1967) are best known. What I call 'Bell's theorem' (as a physical, and not just mathematical, theorem) states, in short, that 'local hidden-variable theories are impossible'. It is essential here to be clear about what 'local' means: unless explicitly stated otherwise, 'local' will mean here '*only invoking not-faster-than-light interactions*' – the relevant meaning in relativity theory. Thus I will use the physics rather than the information-theoretic jargon, where 'local' has different meanings and is usually equated with 'satisfying a Bell inequality' or with 'jointly satisfying the Bell-premises' (namely Eqs. 11.1a, 11.1b, and 11.1c below), or with other conditions (Brunner et al. 2014). Of course, in the context of physics the important and more general question is: *can quantum mechanics be derived from a local theory – local in the sense defined above*. But according to the received wisdom condensed in Bell's theorem, this is not possible.

But one cannot escape from following observation: hundred years after quantum mechanics' birth, and fifty years after's Bell's discovery, a large part of physicists remains skeptical about the most extraordinary claim of the standard interpretation of quantum mechanics and of Bell's theorem; namely that nature is either a-causal (quantum events arise out of nothing, have no causes) or nonlocal (superluminal interactions exist), or both. This is such a counterintuitive and from other respects unlikely prospect, that attempts to adopt a more classical picture deserve full attention – especially in view of the lasting problem of unifying quantum mechanics and relativity theory. Here we will explore such a potential revision of Bell's theorem, not of course of its mathematics, but of the physical validity of its premises (Eqs. 11.1a, 11.1b, and 11.1c below). We are well aware that our arguments contain a speculative element; but the good news is that they can be tested on model systems. And we predict that in the process a wealth of new insights on the real Bell experiment will be gained – as a minimal pay-off.

Now, recent experiments go against the possibility of local HVTs, since they are generally believed to have closed the most relevant loopholes for such theories (e.g. Giustina et al. 2015; Hensen et al. 2015; Shalm et al. 2015), in these references one finds the rich history of Bell tests). These experiments use a sophisticated set-up to close e.g. the detection, locality and freedom-of-choice loopholes, notably by quickly and randomly varying analyzer settings, more generally by imposing spacelike separation between relevant events. One supplementary merit of these articles (e.g. Giustina et al. 2015; Hensen et al. 2015; Shalm et al. 2015) is that they have sharply defined under which conditions the Bell inequality may be assumed to hold, therefore under which precise assumptions local HVTs would be eliminated. In particular, the authors Giustina et al. (2015) specify in detail how their experiment would close the freedom-of-choice loophole: "This loophole can be closed only under specific assumptions about the origin of [the HV] λ. Under the assumption that λ is created with the particles to be measured, an experiment in which the settings are generated independently at the measurement stations and spacelike separated from the creation of the particles closes the loophole."

Here I will investigate the admissibility of local HVTs that do *not* assume that the HVs λ are 'created with the particle pair'. In particular HV models will be studied that include variables describing, not the particle pair, but a background field in which the Bell particles propagate. As is implicit in the above quote from Giustina et al. (2015), there is an argument why for such theories the freedom-of-choice loophole cannot be closed. Indeed, in a slightly more precise wording, experiments can 'close the freedom-of-choice loophole' in that they 'impose measurement independence' (MI), or the stochastic independence between the HVs λ and the left and right analyzer variables (a,b). As recalled in Sect. 11.2, all Bell inequalities assume MI, the condition (11.1c) below. Under the precise assumptions of the above quote, MI is valid (unless the universe would be superdeterministic or nonlocal). But assume now more generally that $\lambda \equiv (\lambda_0, \lambda_1, \lambda_2)$, where λ_1 are properties of a background field in the space-time neighborhood of the left analyzer setting (at angle a), λ_2 properties of the field close to the right analyzer setting (b), and λ_0 all other relevant properties. Then the conditional probability $\rho(\lambda|\,a, b) \equiv (\lambda_0, \lambda_1, \lambda_2|\,a, b)$ will in general be different from the unconditional $\rho(\lambda_0, \lambda_1, \lambda_2)$, simply because λ_1 can interact with analyzer 'a' and λ_2 with analyzer 'b'. Examples of this type of correlations in known systems, hydrodynamic pilot-wave systems and spin-lattices, will be given in Sect. 11.3 and the Appendix. In Sect. 11.4 it will be argued that *in the presence of a resonant background or pilot-wave* MI-violation does not necessarily imply violation of free will or nonlocality. This suffices to reach our main conclusion, namely that in a pilot-wave system one of the premises of the Bell-inequality, namely MI, does not necessarily hold, and that therefore the Bell inequality could *possibly* be violated in such a system. This result can thus be seen as corroborating a series of articles (see in particular Hall 2010, 2011, 2015; 't Hooft 2015; Khrennikov 2008; Vervoort 2000, 2013, 2016; Nieuwenhuizen 2009; Brans 1988; Kupczynski 1986; Khrennikov 2011; 't Hooft 2016; 't Hooft 2018), which have contested the physical validity of MI and therefore of the no-go theorem. As far as I know Brans (1988) was the first to construct an explicit model violating MI and reproducing the quantum correlation of the Bell experiment.

In Sect. 11.4 it will be argued that in a resonant background a second premise of the Bell inequality, namely outcome independence (OI, 11.1a) can also be violated in a physically acceptable manner, i.e. compatible with locality. All depends on what the hidden-variables are supposed to represent. A background-based or pilot-wave toy model will be given that violates both MI and OI and that maximally violates the Bell inequality, while yet being non-signaling (Sect. 11.4).

To physically justify the model, it will be shown in Sect. 11.3 that the types of correlations needed (violation of MI and OI) exist in well-studied systems involving a resonant background. Indeed, this work is inspired by fluid-mechanical systems that were discovered about a decade ago by Couder, Fort and collaborators (2005;

Eddi et al. 2012; Perrard et al. 2014), and since then investigated in great detail also by other teams, notably by Bush and collaborators (Bush 2015a, b; Harris and Bush 2014) and Gilet and collaborators (Sampara and Gilet 2016; Gilet 2016). These systems, oil droplets walking over a vibrating fluid film, exhibit a remarkable series of quantum-like features, all induced by a pilot-wave. Section 11.3 summarizes the essential probabilistic properties of these hydrodynamic systems; it will be shown that the long-range correlations we need massively arise here, due to resonant interaction of the droplets with a pilot-wave. In the Appendix Ising spin-lattices are investigated, which are also massively correlated through a stochastic background (namely a collection of spins in Boltzmann equilibrium). Even if spin-lattices are static and therefore not realistic as an analogue for pilot-wave HVTs, they can illustrate at least some properties of the toy model, in particular violation of MI and of the Bell inequality. It is hoped they could serve as a 1st-order approximation of realistic HVTs. Note that spin-lattices are cellular automata, and that they thus have a simple conceptual link with 't Hooft's HVT, the Cellular Automaton Interpretation of quantum mechanics ('t Hooft 2015, 2016). 't Hooft also contests the general validity of MI.

The conclusions we arrive at can be tested, by experimental means that seem well within current reach. Indeed, any *classical* background-based system, in particular the hydrodynamic system of Couder et al. (2005), Eddi et al. (2012), Perrard et al. (2014), Bush (2015a, b), and Harris and Bush (2014), that would violate a Bell inequality would corroborate our model. Since the detailed correlations in the hydrodynamic system are not known, we cannot propose a detailed experiment; but a not yet fully specified class of hydrodynamic Bell-tests can be proposed (Sect. 11.5). Finally, Sect. 11.6 discusses candidate theories for the generic model: it was recently argued (e.g. Bush 2015a; Groessing et al. 2014; Couder and Fort 2012) that the surprising quantum/hydrodynamic analogies incite one to revisit sub-quantum theories as de Broglie's pilot-wave theory (de Broglie 1987) and its modern variants, notably stochastic electrodynamics (de la Peña et al. 2015), as well as other theories involving a background or pilot wave (Brady and Anderson 2014; Groessing et al. 2014; Sbitnev 2015). Preliminary ideas of the present work were presented in Vervoort (2016), but here a different and substantially elaborated model is given, as well as a comparison with spin-lattices.

11.2 Premises of Bell-Inequalities

Let us first recall the assumptions on which Bell-type inequalities are based. In a Bell-type experiment one measures correlations $P(\sigma_1, \sigma_2| a, b)$, i.e. joint probabilities for finding an outcome with value 'σ_1' (± 1) on the left particle and 'σ_2' on the right particle, given that the value of the analyzer variable on the left is 'a', and 'b' on the right (the analyzer variables themselves will sometimes, when necessary for clarity, be denoted by x and y). A HV model assumes that these correlations can be explained by HVs λ with a distribution ρ. In the most general stochastic setting, of which the deterministic case is but a special instance, the (minimal) premises of the Bell inequality are the following conditions, often termed 'outcome independence' (OI), 'parameter independence' (PI) (or 'no-signaling') and 'measurement independence' (MI) (e.g. Hall 2011):

$$\mathrm{P}(\sigma_1|\sigma_2, \mathrm{a}, \mathrm{b}, \lambda) = \mathrm{P}(\sigma_1|\mathrm{a}, \mathrm{b}, \lambda) \qquad (\mathrm{OI}), \tag{11.1a}$$

$$\mathrm{P}(\sigma_2|\mathrm{a}, \mathrm{b}, \lambda) = \mathrm{P}(\sigma_2|\mathrm{b}, \lambda) \text{ and similarly for } \sigma_1 \qquad (\mathrm{PI}), \tag{11.1b}$$

$$\rho(\lambda|\mathrm{a}, \mathrm{b}) = \rho\left(\lambda|\mathrm{a}', \mathrm{b}'\right) = \rho(\lambda). \qquad (\mathrm{MI}) \tag{11.1c}$$

These conditions of stochastic independence are supposed to hold for all relevant values of the variables $(\lambda, \sigma_1, \sigma_2, \mathrm{x}, \mathrm{y})$ appearing in the model or theory. Note that the conjunction of OI and PI is equivalent to the well-known Clauser-Horne factorability condition (Clauser and Horne 1974):

$$\mathrm{P}(\sigma_1, \sigma_2|\mathrm{a}, \mathrm{b}, \lambda) = \mathrm{P}(\sigma_1|\mathrm{a}, \lambda).\mathrm{P}(\sigma_2|\mathrm{b}, \lambda). \tag{11.2}$$

Assuming the validity of (11.1a, 11.1b, and 11.1c) one finds, using standard rules of probability calculus:

$$\mathrm{P}(\sigma_1, \sigma_2|\mathrm{a}, \mathrm{b}) = \int \mathrm{P}(\sigma_1|\mathrm{a}, \lambda).\mathrm{P}(\sigma_2|\mathrm{b}, \lambda).\rho(\lambda).\mathrm{d}\lambda. \tag{11.3}$$

The form (11.3) leads without further assumptions to a Bell inequality:

$$\mathrm{X}_{\mathrm{BI}} = |\, \mathrm{M}(\mathrm{a}, \mathrm{b}) + \mathrm{M}\left(\mathrm{a}', \mathrm{b}\right) + \mathrm{M}\left(\mathrm{a}, \mathrm{b}'\right) - \mathrm{M}\left(\mathrm{a}', \mathrm{b}'\right) | \leq 2 \forall \left(\mathrm{a}, \mathrm{a}', \mathrm{b}, \mathrm{b}'\right), \tag{11.4}$$

where the average product

$$\mathrm{M}(\mathrm{x}, \mathrm{y}) = <\sigma_1.\sigma_2>_{\mathrm{x},\mathrm{y}} = \sum_{\sigma_1\sigma_2} \sigma_1.\sigma_2 P(\sigma_1, \sigma_2|x, y). \tag{11.5}$$

OI, PI and MI are generally believed to describe *any* reasonable HVT. Jointly they are sufficient conditions for a Bell inequality to hold for the given model or theory; they are however not necessary conditions (e.g. MI may be violated in a model while the Bell inequality is still satisfied). OI and PI conjoined or the factorability condition (11.2) are often termed the 'locality condition' and are assumed to necessarily follow from Einstein locality (Bell 2004). However, in Sect. 11.4 it will be argued that OI is not necessarily a good characterization of locality in pilot-wave systems. An even more subtle premise is MI. The usual reasoning to justify this assumption goes as follows: in a local system λ must be independent of (a,b); if not (a,b) would depend on λ (by Bayes' rule); but that is impossible because (a,b) can be freely or randomly chosen and such free variables cannot be determined by variables that determine the particle outcomes – unless one accepts a conspiratorial (superdeterministic) or a nonlocal world. Based on this classic argument (Bell 2004; Hall 2010, 2011, 2015) one often calls MI 'freedom-of-choice'. But as mentioned above, the general validity of MI may be questioned (Hall 2010, 2011, 2015; 't Hooft 2015; Khrennikov 2008; Vervoort 2000, 2013, 2016; Nieuwenhuizen 2009; Brans 1988; Kupczynski 1986; Khrennikov 2011; 't Hooft 2016, 2018). In our view, the most cogent argument against MI is the fact that in a truly deterministic world it is not applicable (Vervoort 2013; 't Hooft 2018).

Another argument against MI and OI is linked to the nature of the HVs λ: do they describe the particle pair or should they be conceived more broadly? And at what time these variables are taken? Some authors specify that the λ belong to the particles, like Bell in his original article (Bell 1964) and Aspect even in his latest review (Aspect 2015). But for most authors the λ are not restricted to particle properties and can or should represent *any* additional information – cf. e.g. the review article (Brunner et al. 2014), and Bell in more recent articles (Bell 1981, 2004). In (1981, p. 56) Bell says: "It is notable that in this argument nothing is said about the locality, or even localizability, of the variable λ. [...] It is assumed only that the outputs [σ_1] and [σ_2], and the particular inputs a and b, are well localized". However we will argue below that if the 'hidden variables' include a pilot-wave MI and OI are not necessarily valid.

The role of a background medium and of massive correlation induced by such a medium can most concretely be illustrated by the hydrodynamic experiments of Refs. (Couder et al. 2005; Eddi et al. 2012; Perrard et al. 2014; Bush 2015a, b; Harris and Bush 2014), so let us turn to these.

11.3 Correlations in the Droplet-Systems of Refs. (Couder et al. 2005; Eddi et al. 2012; Perrard et al. 2014; Bush 2015a, b; Harris and Bush 2014)

In a series of experiments the groups of Couder, Fort, Bush, Gilet and collaborators have succeeded in creating oil droplets that hover over a vertically vibrating oil film. Under precise experimental conditions the droplets rapidly bounce on the oil surface, while being propelled by the surface wave they create. With such 'walking' droplets an impressive series of experiments can be performed exhibiting properties that mimic quantum phenomena, such as single particle diffraction and interference, quantization of angular momentum (the droplets can only rotate on a discrete series of radiuses), a form of tunneling, etc. (Couder et al. 2005; Eddi et al. 2012; Perrard et al. 2014; Bush 2015a, b; Harris and Bush 2014; Sampara and Gilet 2016; Gilet 2016). When the oil bath is rotated, two droplets attract each other via the surface wave they generate. An increasing rotation speed lifts the degeneracy between states, which is the analogue of Zeeman splitting (Eddi et al. 2012). Moreover these experimental analogies are backed-up by certain intriguing formal analogies[1] (e.g. Gilet 2016). For our purpose, we do not need to take into account any of the detailed mechanisms involved (for in-depth Newtonian modeling of the system, see e.g. Bush 2015b). We only need to note following general probabilistic properties (P1–P4) of the droplet-systems:

P1. *The stable 'walking' regime is a probabilistic regime.* As illustrated by the complex phase diagram of possible movements (Bush 2015b), the walking regime occurs only in well-defined experimental conditions, i.e. for precise values of the control parameters of the system. These essentially are the frequency (f) and amplitude (A) of the external vibration, the mass and size of the droplet (m and r), the geometrical parameters of the oil film and bath ($\{d_i\}$), and the viscosities of film and droplet (μ_f and μ_d). If these parameters $\{f, A, r, m, \{d_i\}, \mu_f, \mu_d\}$ lie within the precise ranges of values documented by the researchers, the droplets walk horizontally; outside these ranges the movement becomes erratic and/or the droplet is captured by the film. In general, the properties measured in this regime are probabilistic: see e.g. the histogram of droplet locations in (Bush 2015b, Fig. 3); the multimodal probability distribution for orbital movement (Harris and Bush 2014, Fig. 10–12); etc. If certain parameters are fine-tuned deterministic trajectories can also be obtained;[2] but in general there are stochastic fluctuations, due to the non-linear (hence potentially chaotic) nature of the system and the presence of uncontrollable environmental parameters, such as air flow, temperature fluctuations etc. Note that the deterministic case is a special case of the probabilistic case, when only probabilities 0 or 1 occur.

P2. *The droplets coherently move in resonance with a periodic pilot-wave.* The droplets' movement is accompanied and induced by a structured pilot-wave, showing high degrees of periodicity and symmetry. The vibrating film gives kinetic energy to the droplet, but the bouncing droplet back-reacts, i.e. periodically hits the film and determines the shape and characteristics of a surface wave on the oil film. This pilot-wave propels the droplets horizontally over the film: while the droplets hit the wave periodically at a determined location just before the crests they receive a constant horizontal momentum. The surface field shows a high degree of symmetry; to good approximation it can be modeled by Huygens – Fresnel theory as a superposition of the circular waves created at the successive impacts.

P3. *In the stable regime the droplet systems are massively correlated, i.e. any system variable is potentially correlated to any other system variable, also at different spacetime points.* This is a consequence of P1 and P2. Take for instance λ_1 to be a field property (e.g. the wave height or velocity) at a certain spacetime

[1] As one example, replacing the de Broglie wavelength in the expression of the radius of the Landau-levels by λ_F, the characteristic wavelength of the droplets' pilot-wave (the so-called Faraday wavelength), leads to radiuses that fit well to those measured on the rotating droplets (Eddi et al. 2012).

[2] John W. M. Bush, private communication.

point, and λ_2 any other field property at any other spacetime point (lying within the boundaries imposed by the experiment). Then λ_1 and λ_2 will in general be correlated, i.e.

$$P\left(\lambda_1 \middle| \lambda_2\right) \neq P\left(\lambda_1\right), \tag{11.6}$$

due to the structure (periodicity and symmetry) of the wave. A simple and illustrative case is when λ_1 and λ_2 are heights of a circular wave at two points that are symmetrical with respect to the center (and taken at identical times); the center is here the impact point of the droplet. If the wave would be perfectly circular, and λ_1 and λ_2 have identical values, the correlation would be perfect ($P(\lambda_1|\lambda_2) = 1 \neq P(\lambda_1)$). Note that this correlation may exist even if λ_1 and λ_2 are taken at spacetime points that are spacelike separated, e.g. when they are simultaneous. Here and in the following symbols as 'λ_1' and 'a' may stand for n-tuples (n-vectors). Note that P3 also holds for the correlation between wave and droplet properties. For instance, if λ_0 is a property of the droplet, say its mass, then in general

$$P\left(\lambda_1 \middle| \lambda_0\right) \neq P\left(\lambda_1\right). \tag{11.7}$$

This simply reflects the fact that the droplet's mass partly determines the characteristics of the surface field, as is well documented in Couder et al. (2005), Eddi et al. (2012); Perrard et al. (2014); Bush (2015a, b), and Harris and Bush (2014). λ_0 might also be a *dynamical* property of the droplet, such as its position or velocity at a given spacetime point, possibly different from the spacetime point at which λ_1 is taken. Obviously (11.7) still holds in general, due to the fact that the droplet coherently follows the structured wave. Or λ_1 and λ_0 might both be particle properties, as in Bush (2015a, Fig. 4a), showing the correlation between droplet speed and location. Note, finally, that P3 also holds for the correlation between e.g. the surface field variables and the control parameters {f, A, r, m, {d_i}, μ_f, μ_d}. Also the latter parameters can be considered stochastic variables: they can be varied in certain experiments (as has been done by the experimenters). Thus we can have stochastic dependencies of the type:

$$P\left(\lambda_1, \lambda_2 \middle| a, b\right) \neq P\left(\lambda_1, \lambda_2\right) \tag{11.8}$$

where (a, b) are e.g. two variables $\in$\{d_i\}, the dimensions of the oil bath. This reflects the fact that the properties of the pilot wave (λ_1, λ_2) strongly depend on the geometry of the bath.

P4. *The ubiquitous correlation of P3 also holds for a 2-droplet system.* Indeed, it has been shown that if two droplets are deposited on a vibrating film they will (always in the stable regime) create a common wave field that strongly correlates their movement. For instance, they can rotate about each other while bouncing in phase or anti-phase (Eddi et al. 2012). In this case identical z-positions of the droplets are perfectly correlated; etc. To use again a jargon from relativity theory, P4 implies in particular that in a 2-droplet system there are correlations between subsystems *that are spacelike separated*. This is not different from the 1-droplet case (cf. e.g. Eq. (11.6)).

Let us now consider a case that has not been performed in the experiments, but that is a straightforward extrapolation of the latter. What would happen if two (identical) droplets could be created in the center of a (roughly symmetric) bath while receiving opposite horizontal momentum? Then one expects[3] that in the walking regime the droplets will move in opposite directions, and that underneath them a surface field with high symmetry will form which again correlates their movements (the simplest assumption is that their movements will be perfectly symmetric; but in realistic cases there will be stochastic deviations). As in Eq. (11.7), the properties of the 2-droplet wave field will again depend on the droplet properties, e.g. the mass λ_0 of both identical droplets. Thus in general

$$P\left(\lambda_1, \lambda_2 \middle| \lambda_0\right) \neq P\left(\lambda_1, \lambda_2\right) \tag{11.9}$$

where λ_1 and λ_2 are again properties of the wave field, for instance the height (λ_1) of the surface wave at some reference point on the average trajectories of the left-moving droplets, and λ_2 the same property for the right-moving droplets. One will also have correlations of following type:

$$P\left(\lambda_1 \middle| \lambda_2, \lambda_0\right) \neq P\left(\lambda_1 \middle| \lambda_0\right). \tag{11.10}$$

In other words λ_0 does not 'screen off' the correlations between λ_1 and λ_2; for fixed λ_0 (say mass) the field variables remain, of course, strongly correlated. This also generally holds when λ_0 stands for a set of control or 'contextual' parameters, say a $\subset\{d_i\}$, the dimensions of the oil bath:

$$P\left(\lambda_1 \middle| \lambda_2, a\right) \neq P\left(\lambda_1 \middle| a\right), \text{ or equivalently,} \tag{11.11a}$$

$$P\left(\lambda_1, \lambda_2 \middle| a\right) \neq P\left(\lambda_1 \middle| a\right) P\left(\lambda_2 \middle| a\right). \tag{11.11b}$$

Fixing some contextual or control parameters does not necessarily decouple field variables, rather to the contrary.

Summarizing P1–P4, in the stable regime there potentially are correlations between any two system variables, even if these variables describe spacelike separated subsystems. P1–P4 are straightforward manifestations of the fact that the

[3] John W. M. Bush, private communication.

system is guided by a structured pilot wave, and that the droplets move coherently with the wave. In the next Section a generic model for a Bell experiment will be proposed in which Bell particles and analyzers interact with a background field or medium. Importantly, for that model I will only rely on probabilities of the type (11.6)–(11.11) that exist in fluid-mechanical systems.

A last remark concerns the 'path-memory' effect in the droplet-systems, studied in detail by Couder et al. (2012). The wavefield as a whole arises out of the superposition of the different waves created along the path of the droplet, so that it contains in a sense a memory of the past. Moreover the trajectory is perturbed by waves scattered by far-away obstacles through a kind of echo-location effect. Couder et al. have termed these non-local effects, but strictly speaking this is a misnomer in the present context of Bell's theorem, where 'non-local effect' means 'superluminal effect'; and of course there is nothing superluminal going on in the droplet systems. 'Delocalization' effect would be a better term. At any rate we do not rely on this path-memory effect in the model of next Section. Most importantly, when considering a hydrodynamic Bell-type experiment one can easily design it in such a way that these delocalized effects cannot come into play (Sect. 11.5).

11.4 Background-Based Hidden Variable Theories. A Toy Model

Goal is to devise a model for a Bell experiment in which the two Bell particles and analyzers interact with an (unknown) background medium or field – say the quantum vacuum, the 'zero-point field' of stochastic electrodynamics (de la Peña et al. 2015), an ether, etc. We do not assume any particular physical properties for this background, except the very general probabilistic features outlined above. In view of their generality it is not excluded that, besides the above fluid systems, other media or fields exhibit exactly these properties.

In a stochastic HV model for a Bell experiment one assumes that the left and right spins, σ_1 and σ_2, are stochastically determined by some λ. Here I assume that the spins are not only determined by the particle properties and the respective analyzer variables, but also by the background. The latter is supposed to interact with the particles, but also with the analyzers. Specifically, in *a background-based HV model* I assume that σ_1 [σ_2] can meaningfully be described by the probability $P(\sigma_1|\lambda_0, \lambda_1, a)$ [$P(\sigma_2|\lambda_0, \lambda_2, b)$]. Here λ_1 (an n-vector) are properties of the background field in the spacetime neighborhood of the left setting event (the analyzer angle assuming a value a); similarly λ_2 are properties of the field close to the right analyzer (b) just before or at measurement. And λ_0 are all other relevant variables that are independent on a and b, e.g. properties 'that are created with the particles to be measured' in the sense of (Giustina et al. 2015) (in the simplest case these are intrinsic properties of the particles, such as mass, but in principle a broader interpretation is possible).

First assume that the factorability condition (11.2) holds, where now $\lambda \equiv (\lambda_0, \lambda_1, \lambda_2)$:

$$P\left(\sigma_1, \sigma_2 \left| \lambda, a, b\right.\right) \equiv P\left(\sigma_1, \sigma_2 \left|\lambda_0, \lambda_1, \lambda_2, a, b\right.\right) = P\left(\sigma_1 \left|\lambda_0, \lambda_1, a\right.\right) P\left(\sigma_2 \left|\lambda_0, \lambda_2, b\right.\right), \tag{11.12}$$

for all values of the variables. So σ_1 only depends on a, λ_0 and λ_1, with which it is in local contact during measurement; similarly on the right. In a local background-based ('BB') model the joint probability $P(\sigma_1, \sigma_2|a, b)$ can then generally be written as follows, using (11.12):

$$P^{BB}\left(\sigma_1, \sigma_2|a, b\right) = \sum_{\lambda_0,\lambda_1,\lambda_2} P\left(\sigma_1, \sigma_2|\lambda_0, \lambda_1, \lambda_2, a, b\right) P\left(\lambda_0, \lambda_1, \lambda_2|a, b\right)$$
$$= \sum_{\lambda_0,\lambda_1,\lambda_2} P\left(\sigma_1|\lambda_0, \lambda_1, a\right) P\left(\sigma_2|\lambda_0, \lambda_2, b\right) P\left(\lambda_0|a, b\right) P\left(\lambda_1, \lambda_2|\lambda_0, a, b\right). \tag{11.13}$$

For λ_0 we assume, by construction, that:

$$P\left(\lambda_0|a, b\right) = P\left(\lambda_0\right). \tag{11.14}$$

Indeed, under the usual assumption of locality and absence of superdeterminism, if λ_0 is for instance an intrinsic particle property 'created at emission', and if 'a' and 'b' are randomly set at spacelike distances from the emission as e.g. in Giustina et al. (2015), Hensen et al. (2015), and Shalm et al. (2015), then (11.14) must hold. Thus (11.13) becomes:

$$P^{BB}\left(\sigma_1, \sigma_2 \left|a, b\right.\right) = \sum_{\lambda_0} P\left(\lambda_0\right) \sum_{\lambda_1,\lambda_2} P\left(\sigma_1 \left|\lambda_0, \lambda_1, a\right.\right) P\left(\sigma_2 \left|\lambda_0, \lambda_2, b\right.\right) \times P\left(\lambda_1, \lambda_2 \left|\lambda_0, a\ , b\right.\right). \tag{11.15}$$

The first key property of a background-based/pilot-wave model is that MI is violated in it:

$$P\left(\lambda_0, \lambda_1, \lambda_2 \left|a, b\right.\right) \neq P\left(\lambda_0, \lambda_1, \lambda_2 \left|a', b'\right.\right), \tag{11.16}$$

for $a \neq a'$ or $b \neq b'$. Since it is assumed that λ_1 directly interacts with the left analyzer at angle a, and λ_2 with the right analyzer, it immediately follows that the probability distribution for (λ_1, λ_2) [and hence for $\lambda \equiv (\lambda_0, \lambda_1, \lambda_2)$] will in general be dependent on (a,b) (cf. Appendix for examples in spin-lattices). More generally (11.16) can be understood as expressing that the field characteristics are determined by (a,b), as happens in hydrodynamic systems: cf. the correlation (11.8). Note that

Eq. (11.16) can well be compatible with locality; one only needs to assume local and localized interactions for Eqs. (11.16) or (11.8) to hold.

It is instructive to first focus on a special case and assume that the analyzer settings are not varying over time, as in a static Bell experiment. Furthermore, let us assume that λ_0 takes only one fixed value. Since MI is violated, it is not a surprise that also the Bell inequality can be violated in the background model. Indeed, recently several toy models have been devised reproducing the quantum statistics via violation of MI only (other models might violate OI and/or PI): see the reviews in Brunner et al. (2014), Hall (2011), Di Lorenzo (2012). Using a result by Di Lorenzo (2012) it is immediate to show that the quantum correlation, P^{QM}, can be recovered as a special instance of a background model that assumes only (11.12)–(11.16). The quantum correlation is:

$$P^{QM}\left(\sigma_1, \sigma_2 \,|a, b\right) = \frac{1}{4}\left[1 - \sigma_1.\sigma_2.\cos\left(a-b\right)\right]. \tag{11.17}$$

When λ_0 has a fixed value Eq. (11.15) reduces to a probability of the type:

$$P^{BB}\left(\sigma_1, \sigma_2 \,|a, b\right) = \sum_{\lambda_1,\lambda_2} P\left(\sigma_1 \,|\lambda_1, a\right) P\left(\sigma_2 \,|\lambda_2, b\right) P\left(\lambda_1, \lambda_2 \,|a\ , b\right). \tag{11.18}$$

In (11.18) all variables may be n-tuples, in particular 3-vectors. Making the substitutions $\lambda_1 \to \overline{\lambda_1}, \lambda_2 \to \overline{\lambda_2}$ and $(a, b) \to \left(\overline{a}, \overline{b}\right)$ and assuming that $\overline{\lambda_1}, \overline{\lambda_2}, \overline{a}$ and $\overline{b}$ are unit vectors, then with the normalized choices of Ref. (Di Lorenzo 2012):

$$\begin{aligned} P\left(\sigma_1 \,\middle|\overline{\lambda_1}, \overline{a}\right) &= \tfrac{1}{2}\left(1 + \sigma_1\overline{\lambda_1}\cdot\overline{a}\right), \\ P\left(\sigma_2 \,\middle|\overline{\lambda_2}, \overline{b}\right) &= \tfrac{1}{2}\left(1 + \sigma_2\overline{\lambda_2}\cdot\overline{b}\right), \\ P\left(\overline{\lambda_1}, \overline{\lambda_2} \,\middle|\overline{a}, \overline{b}\right) &= \tfrac{1}{4}d\overline{\lambda_1}d\overline{\lambda_2} \sum_{\overline{p}=\pm\overline{a},\pm\overline{b}} \delta\left(\overline{\lambda_1} - \overline{p}\right)\delta\left(\overline{\lambda_2} + \overline{p}\right), \end{aligned} \tag{11.19}$$

(11.18) reduces to the quantum correlation (11.17), as a simple calculation shows. Note this model is 'local' in the sense that OI and PI are obviously satisfied; yet there appears to be a hidden nonlocality in it. Indeed, the expression for $P\left(\overline{\lambda_1}, \overline{\lambda_2} \,\middle|\overline{a}, \overline{b}\right)$ in (11.19) can only be considered local and non-superdeterministic if the Bell experiment is static: it assumes that $\overline{\lambda_1}$ depends on $\overline{b}$ and $\overline{\lambda_2}$ on $\overline{a}$, which is only conceivable if a delocalized (but subluminal) influence is established between the left and right wings. (In the droplet systems this can occur, since the global wave field can be shaped by far-away boundary conditions.) But such a mechanism cannot work in a dynamic experiment, where such long-range interactions cannot have an effect, precluding the correlation (11.19). And indeed, it is straightforward to show that model (11.19) is 'signaling' (cf. below; not all of the conditions (11.28) are satisfied). Interestingly, several essential properties of toy model (11.19) will be seen to exist in spin-lattices.

Note that the model (11.19) not only violates MI; it does so in such a manner that

$$P\left(\overline{\lambda_1}, \overline{\lambda_2} \middle| \overline{a}, \overline{b}\right) \neq P\left(\overline{\lambda_1} \middle| \overline{a}\right) P\left(\overline{\lambda_2} \middle| \overline{b}\right). \tag{11.20}$$

This is a property of non-factorability, or rather *'non-screening-off'* at the hidden-variable level: $(\overline{a}, \overline{b})$ do not screen-off the correlations between $\overline{\lambda_1}$ and $\overline{\lambda_2}$, in other words in an ensemble with fixed analyzer variables the HVs are not decoupled. If the equality sign holds in Eq. (11.20) one immediately proves that the correlation (11.19) satisfies a Bell inequality.

Beyond model (11.19), condition (11.20) will appear to be a general property of background-based/pilot-wave models. A first important hint that such models are physically allowed is that (11.20) is ubiquitous in physical systems that are compatible with free will and locality. Indeed, (11.20) is a type of correlation that exists in the droplet-systems: (11.20) is a special case of (11.11) [as can be seen by replacing in (11.11b) $\lambda_1 \rightarrow \overline{\lambda_1}, \lambda_2 \rightarrow \overline{\lambda_2}$ and $a \rightarrow (\overline{a}, \overline{b})$]. Of course, in the hydrodynamic experiments the control parameters (a,b) can be freely set, as in the Bell experiment. Therefore (11.11) and (11.20) are surely compatible with free will.

In the generic model (11.15) the non-screening-off condition is:

$$\begin{aligned} P\left(\lambda_1, \lambda_2 \middle| \lambda_0, a, b\right) &= P\left(\lambda_1 \middle| \lambda_0, a, b\right) P\left(\lambda_2 \middle| \lambda_0, \lambda_1, a, b\right) \\ = P\left(\lambda_1 \middle| \lambda_0, a\right) P\left(\lambda_2 \middle| \lambda_0, \lambda_1, b\right) &\neq P\left(\lambda_1 \middle| \lambda_0, a\right) P\left(\lambda_2 \middle| \lambda_0, b\right). \end{aligned} \tag{11.21}$$

The first equality follows from the product rule of probability calculus; the second from a locality assumption that λ_1 only depends on a (not on b), and similarly in the right wing; and the last inequality expresses the fact that λ_1 and λ_2 are conditionally correlated. If the equality sign holds in (11.21), (11.15) is of the Bell-Clauser-Horne type and satisfies a Bell inequality. A correlation of the type (11.21) exists in droplet-systems, but also in spin-lattices (cf. Appendix).

Let us now turn to the validity of OI in pilot-wave systems; also this assumption is not so obvious as is usually believed. First, note that the decorrelations (11.1a) and (11.2) are assumptions which are surely allowed by probability theory, but for which the literature ultimately provides only an intuitive justification (cf. e.g. Bell 1964, 1981, 2004; Brunner et al. 2014; Hall 2010, 2011, 2015). In the end (11.1a) and (11.2) are just probabilistic assumptions, and within physics no proof exists that OI could not be violated in supercorrelated systems as the droplet-systems. Inspecting the definition (11.1a) of OI, it becomes clear that all depends on what one denotes by the HVs 'λ':

$$P\left(\sigma_1 \middle| \sigma_2, a, b, \lambda\right) = P\left(\sigma_1 \middle| a, b, \lambda\right). \qquad \text{(OI)} \tag{11.1a}$$

Consider for instance a Bell experiment on the droplet systems (involving two correlated droplets, cf. Sect. 11.5) and take $\lambda = \lambda_0$, some intrinsic droplet property as mass, and σ_1 and σ_2 some dynamical droplet properties taken at measurement (say the z-component of their position above a reference plane), then it is quite obvious that in general OI will *not* be valid. Indeed (a,b, λ_0) are contextual parameters influencing the pilot-wave properties (λ_1 and λ_2); which are in general correlated (Sect. 11.3). Since the droplets (σ_1 and σ_2) resonantly follow the pilot-wave (property P2, Sect. 11.3), also the latter properties will be correlated in general. E.g. $P(\sigma_1 | a, b, \lambda_0)$ may be $= 0.5$ while $P(\sigma_1 | \sigma_2, a, b, \lambda_0)$ may be ≈ 1 if $\sigma_1 \approx \sigma_2$, due to the resonant movement in a symmetric surface field. $P(\sigma_1 | \sigma_2, a, b, \lambda_0) = 1$ may hold in the ideal case of a perfectly symmetric surface field. It would thus appear that OI is questionable, and that the core of the locality condition (11.2) in background-based systems is not OI but PI, since PI must necessarily be satisfied.

Accepting violation of OI, we have in general, instead of (11.12):

$$P\left(\sigma_1, \sigma_2 \middle| \lambda, a, b\right) \equiv P\Big(\sigma_1, \sigma_2 \Big| \lambda_0, \lambda_1, \lambda_2, a, b\Big)$$
$$= P\left(\sigma_1 \middle| \lambda_0, \lambda_1, a\right) P\Big(\sigma_2 \Big| \sigma_1, \lambda_0, \lambda_2, b\Big) \neq P\left(\sigma_1 \middle| \lambda_0, \lambda_1, a\right) P\Big(\sigma_2 \Big| \lambda_0, \lambda_2, b\Big). \tag{11.22}$$

A next question is then: can one construct a background-based toy model (i.e. satisfying (11.15), (11.21) and (11.22)) that violates the Bell inequality while being non-signaling? This is indeed possible, as will now be shown. If the arguments above are correct, such a model would be compatible with locality (not involve superluminal influences) and free will. We have now:

$$P^{BB}\left(\sigma_1, \sigma_2 | a, b\right) = \sum_{\lambda_0} P\left(\lambda_0\right) \sum_{\lambda_1, \lambda_2} P\left(\sigma_1 | \lambda_0, \lambda_1, a\right) P\left(\sigma_2 | \sigma_1, \lambda_0, \lambda_2, b\right)$$
$$\times P\left(\lambda_1 | \lambda_0, a\right) P\left(\lambda_2 | \lambda_0, \lambda_1, \ b\right), \tag{11.23}$$

where the probabilities should satisfy the following normalization conditions:

$$\sum_{\lambda_0} P\left(\lambda_0\right) = 1, \tag{11.24a}$$

$$\sum_{\sigma_1} P\left(\sigma_1 | \lambda_0, \lambda_1, a\right) = 1, \qquad \forall\left(\lambda_0, \lambda_1, \mathrm{a}\right), \tag{11.24b}$$

$$\sum_{\sigma_2} P\left(\sigma_2 | \sigma_1, \lambda_0, \lambda_2, b\right) = 1, \qquad \forall \sigma_1 \left(\lambda_0, \lambda_2, \mathrm{b}\right), \tag{11.24c}$$

$$\sum_{\lambda_1} P(\lambda_1 | \lambda_0, a) = 1, \qquad \forall (\lambda_0, \mathrm{a}), \tag{11.24d}$$

$$\sum_{\lambda_2} P(\lambda_2 | \lambda_0, \lambda_1, b) = 1, \qquad \forall (\lambda_0, \lambda_1, \mathrm{b}). \tag{11.24e}$$

For the proof it is sufficient to make judicious choices for the probabilities in (11.23 and 11.24); in particular let us assume that λ_0 has one fixed value and that λ_1 and λ_2 only assume two values, which can be set to 1 and 2:

$$\lambda_1, \lambda_2 = 1, 2. \tag{11.25}$$

Equation (11.23) can then be written without loss of generality:

$$P^{BB}(\sigma_1, \sigma_2 | a, b) = \sum_{\lambda_1, \lambda_2} P(\sigma_1 | \lambda_1, a) \, P(\sigma_2 | \sigma_1, \lambda_2, b) \, P(\lambda_1 | a) \, P(\lambda_2 | \lambda_1, \; b). \tag{11.26}$$

Make now following normalized choices, satisfying (11.24):

$$\begin{aligned}
&P(\lambda_1 = 1 | a) = 0.5 = P\left(\lambda_1 = 1 \middle| a'\right) \\
&P(\sigma_1 = +1 | \lambda_1 = 1, a) = 1 = P\left(\sigma_1 = +1 \middle| \lambda_1 = 2, a'\right); \\
&P\left(\sigma_1 = +1 \middle| \lambda_1 = 1, a'\right) = 0 = P(\sigma_1 = +1 | \lambda_1 = 2, a) \\
&P(\sigma_2 = +1 | \sigma_1 = +1, \lambda_2 = 1, b) = 1 = P(\sigma_2 = +1 | \sigma_1 = +1, \lambda_2 = 2, b) \\
&P\left(\sigma_2 = +1 \middle| \sigma_1 = +1, \lambda_2 = 1, b'\right) = 0; \; P\left(\sigma_2 = +1 \middle| \sigma_1 = +1, \lambda_2 = 2, b'\right) = 1 \\
&P(\sigma_2 = +1 | \sigma_1 = -1, \lambda_2 = 1, b) = 0; \; P(\sigma_2 = +1 | \sigma_1 = -1, \lambda_2 = 2, b) = 1 \\
&P\left(\sigma_2 = +1 \middle| \sigma_1 = -1, \lambda_2 = 1, b'\right) = 0; \; P\left(\sigma_2 = +1 \middle| \sigma_1 = -1, \lambda_2 = 2, b'\right) = 1 \\
&P(\lambda_2 = 1 | \lambda_1 = 1, b) = 1 = P(\lambda_2 = 1 | \lambda_1 = 2, b); \; P\left(\lambda_2 = 1 \middle| \lambda_1 = 1, b'\right) = 0; \\
&P\left(\lambda_2 = 1 \middle| \lambda_1 = 2, b'\right) = 1.
\end{aligned} \tag{11.27}$$

These choices (model BB-1) are summarized in Table 11.1:

For the analyzer variables it is assumed P(x) = P(y) = 0.5 and P(x, y) = 0.25, $\forall$ (x, y), as is usual in Bell-experiments. This model is semi-deterministic since besides the analyzer variables (x,y) and λ_1 all other variables have only values with a conditional probability 0 or 1. Thus if (x,y) = (a,b), Table 11.1 shows that σ_1 and σ_2 have equal sign, therefore M(a, b) = $<\sigma_1\sigma_2>_{a,\,b}$ = + 1. Similarly M(a',b) = M(a,b') = +1 and M(a',b') = −1, implying X_{BI} = 4: model BB-1 maximally violates the Bell inequality.

Table 11.1 Background-based model BB-1 assumes that the analyzer variables (x and y) are equiprobable and independent (P(x, y) = P(x)P(y) = 0.25 ∀ (x, y)) and that $P(\lambda_1 = 1 \mid x) = 0.5, \forall x$; the other variables have values with a conditional probability 0 or 1 (cf. 11.27)

x	y	λ_1	σ_1	λ_2	σ_2
a	b	1	+1	1	+1
		2	−1	1	−1
a′	b	1	−1	1	−1
		2	+1	1	+1
a	b′	1	+1	2	+1
		2	−1	1	−1
a′	b′	1	−1	2	+1
		2	+1	1	−1

Note, and this is essential, that the model does not allow superluminal signaling.[4] Indeed, Table 11.1 shows there is no dependence between the relevant left- and right-wing variables, namely between y and σ_1; y and λ_1; y and (σ_1, λ_1); and similarly for x. If e.g. y and σ_1 would be dependent, then measuring (many times) σ_1 would inform Alice about Bob's setting choice, allowing superluminal signaling. In detail, as can be seen in Table 11.1, model BB-1 satisfies following non-signaling conditions (these are not all independent, but that is not important here):

$$\begin{aligned}
P\left(\sigma_1 \middle| x, b\right) &= P\left(\sigma_1 \middle| x, b'\right) && \forall (\sigma_1, x)\\
P\left(\lambda_1 \middle| x, b\right) &= P\left(\lambda_1 \middle| x, b'\right) && \forall (\lambda_1, x)\\
P\left(\sigma_1, \lambda_1 \middle| x, b\right) &= P\left(\sigma_1, \lambda_1 \middle| x, b'\right) && \forall (\sigma_1, \lambda_1, x)\\
P\left(\sigma_2 \middle| a, y\right) &= P\left(\sigma_2 \middle| a', y\right) && \forall (\sigma_2, y)\\
P\left(\lambda_2 \middle| a, y\right) &= P\left(\lambda_2 \middle| a', y\right) && \forall (\lambda_2, y)\\
P\left(\sigma_2, \lambda_2 \middle| a, y\right) &= P\left(\sigma_2, \lambda_2 \middle| a', y\right) && \forall (\sigma_2, \lambda_2, y).
\end{aligned} \tag{11.28}$$

Now, a basic assumption of the model is (11.21), non-screened-off MI-violation, which is for fixed λ_0:

$$P\left(\lambda_1, \lambda_2 \middle| a, b\right) = P\left(\lambda_1 \middle| a\right) P\left(\lambda_2 \middle| \lambda_1, b\right) \neq P\left(\lambda_1 \middle| a\right) P\left(\lambda_2 \middle| b\right). \tag{11.29}$$

The reader unaware of (11.28) could be tempted to conclude that (11.29) necessarily is nonlocal (allows superluminal signaling), based on following erroneous reasoning: (11.29) shows that λ_2 depends on b and that λ_1 depends on λ_2; hence λ_1 must depend on b, which is nonlocal. But probabilistic dependence is not necessarily

[4] A related model presented in Vervoort (2016) is local in the sense of Clauser-Horne, but is signaling.

transitive; and Table 11.1 (satisfying in particular the second of the conditions (11.28)) indeed proves that λ_1 does *not* depend on b. The reason why dependence is not transitive here is that λ_1 and λ_2 do not cause each other; they are caused by the action of their respective analyzer (a[b] partly determining λ_1 [λ_2]) *and* by a common cause, namely the action of the particle pair simultaneously creating (or partly influencing) λ_1 and λ_2. In sum there is no direct causal path from λ_1 to b.

More generally, going beyond model BB-1, it seems possible to physically justify our basic assumptions (11.21) and (11.29) (MI-violation) and (11.22) (OI-violation) by reference to the droplet-systems, also in a dynamic Bell-experiment. As already stated, these types of correlations are encountered in the hydrodynamic systems: e.g. (11.21) is a correlation of the type (11.10) and (11.11) (replace e.g. a in (11.11) by (a, b, λ_0)). In the droplet systems (11.21) can be interpreted in a simple manner: in an ensemble with fixed droplet masses (λ_0) and fixed control parameters (a,b), there exists in general a correlation between field properties (λ_1, λ_2) – as manifestly follows from the experiments (Couder et al. 2005; Eddi et al. 2012; Perrard et al. 2014; Bush 2015a, b; Harris and Bush 2014) (property P4, Sect. 11.3). We thus can explain (11.21) and (11.29) in a real Bell experiment by the hypothesis that each particle pair leaving the source generates, or is accompanied by, a structured background wave showing high symmetry and periodicity; hence λ_1 and λ_2, properties of the background wave taken at measurement at (symmetric) points close to the detectors, can be strongly correlated. Likewise (11.22) can also be justified by the fact that the particles, as the droplets, resonantly move in a structured background field. Finally, it seems that there is no reason why varying analyzer settings, as in the advanced experiments (Giustina et al. 2015; Hensen et al. 2015; Shalm et al. 2015), would have to destroy the correlations we need. To maintain (11.21) and (11.22), one simply needs to assume that the switching does not totally disrupt the structure in the pilot-wave, in other words the long-range correlations. A simple physical picture in which this happens is when the analyzers only slightly influence the properties λ_1, λ_2 of the pilot-wave (on their respective sides) compared to the particles themselves (λ_0); more generally (in model BB-1 the dependence between λ_2 and b is strong) the analyzers should not totally randomize the correlation between λ_1 and λ_2. This seems a harmless assumption not violating any known physical law. In the droplet systems the droplet properties (λ_0) strongly determine the pilot-wave; a small object in the bath, characterized by a (varying) parameter 'a', can well be assumed to have a smaller influence – small enough for sufficient correlation between λ_1 and λ_2 to persist.[5]

[5] Note that some influence of a on λ_1 and on σ_1 and/or of b on λ_2 and on σ_2 is still required, else the Bell inequality is always satisfied. A corollary of above analysis is that the background system envisaged here might need to be nonlinear, in the sense that tiny influences of one parameter (a) on another (λ_1) cause a macroscopically detectable difference in outcome in a third quantity (σ_1). Interestingly, this is again observed in the droplet systems, which exhibit such exponential sensitivity on initial conditions, as shown in Perrard et al. (2014), Bush (2015b). The importance of non-linearity in the framework of local HVTs was already noted in Vervoort (2000).

Thus a background or pilot-wave model based on (11.21) and (11.22) can potentially violate a Bell inequality; we have argued it is compatible with locality (subluminal interactions) and free will. Clearly, this is a 'no no-go' claim based on a toy model, not a proof that there exist real background systems violating a Bell inequality. In the spirit of recent work by Popescu and others (Popescu 2014; Gallego et al. 2011), model BB-1 is "even more nonlocal than quantum mechanics, yet fully consistent with relativity". Following these authors, an interesting question is whether there are physical principles that would *limit* the value of X_{BI} in more realistic background models. This question may be more easily answered here, since background models have a physical interpretation.

At this point a rather obvious question arises, namely whether the above model involving "particles + a pilot wave", can be generalized to just "particles as (stochastic) classical fields". Indeed, in the background model I considered the interaction of Bell particles with a stochastic background, where in the simplest interpretation λ_0 is a property of the particle and $\lambda_{1(2)}$ of the field. But can one not suppose that *all* properties λ_0, λ_1, λ_2 describe a stochastic classical field, and *that the particles are singularities of the field*, λ_0 describing these singularities ? (One will note the analogy of this picture with quantum field theories, in which particles are excited states of the delocalized vacuum state.) It seems that also in this picture we can have the essential correlations (11.21) and (11.22). Then one would come to the almost discomfortingly simple conclusion that as soon as the HVs λ are associated with extended fields, Bell's theorem is in jeopardy – under the conditions discussed for the background model. But is this not a perfect prelude to quantum field theories ?

In the Appendix it is shown that some of the probabilistic properties considered above actually exist in Ising spin-lattices (for a fuller treatment see Vervoort). Just as droplet-systems, spin-lattices are highly correlated through a stochastic background (namely the spins in between the 'test' and 'control' spins in Fig. 11.2). They are not a dynamic system as a Bell experiment and are not realistic templates for pilot-wave HVTs, but they nevertheless allow to nicely illustrate several properties of the generic background model, in particular non-screened-off MI-violation. They are 'local' in the sense of satisfying OI and PI (the Clauser-Horne factorability), but they are signaling, and thus nonlocal in this sense. Maybe their description can be generalized to a dynamical situation; it is well known that related lattice-gas models exist for the Navier-Stokes equation (Frisch et al. 1986).

11.5 Hydrodynamic Bell Experiments

In view of the lasting mysteries surrounding quantum mechanics and Bell's theorem, it seems that the droplet-systems, as macroscopic pilot-wave systems, offer a unique opportunity for advancing the debate on several fronts. Above it was argued that in such systems MI and OI are questionable and may not hold. Note that MI, OI and PI are sufficient but not necessary conditions for the Bell inequality; so

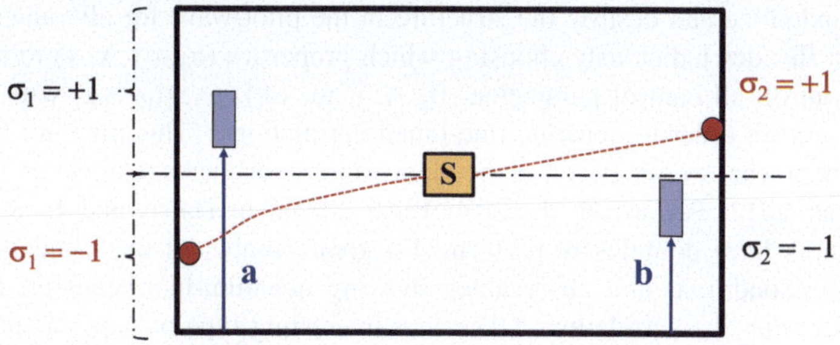

Fig. 11.1 Conceptual scheme of a Bell-type experiment in droplet-systems (S = source).

even if for instance MI would not be valid in some pilot-wave system there is no guarantee that a Bell inequality is violated. All depends on the system in question (including all boundary and measurement conditions) and on the precise strength of the correlations (11.23) it generates. Although we do not know the numerical values of these correlations for the droplet systems (Couder et al. 2005; Eddi et al. 2012; Perrard et al. 2014; Bush 2015a, b; Harris and Bush 2014), the walkers are obviously excellent candidates to test the generic model. Let us therefore present some guidelines for such experiments. These seem particularly relevant in view of the difficulty to devise genuinely loophole-free experiments on quantum systems (Giustina et al. 2015; Hensen et al. 2015; Shalm et al. 2015; Santos 2004): this should be much easier in the macroscopic realm. In order to close e.g. the locality loophole one just needs to consider the sonic speed in the medium, not the light speed.

Clearly, Bell's reasoning leading to the Bell inequality does nowhere rely on the fact that σ_1 and σ_2 in (11.1) are spins; *any* bi-partite, classical, local system on which free-willed experimenters perform a Bell experiment should satisfy a Bell inequality according to (Bell 1964, 1981, 2004). Thus σ_1 and σ_2 may – a priori – represent *any* property of the droplets. This greatly opens the parameter space of the experiments that can be envisaged. A few general remarks on such a hydrodynamic Bell experiment can be made without specifying[6] (σ_1, σ_2, a, b); a conceptual scheme of such a test is given in Fig. 11.1.

The first challenge is to generate pairs of correlated droplets that move roughly in opposite directions (but this should be possible: see P4 in Sect. 11.3). Then, in order to implement the model of Sect. 11.4, some measurement devices characterized by parameters x (left) and y (right) should interact with the fluid bath, thus determining the pilot-wave properties in their neighborhood and by the same token some dynamical droplet property (σ). The interaction should not be so strong as to

[6] There is some incentive to choose for σ the hydrodynamic equivalent of spin, which might suggest to use a rotating bath, or rotating or magnetic droplets. Indeed, the existence of hydrodynamic 'spin states' has already been suggested (Bush 2015a). But it is wise to start with the much simpler experiment suggested below.

totally randomize and destroy the structure in the pilot-wave for all values of the variables. Besides judiciously choosing which properties (σ_1, σ_2, x, y) to measure and to control, all control parameters {f, A, r, m, {d_i}, μ_f, μ_d,...} will have to be optimized in order to generate fine-tuned correlations. (This may not be more difficult than what was achieved in the pioneering experiments (Couder et al. 2005; Eddi et al. 2012; Perrard et al. 2014; Bush 2015a, b; Harris and Bush 2014): their authors have doubtlessly performed a great number of experiments before finding the conditions and observables showing quantum-like behavior, in other words quantum-like correlations.) The first interesting type of experiments could be static: then one has to determine on an ensemble of droplet-pairs the average products M(x,y) in (11.5) in four consecutive experiments corresponding to the four combinations (x,y). Of course, in such a static experiment one has, besides those discussed above, an additional resource for violating a Bell inequality, namely the delocalized effects that exist in the droplet-systems (cf. last paragraph of Sect. 11.3).

In the ideal, dynamic test one activates (x,y) just before (σ_1,σ_2) are measured, in such a manner that x[y] cannot influence $\sigma_2[\sigma_1]$ in time (taking the droplet and sound speeds in the background medium into account); this to close the locality loophole as in experiments (Giustina et al. 2015; Hensen et al. 2015; Shalm et al. 2015). This can be achieved by measuring for each droplet pair the (simultaneous) time of emission from a source (T_0), by activating (x,y) at time $T_0 + T_1$, and by (non-intrusively) measuring (σ_1,σ_2) at $T_0 + T_1 + \varepsilon$ where ε is small enough to prevent that the perturbation induced by (x,y) can travel to the other side.

In Fig. 11.1 I took for simplicity $\sigma = +1$ if the droplet hits the side-wall above the symmetry plane of the bath; as an example a and b can be positions of some small objects, e.g. pins pushed in the bath, ideally during the 'flight' of the droplets just before measurement. A more realistic case may be to take (σ_1,σ_2) as the instantaneous deflections of the droplets at $T_0 + T_1 + \varepsilon$ (+1[−1] if the angle with respect to a reference line is positive [negative], again much as in Fig. 11.1). But again, Fig. 11.1 should merely be seen as a conceptual scheme.

Model BB-1 can be a source of inspiration: it suggests to choose (b, b') in such a manner that when y = b, λ_2 is not dependent on λ_1, while when y = b', λ_2 is (perfectly) correlated with λ_1. This suggests to take as measurement system one that can disrupt the correlation between λ_1 and λ_2 when y has a specific value (b). Model BB-1 also suggests performing the experiment in the stochastic regime, in which the droplets follow stochastic trajectories that are not straight lines. Then violation of OI is a possible resource for violating a Bell inequality (it is well-known that in deterministic systems OI is always satisfied (Hall 2011)).

Even independently of our model, comparing the static and dynamic case will allow to investigate a wealth of interesting questions related to the quantum version of the Bell experiment. In general, the role of loopholes can be studied much more easily in the fluid system. For instance, if the Bell inequality would be violated in the static experiment but not in the dynamic one, interesting information on the precise dynamics of some delocalized effect could be obtained. But also the relevance and meaning of the still quite mysterious detection, fair sampling and coincidence loopholes (Santos 2004) can be studied much more easily, simply because the particles can be non-intrusively observed and 'killed in flight'. We emphasize this is a phenomenal advantage over the photon experiments, where one largely moves in the dark; hence surprises are not excluded. Finally, note that all these loopholes correspond to what could be called, with caution, 'generalized non-local effects'. In nonlinear, chaotic systems as the droplet systems such effects, *even if arbitrarily weak*, could be enhanced and lead to violation of the Bell inequality, as was shown in Vervoort (2000).

11.6 Existing Attempts at Theories

A natural question is whether the background-based theories envisaged here are already represented by existing (be it not yet finalized) theories, aiming at completing quantum mechanics via hidden background variables. The first candidates that come to mind are Madelung's hydrodynamic interpretation of quantum mechanics (Madelung 1927), and, especially, Louis de Broglie's pilot-wave theory (de Broglie 1987). Further recent and noteworthy attempts, inspired by the droplet systems, are Brady and Anderson (2014), Groessing et al. (2014), and Sbitnev (2015). Maybe the most elaborated theory is stochastic electrodynamics (de la Peña et al. 2015), a variant of de Broglie's pilot-wave theory. Even if these theories are not final, it has recently been argued that in view of new experimental data (Couder et al. 2005; Eddi et al. 2012; Perrard et al. 2014; Bush 2015a, b; Harris and Bush 2014) they deserve renewed attention (Bush 2015a; Couder and Fort 2012). In de Broglie's theory a quantum particle resonantly interacts with its pilot-wave at the Compton frequency $\omega_C = mc^2/\hbar$ (the Zitterbewegung); at the same time it is guided by a monochromatic pilot-wave in real space characterized by the de Broglie wavelength $\lambda_B = h/p$ (Bush 2015a). Qualitatively this composed dynamics quite remarkably corresponds to the movement of the droplets (where the role of the de Broglie wavelength is taken by the Faraday wavelength); it is also reflected in the generic background model. In Madelung's theory as well as in variants of Bohm's theory (Bohm and Vigier 1954) particles are dragged by a 'Madelung fluid' while undergoing Brownian deviations around the average streamlines; the latter provide the hidden background variables. Stochastic electrodynamics, investigated since the 1960s by several researchers starting from Marshall, Boyer, de la Pena, Cetto and others (cf. review de la Peña et al. 2015), allows to recover several quantum features and equations based on one main ingredient: a stochastic 'zero-point field' (ZPF), resonantly interacting with particles and exchanging energy with them. Thus these theories involve a background field in the sense of the model of Sect. 11.4. Moreover it seems they might allow for the long-range correlations this model invokes. Indeed, in stochastic electrodynamics entanglement and *apparent* non-locality arise through '*common resonance modes*' of the particles (de la Peña et al. 2015, p. 248), quite similarly as in the model of Sect. 11.4.

11.7 Conclusion

More than fifty years after its discovery, the interpretation of Bell's theorem remains highly debated, for the simple reason it leaves us the choice between two utterly unpleasing worldviews, involving either a-causality (things happen without any reason, 'out of the blue') or superluminal interactions. From one point of view, it is almost curious that one interpretation – termed 'superdeterminism', even if 'determinism' (the view that all events have causes) suffices as a term (Vervoort 2013) –, has so few adepts, since it allows to solve the problem. According to this deterministic model Bell's theorem should be interpreted differently – namely as proving that MI (measurement independence) is not valid at the quantum level. Technically, determinism taken seriously amounts to non-validity of MI. We have argued here that hydrodynamic pilot-wave systems (Couder et al. 2005; Eddi et al. 2012; Perrard et al. 2014; Bush 2015a, b; Harris and Bush 2014) offer a unique opportunity to test whether MI and the Bell inequality are valid at the macroscopic scale. In these systems there may be an additional ground why MI is violated, in a manner that is clearly compatible with free will (Sect. 11.4).

More precisely, we provided a background toy model violating MI and OI (outcome independence) and maximally violating the Bell inequality, while yet being non-signaling. We argued that all types of correlations needed in this model arise in the hydrodynamic pilot-wave systems of refs. (Couder et al. 2005; Eddi et al. 2012; Perrard et al. 2014; Bush 2015a, b; Harris and Bush 2014). This suggests that this toy model is compatible with locality (subluminal interactions) and free will.

As stated in Ref. (Giustina et al. 2015), MI can only be imposed, and therefore the freedom-of-choice loophole closed, under the assumption 'that λ is created with the particles to be measured'. In the toy model of Sect. 11.4 the λ do not only describe (particle) properties at emission, but also a resonant background in the spacetime neighborhood of the detection events. Of course, one could say that such pilot-wave HVTs rely on another type of nonlocality, or rather holism, since they invoke delocalized fields. But the difference is clear: this is physics as usual, and not spooky action-at-a-distance. In the intuitively simplest picture, directly inspired by the droplet-systems, entanglement and *apparent* nonlocality can arise in such theories when particles coherently move in resonance with a periodic wave (or maybe, by extension, when they are singularities of such a field).

Finally, we emphasized that hydrodynamic Bell experiments could provide decisive new insights in Bell's theorem, also independently of our model. Indeed, there is a continuing debate on the relevance of loopholes invalidating Bell's theorem. It is much easier to experimentally close or open these loopholes in the hydrodynamic systems, so that their interpretation can be studied in detail.

Acknowledgements I would like to thank John Bush, Marc Fleury, Lorenzo Maccone, Ward Struyve, and especially Scott Glancy for highly instructive discussions. The article is dedicated to a mentor since a decade, Mario Bunge.

Appendix: Bell-Type Experiment on Spin-Lattices

Let us suppose Alice and Bob share an ensemble of identical spin-lattices containing 10 spins (all $\sigma_i = \pm 1$), all at the same temperature T (Fig. 11.2). Also suppose Alice and Bob perform a Bell-type experiment on this ensemble by measuring σ_1, σ_a (Alice's wing) and σ_2, σ_b (Bob's wing).

The Hamiltonian is the classical spin-1/2 Ising Hamiltonian:

$$H(\theta) = -\sum_{i,j} J_{ij}\sigma_i\sigma_j - \sum_i h_i\sigma_i. \tag{11.30}$$

Here θ is a 10-spin configuration $(\sigma_a, \sigma_b, \sigma_1, \ldots \sigma_8)$, the h_i are local magnetic fields, and the J_{ij} are the interaction constants between σ_i and σ_j, as usual assumed to be zero beyond nearest neighbors. Thus the (Coulomb) interaction between the spins is local and localized; at the same time it is 'transmitting' in that the elements of the lattices 'feel' each other over long distances. Concomitantly spin-lattices are massively correlated: every spin is correlated with every other spin in the lattice (Yeomans 1992). This implies in particular that the system is signaling: not every condition in (11.28) is satisfied. Finally it is assumed that the probability of a given spin configuration is the usual Boltzmann probability:

$$P(\theta) = e^{-\beta H(\theta)}/Z, \text{ with } Z = \sum_\theta e^{-\beta H(\theta)}, \text{ the partition function.} \tag{11.31}$$

It is then straightforward to calculate the Bell probabilities $P(\sigma_1, \sigma_2 | \sigma_a, \sigma_b)$ and the inequality (11.5) where the role of a (b) is now taken by σ_a (σ_b) (take $a \equiv b \equiv +1$ and $a' \equiv b' \equiv -1$). The role of the 'background' HVs is assumed by the intermediate spins: $\lambda \equiv \sigma_\lambda \equiv (\sigma_3, \sigma_4, \ldots \sigma_8)$.

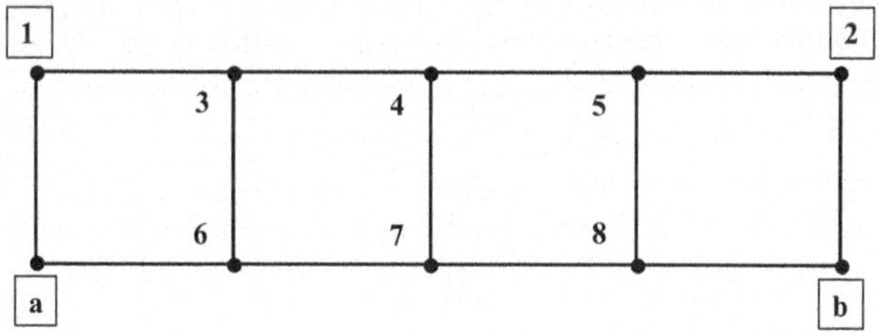

Fig. 11.2 10 spins on a lattice

As an example, if one assumes that all J_{ij} are equal (all $J_{ij} = J$) and all $h_i = 0$ then one obtains (the sums $\sum_{i,j}$ run over the 13 first-neighbour pairs):

$$
\begin{aligned}
Z\,P(\sigma_1, \sigma_2, \sigma_a, \sigma_b) &= \sum_{\sigma_3\sigma_4\ldots\sigma_8} e^{-\beta H(\theta)} = \sum_{\sigma_3\sigma_4\ldots\sigma_8} e^{\beta J \sum_{i,j} \sigma_i\sigma_j} \\
&= \sum_{\sigma_3\sigma_4\ldots\sigma_8} \prod_{i,j} e^{\beta J \sigma_i\sigma_j} \\
&= \sum_{\sigma_3\sigma_4\ldots\sigma_8} \prod_{i,j} \left[\cosh\left(\beta J \sigma_i\sigma_j\right) + \sinh\left(\beta J \sigma_i\sigma_j\right)\right] \\
&= \sum_{\sigma_3\sigma_4\ldots\sigma_8} \prod_{i,j} \left[\cosh\left(\beta J\right) + \sigma_i\sigma_j \sinh\left(\beta J\right)\right] \\
&= \sum_{\sigma_3\sigma_4\ldots\sigma_8} \left(\cosh\left(\beta J\right)\right)^{13} \prod_{i,j} \left[1 + \sigma_i\sigma_j \tanh\left(\beta J\right)\right] \\
&= \sum_{\sigma_3\sigma_4\ldots\sigma_8} \alpha \prod_{i,j} \left[1 + K\sigma_i\sigma_j\right] \\
&= \alpha\left(1 + K\sigma_1\sigma_a\right)\left(1 + K\sigma_2\sigma_b\right) \\
&\quad \sum_{\sigma_3\sigma_4\ldots\sigma_8} \left(1 + K\sigma_1\sigma_3\right)\left(1 + K\sigma_a\sigma_6\right)\ldots\left(1 + K\sigma_5\sigma_2\right) \\
&\quad \times \left(1 + K\sigma_8\sigma_b\right) \\
&= \alpha\left(1 + K\sigma_1\sigma_a\right)\left(1 + K\sigma_2\sigma_b\right) \times \\
&\quad \sum_{\sigma_3\sigma_4\ldots\sigma_8} \left[1 + K\left(\sigma_1\sigma_3 + \sigma_a\sigma_6 + \ldots\right)\right. \\
&\quad + K^2\left(\sigma_1\sigma_3\sigma_a\sigma_6 + \sigma_1\sigma_3^2\sigma_6 + \ldots\right) + \ldots \\
&\quad \left. + K^{11}\sigma_1\sigma_a\sigma_3^3\sigma_4^3\sigma_5^3\sigma_6^3\sigma_7^3\sigma_8^3\sigma_2\sigma_b\right].
\end{aligned}
\tag{11.32}
$$

Here $\alpha \equiv (\cosh(\beta J))^{13}$ and $K \equiv \tanh(\beta J)$. The only non-zero terms are those in which all σ_i appearing as indices (namely $\sigma_3, \sigma_4, \ldots, \sigma_8$) are squared. The lowest-order terms in which this occurs are $K^3\ \sigma_1\sigma_3^2\sigma_6^2\sigma_a$ and $K^3\ \sigma_2\sigma_5^2\sigma_8^2\sigma_b$. These terms correspond to a path linking the nodes 1-3-6-a and 2-5-8-b respectively (cf. Fig. 11.2); the power of K corresponds to the number of segments in the path. To retain all non-zero terms, we thus have to count 1) all direct (i.e. not self-intersecting) paths linking nodes 1 and 2, 1 and a (and 2 and b), 1 and b (and 2 and a) and a and b; 2) all closed loops (such as 3-4-7-6-3); and all products of such paths that have no segments in common (such as 1-3-6-a and 4-5-8-7-4). This leads to:

$$
\begin{aligned}
Z\,P\left(\sigma_1,\sigma_2,\sigma_a,\sigma_b\right) = \alpha\left(1+K\sigma_1\sigma_a\right)\left(1+K\sigma_2\sigma_b\right)2^6 \times \\
\times \left\{1+\left(K^3+K^5+2K^7\right)\left(\sigma_1\sigma_a+\sigma_2\sigma_b\right)\right. \\
+\left(K^4+3K^6\right)\left(\sigma_1\sigma_2+\sigma_a\sigma_b\right)+ \\
+\left(K^6+3K^8\right)\sigma_1\sigma_2\sigma_a\sigma_b+\left(3K^5+K^7\right)\left(\sigma_1\sigma_b+\sigma_2\sigma_a\right) \\
\left.+2K^4+K^6\right\}.
\end{aligned}
\tag{11.33}
$$

Using the same procedure we find:

$$
\begin{aligned}
Z\,P\left(\sigma_a,\sigma_b\right) = \sum_{\sigma_1\sigma_2\ldots\sigma_8} e^{\beta J \sum_{i,j}\sigma_i\sigma_j} = \alpha \sum_{\sigma_1\sigma_2\ldots\sigma_8}\prod_{i,j}\left[1+K\sigma_i\sigma_j\right] \\
= \alpha 2^8\left[1+\sigma_a\sigma_b\left(K^4+10K^6+5K^8\right)+4K^4+3K^6+5K^8+3K^{10}\right].
\end{aligned}
\tag{11.34}
$$

Dividing (11.32) by (11.34) we obtain $P(\sigma_1, \sigma_2 \mid \sigma_a, \sigma_b)$. For instance,

$$
P(+,+|+,+) = \frac{(1+K)^2\left[2K^3+4K^4+8K^5+8K^6+6K^7+3K^8\right]}{2^2\left[1+5K^4+13K^6+10K^8+3K^{10}\right]}.
\tag{11.35}
$$

This implies that for a lattice with homogeneous interactions $J_{ij} = J = 1$ and $\beta = 1$, $P(+,+ \mid +,+) = 0.95$. X_{BI} can then numerically be evaluated via (11.4 and 11.5). E.g., for $J = 1 = \beta$, $X_{BI} = -0.667$, a value that satisfies the Bell inequality. In the weak-interaction limit $K << 1$ one finds $X_{BI} \approx -2.\ K^2$, satisfying the Bell inequality. In the strong-interaction limit $K >> 1$ one obtains $X_{BI} \approx 1$, again satisfying (11.4).

Importantly, while above outcomes are obtained for all $J_{ij} = J$ and all $h_i = 0$, numerical simulation shows that the result is more interesting if one varies the J_{ij} and h_i ($\neq 0$) over the lattice. In that case *the BI can be strongly violated*, for a broad interval of values for β, h_i, J_{ij}. For instance, for $\beta = 1$, $h_i \in \{-1, 1, 3\}$, $J_{ij} \in \{1, 2, 3, 4\}$ and keeping left-right symmetry in the lattice, one finds that $X_{BI} = 2.87$ at its local maximum (Vervoort). This exceeds $2\sqrt{2} \approx 2.83$, the Tsirelson bound and the value for the singlet state in the original Bell experiment. In the smallest hexagonal lattice one finds, by an amusing coincidence, $X_{BI}^{max} = 2.82843 = 2\sqrt{2} + 3.10^{-6}$.

Even if some lattices violate the Bell inequality, it is possible to prove that they satisfy OI and PI (Vervoort) (as also follows from the theory of Markov fields Kindermann and Snell 1980). *At the same time all lattices violate MI*; thus MI-violation is the resource for violation of the Bell inequality. In Vervoort, it is further shown that non-screening-off properties as (11.9) and (11.21) hold in the lattice of

Fig. 11.2, e.g. if one takes $\lambda_1 \equiv (\sigma_3, \sigma_6)$, $\lambda_2 \equiv (\sigma_5, \sigma_8)$ and $\lambda_0 \equiv \sigma_4$. This actually follows directly from the theory of Markov random fields.

Clearly, to make this study fully relevant one would have to go beyond a static Bell experiment; it is an open question whether Ising-like (lattice-gas) models can be adapted to the dynamical phenomena considered in Sect. 11.4, notably to describe a pilot-wave.[1] Still, classic spin-lattices already exhibit several intriguing similarities with the generic model of Sect. 11.4 and can illustrate some of the abstract correlations considered there, notably 'non-screened-off MI-violation' leading to violation of the Bell inequality.

This chapter is a reprint of the article: L. Vervoort, "Are Hidden-Variable Theories for Pilot-Wave Systems Possible?", Foundations of Physics (2018), Vol 48–7, pp. 803–826, cf. https://link.springer.com/article/10.1007/s10701-018-0184-x

References

Aspect, A. 2015. Viewpoint: Closing the door on Einstein and Bohr's quantum debate. *APS Physics* 8: 123.

Bell, J.S. 1964. On the Einstein-Podolsky-Rosen paradox. *Physics* 1: 195–200.

———. 1981. Bertlmann's Socks and the nature of reality. *Journal de Physique* 42 (Complément C2): C2-41–C2-62.

———. 2004. *Speakable and unspeakable in quantum mechanics*. Cambridge: Cambridge University Press.

Bohm, D., and J.P. Vigier. 1954. Model of the causal interpretation of quantum theory in terms of a fluid with irregular fluctuations. *Physics Review* 96: 208.

R. Brady, R. Anderson. 2014. arXiv:1401.4356 [quant-ph].

Brans, C. 1988. Bell's theorem does not eliminate fully causal hidden variables. *International Journal of Theoretical Physics* 27: 219–226.

Brunner, N., et al. 2014. Bell nonlocality. *Reviews of Modern Physics* 86: 419.

Bush, J.W.M. 2015a. The new wave of pilot-wave theory. *Physics Today* 68 (8): 47–53.

———. 2015b. Pilot-wave hydrodynamics. *Annual Review of Fluid Mechanics* 47: 269–292.

Clauser, J., and M. Horne. 1974. Experimental consequences of objective local theories. *Physical Review D* 10: 526–535.

Couder, Y., and E. Fort. 2012. Probabilities and trajectories in a classical wave-particle duality. *Journal of Physics Conference Series* 361: 012001.

Couder, Y., S. Protière, E. Fort, and A. Boudaoud. 2005. Dynamical phenomena: Walking and orbiting droplets. *Nature* 437: 7056.

de Broglie, L. 1987. Interpretation of quantum mechanics by the double solution theory. *Ann. Fond. Louis de Broglie* 12: 1–23.

de la Peña, L., A.M. Cetto, and A. Valdés Hernández. 2015. *The emerging quantum: The physics behind quantum mechanics*. Springer.

[1] The Ising Hamiltonian could also be assumed for moving particles, as in lattices gases, which can be described by Ising-like Hamiltonians and which can simulate the Navier-Stokes equation. However the Boltzmann probability (11.31) is only valid in equilibrium. Note also that the Ising Hamiltonian can be seen as a 1st-order approximation (it consists of the first two terms of a Taylor expansion in the spins).

Di Lorenzo, A. 2012. A simple model for the spin-singlet: mathematical equivalence of non-locality, slave will, and conspiracy. *Journal of Physics A: Mathematical and Theoretical* 45: 265302.
Eddi, A., J. Moukhtar, S. Perrard, E. Fort, and Y. Couder. 2012. Level-splitting at a macroscopic scale. *Physical Review Letters* 108: 264503.
Frisch, U., et al. 1986. Lattice-gas automata for the Navier-Stokes equation. *Physical Review Letters* 56: 1505–1508.
Gallego, R., L.E. Würflinger, A. Acín, and M. Navascués. 2011. Quantum correlations require multipartite information principles. *Physical Review Letters* 107: 210403.
Gilet, T. 2016. Quantum-like statistics of deterministic wave-particle interactions in a circular cavity. *Physical Review E* 93: 042202.
Giustina, M., et al. 2015. Significant-loophole-free test of Bell's theorem with entangled photons. *Physical Review Letters* 115: 250401.
Groessing, G., S. Fussy, J. Mesa Pascasio, and H. Schwabl. 2014. *arXiv*:1403.3295 [quant-ph].
Hall, M.J.W. 2010. Local deterministic model of singlet state correlations based on relaxing measurement independence. *Physical Review Letters* 105: 250404.
———. 2011. Relaxed Bell inequalities and Kochen-Specker theorems. *Physical Review A* 84: 022102.
———. 2015. The significance of measurement independence for Bell inequalities and locality. *arXiv*:1511.00729 [quant-ph].
Harris, D.M., and J.W.M. Bush. 2014. Droplets walking in a rotating frame: From quantized orbits to multimodal statistics. *Journal of Fluid Mechanics* 739: 444–464.
Hensen, B., et al. 2015. Loophole-free Bell inequality violation using electron spins separated by 1.3 kilometers. *Nature* 526: 682.
Khrennikov, A. 2008. *Interpretations of probability*. Berlin: de Gruyter.
———. 2011. *Journal of Physics: Conference Series* 306: 012021.
Kindermann, R., and J.L. Snell. 1980. *Markov random fields and their applications*. American Mathematical Society.
Kochen, S., and E.P. Specker. 1967. The problem of hidden variables in quantum mechanics. *Journal of Mathematics and Mechanics* 17: 59–87.
Kupczynski, M. 1986. *Physics Letters A* 116: 417–422.
Madelung, E. 1927. Quantentheorie in hydrodynamischer form. *Zeitschrift für Physik* 40 (3–4): 322–326.
Nieuwenhuizen, Th.M. 2009. Where Bell went wrong. *AIP Conference Proceedings* 1101: 127.
Perrard, S., et al. 2014. Chaos driven by interfering memory. *Physical Review Letters* 113: 104101.
Popescu, S. 2014. Nonlocality beyond quantum mechanics. *Nature Physics* 10: 264–270.
Sampara, N., and T. Gilet. 2016. Two-frequency forcing of droplet rebounds on a liquid bath. *Physical Review E* 94: 053112.
Santos, E. 2004. The failure to perform a loophole-free test of Bell's inequality supports local realism. *Foundations of Physics* 34: 1643.
V. Sbitnev. 2015. *arXiv*:1504.07497 [quant-ph].
Shalm, L., et al. 2015. Strong loophole-free test of local realism. *Physical Review Letters* 115: 250402.
't Hooft, G. 2015. Models on the boundary between classical and quantum mechanics. *Philosophical Transactions of the Royal Society A* 373: 20140236.
———.. 2016. The cellular automaton interpretation of quantum mechanics. In *Fundamental theories of physics*, Vol. 185. Springer, *arXiv*:1405.1548 [quant-ph].
———.. 2018. Free will in the theory of everything. *arXiv*:1709.02874 [quant-ph].
Vervoort, L. 2000. Bell's theorem and nonlinear systems. *Europhysics Letters* 50: 142.
———. 2013. Bell's theorem: Two neglected solutions. *Foundations of Physics* 43: 769–791.
———. 2016. No-go theorems face background-based theories for quantum mechanics. *Foundations of Physics* 46 (4): 458–472.
L. Vervoort, to be published.
Yeomans, J.M. 1992. *Statistical mechanics of phase transitions*. Oxford: Oxford Science Publications.

Chapter 12
The Wave-Memory Interpretation of Quantum Mechanics

Paulo Castro

Abstract Present quantum mechanics has been a source of puzzlement since its inception in 1927. The theory has been undoubtedly successful predicting results in a large number of experimental situations. However, its heavy epistemological legacy, coming from the initial ideas of Niels Bohr about complementarity and about the possibility for physical description, has turned quantum mechanics in a kind of ontological black box. This work is an attempt to overcome such epistemological opacity, proposing an interpretation that will present Bohrian quantum mechanics formalism and the pilot-wave framework as two complementary and mutually non-exclusive theoretical descriptions of quantum phenomena. The two theories will, hopefully, become the foundations of a new and more complete approach to quantum reality. One that will, simultaneously, serve as a highly efficient predictive formalism and also as an intelligible description about quantum phenomena in four-dimensional spacetime. I will start from the pilot-wave framework, proposing that quantum waves are memory carrying structures that encode probabilistic distributions of the fundamental behaviors of corpuscles. This nomological information, as I will call it, will have a necessitarian weight upon all corpuscles behaviors, in the sense that the only behaviors allowed in Nature are the ones that can be encoded or have been encoded in the quantum wave carriers. Such a picture of quantum physics will imply that corpuscles exchange nomological information with the carrier waves, by means of the guidage or pilot-wave effect. Secondly, I will suggest that the actual Bohrian quantum mechanics formalism, the one used in the Copenhagen School interpretation, describes not the actual states and properties of quantum corpuscular entities, but the nomological information about those states and properties. As such Bohrian quantum mechanics is a description about not the actual physical observable dynamics, but about the dynamics of the nomological information encoded in the carrier quantum waves. The adoption and consistent

---

P. Castro (✉)
Centre for Philosophy of Sciences of the University of Lisbon (CFCUL), University of Lisbon, Lisbon, Portugal
e-mail: jpcastro@ciencias.ulisboa.pt

P. Castro et al. (eds.), *Advances in Pilot Wave Theory*, Boston Studies in the Philosophy and History of Science 344,
https://doi.org/10.1007/978-3-031-49861-9_12

interplay of both theoretical descriptions, pilot-wave mode and Bohrian mode, will serve epistemological completeness in the study of quantum phenomena.

Keywords Philosophy of quantum mechanics · Pilot-wave theory · Copenhagen School interpretation · Bohrian quantum mechanics · Nomological information · Quantum wave carriers

12.1 Introduction

Quantum mechanics has been a source of astonishment and puzzlement since its very early inception around 1927, when the corpus of the actual theory began to build up from the now legendary fifth Solvay conference (Bacciagaluppi and Valentini 2009). Niels Bohr's philosophical insights at the time were embraced, due to his profound charisma and influence in the physics community. These insights developed after Heisenberg's realization that the uncertainty relations, in the undulatory formalism, would have an epistemological impact on the classical causal description, involving position and momentum (Bohr 1985). From this fact Bohr rationalized quantum mechanics as a theory that would include the ultimate limits to the human cognition of reality, grounded on his Principle of Complementarity. Although not present even in the late ideas of Bohr, certainly holding a much more cautious stand, a stronger idealistic understanding of quantum phenomena sparked from the concept of limited cognition of reality. It kept feeding the imagination of physicists, mathematicians, and philosophers of Science, over decades. The cultural academic environment has progressively grown used to accept the causal influence of a kind of platonic potential realm, where a wave-like structure metaphysically exists to rule the "creation into the world" of potential atomic corpuscles, brought into existence by acts of conscious observation.

This highly idealistic way of thinking has seemingly shed, upon contemporary reasoning about physics, the idea that mathematical formalism is to be thought as more important than the reality it describes and, surprisingly, even more important than the intelligibility it was supposed to be grounded on. It is rather remarkable that already at the Solvay conference of 1927, Louis de Broglie was trying to counteract this same opacity brought about by the apparent lack of intelligibility quantum wave-corpuscle contradictory experimental results present. At the same event, standing on previous work on the double-solution theory, de Broglie suggested what is now called the Pilot wave model (Bacciagaluppi and Valentini 2009), where wave and corpuscle would both exist at all times, interacting trough what he thought to be a nonlinear guidage effect. The wave would guide the corpuscle, and this would generate the former, behaving as a moving oscillator. De Broglie, deeply influenced by the ideas of Albert Einstein, was aiming at a deterministic theory of quantum mechanics, which of course, might not actually be the case. He was also unaware of the subtill difficulties that the entanglement phenomena and the tunnel effect pose to our understanding of spacetime at the quantum scale.

However, he was surprisingly prescient into the fact that a strong relation does exist in Nature between corpuscle behavior and wavelike properties. It so happens that while Bohr explored this fact pursuing an epistemological train of thoughts, de Broglie tried to make sense of it from a more realistic predisposition, aiming at an ontological proposal about how the world might be. It is my humble understanding that quantum mechanics, although fitting phenomena description in a remarkable way, still seems a theory overweighted by epistemological considerations about the limitations of human knowledge, or else about the overexaggerated demand that all presently unobservable aspects of reality are to be simply dismissed as non-existent. As such, perhaps it is time to repursue de Broglie's ontological way and try to imagine what is implying such possible epistemological limitations, if really there are such. It is surely a very difficult task, but such an enterprise my hopefully bring a theory that will be epistemologically and ontologically weighted in a more balanced sense. It will most probably be a theory that while preserving some of the previous conundrums and perplexities of former attempts, allows framing them in a more intelligible way, pointing to new solutions and unexpected experimental and technological applications. It is a very disturbing sensation to address theoretically an issue without properly understanding it or being taught that one's understanding capability about it is to be somehow limited, as an a priori assumption held by the theory itself. I propose that such philosophical predisposition is inappropriate when it comes to scientific progress and that one must always do the upmost effort to extend human intelligibility about reality. Even when the generalized present conviction seems to dismiss you from such effort.

In what follows I will try to present a possible intelligible framework for quantum phenomena. Of course, it is just a very limited and initial attempt, only devised to suggest a possible alternative way of thinking. I have hence called it an interpretation because it is just that. This attempt will also feed from the spirit of the present time, while concerning the use of an alternative concept of information, which I have qualified as nomological. What will be proposed will, expectedly and rather unavoidably, have embedded its own failures and perplexities in a mere attempt to understand reality. Perhaps too ambitiously, the present paper will have as its ultimate goal to serve a possible theoretical unification between two apparently irreconcilable descriptions: the pilot wave description and the description using what I have called the Bohrian formalism, to be meant here as the one currently serving the Copenhagen Scholl interpretation. This unification attempt will be done by means of a speculation about quantum reality, involving the idea that there is a kind of natural information encoded in quantum waves affecting the corpuscles behaviors. I have called this kind of natural encoded knowledge the "nomological information", identifying their carriers as natural memory storing structures. This is the reason why I have called my proposal the "wave-memory interpretation of quantum mechanics". I will use the words "information", "encoded" and "memory" in a non-rigorous way, albeit adopting enough semantic charge to expose the present ideas. For instants, although the word "encoded" presupposes the existence of a code, I cannot precisely say what that code precisely is in the actual state of my reasoning. The sense hence taken is only that there is a natural structure able to contain information about other physical entities.

In Sect. 12.2, I will try to provide motivation for the nomological information hypothesis, starting from the Pilot-wave theory framework. In Sect. 12.3, I will take a more philosophical stand, dealing with some "law-like" or necessitarian aspects of the interpretation, introducing an ontology that tries to capture its main features. Finally, in Sect. 12.4, I will take a rather small part of the Bohrian formalism approach to offer motivation for a possible encoding schema for the nomological information. To accomplish this, I will apply the canonical procedure of wave packaging, first to infinitely spread plane waves and, secondly to space limited Morlet wavelet solitons, both considered as information wave carriers. From the wavelet packaging process, following a previous author (Croca 2003), I will derive a new generalized position-momentum uncertainty relation, which can be interpreted as a measure of nomological order in any physical situation.

12.2 Encoding the Behavior of the Corpuscle in Pilot Wave Phenomena

In recent years a new domain of experimental work developed in fluid dynamics physics: the so-called Hydrodynamic Quantum Analogs (HQA) research domain, with a small world spread community of experimentalists and theoretical physicists, presently highly motivated by John Bush at the MIT in the United States (Bush and Oza 2021). The movement started in 2005 with the pivotal work by Yves Couder, S. Protière, Emmanuel Fort and A. Boudaoud at Paris, France (Couder et al. 2005; Couder and Fort 2006). The French team found out that a millimetric oil droplet could self-propel, bouncing on the surface of a vibrating liquid bath. The sequence of bouncing positions would form a trajectory, each bounce resulting from the resonant interaction between the droplet and the quasi-monochromatic wave field produced by the droplet itself, that have propagated on the oil surface. This physical scenario hence constituted a pilot wave phenomenon, where a generated wave, taking the droplet as the oscillatory corpuscle, would in turn guide the droplet's trajectory. It showed, once and for all, that pilot wave phenomena and its dynamics do exist, even if at a macroscopic level and with strong deterministic characteristics. The oil bath would have to be acted on from below by a vibrating force, producing an amplitude acceleration below the Faraday threshold. That is, the acceleration value above which pilot wave phenomena will no longer exist, giving way to standing Faraday waves on the oil surface (Bush and Oza 2021). Over the years the HQA community has put the analog approach to the test, trying to reproduce and study, either experimentally or with the use of computational models, several situations resembling the quantum reality cases. From the tunneling effect (Eddi et al. 2009; Tadrist et al. 2020) to the quantum corral effect (Harris et al. 2013; Cristea-Platon et al. 2018), to Landau levels (Fort et al. 2010; Oza et al. 2014), Friedel oscillations

(Sáenz et al. 2020), diffraction from slits (Couder and Fort 2006; Pucci et al. 2018; Ellegaard and Levinsen 2020), spin states (Labousse et al. 2016; Oza et al. 2018), Zeeman splitting (Eddi et al. 2012), the quantum harmonic oscillator (Kurianski et al. 2017; Perrard and Labousse 2018), or the Hong-Ou-Mandel effect (Valani et al. 2018). Besides the strong interest that such analogies invoke, the most striking revelation coming from the HQA field is that there seems to be a relation between the resulting trajectories stability and what one can call the path memory encoded in the wavefield. This dependence had already been noted by Antonin Eddi and co-workers in 2011 (Eddi et al. 2011), with Stéphane Perrard and et al. even comparing the droplet- wavefield system to a natural Turing machine (Perrard et al. 2016). The overall pilot wavefield, guiding the corpuscle, is composed of several prior waves, generated by the corpuscle prior bounces. This superposition of waves produces a complicated field that in turn affects the corpuscle next bouncing, acting as a "natural path memory storage system", and holding kinematic information about the droplet's behavior. It should be mentioned that the storage process is correlated with the dynamic evolution of the field. Another very revealing fact about this pilot-wave phenomenon is that the decaying time of the prior waves define the wavefield memory storage capacity, experimentally controlled by the acceleration impressed on the bath. This in so much that if, on one hand, the decaying time happens to be very rapid (low acceleration), the pilot wave effect will be less stringent. If, on the other hand, the decaying time is extended (using high acceleration forcing on the oil bath, either near or above the Faraday threshold), the pilot wave phenomena will give way to other kind of effects, including some analogous to quantum chaos, where a droplet can bounce unpredictably over the bath (Bush and Oza 2021). This suggests that the memory encoding physical structure has a storing capability limit with a strong effect on the droplet's behavior. HQA experimental and research domain is a very rich one, full of insights and surprises that deserve a much wider description than the one presented here. However, the embedded "nomological" memory (nomological, from the Greek: *nomos*, law, and *logos*, reason, order) seemingly points to a new way of thinking about the ontological status of a quantum wave, one that I will use in the interpretation I intend to propose further on.

Hydrodynamic pilot wave phenomena appear to vindicate Louis de Broglie's original proposals (de Broglie 1960; de Broglie 1969), where the quantum corpuscle acts as a moving oscillator in phase with the quantum wave, guided in turn by what de Broglie called the phase wave of the corpuscle. De Broglie arrived at such a unifying picture of waves and quanta, by means of a law-like equivalence between two principles involving stationary values. In his own words, and from his 1924 Doctoral Thesis: "The Fermat principle applied to the phase wave is identical to the Maupertuis principle applied to the moving entity; the dynamically possible trajectories of the moving entity are identical to the possible rays of the waves" (de Broglie 2021, p. 38). In the course of reasoning following this theoretical equivalence, de Broglie identified a formal connection between the corpuscle's momentum and his "phase wave", as he put it, hence performing a major unification in physics between moving atomic entities and quantum waves. Already in the Doctoral Thesis he proposed the following relation between the quantum wave phase ϕ and the relativistic momentum-energy vector J (de Broglie 2021, p. 38):

$$d\phi = \frac{2\pi}{h} J_i dx^i \tag{12.1}$$

Which for the simplest case, gives what de Broglie later called the "guidage" relation:

$$p = \hbar \nabla \phi \tag{12.2}$$

For this relation between the kinematic behavior of corpuscles and the undulatory properties of matter, Louis de Broglie received the Nobel prize in 1929.

After the 1927 Solvay conference, Louis de Broglie proposals were generally dismissed by the scientific community. This was to go on until American physicist David Bohm appeared with his early 1952 papers about the hidden variable theoretical formulation of quantum mechanics (Bohm 1952a, b), which endured until 1993, when Bohm forwarded his "implicate order" philosophical interpretation of pilot wave theory (Bohm and Hiley 1995, p. 30), still praised today in the Physics community. In Bohm's approach, the action S would still be related with the wave's phase, but the guiding process would be primarily due to the action of a quantum force originating from a quantum potential. To express the wave-corpuscle guiding dependence, Bohm considered the classical Hamilton-Jacobi equation with an extra term, corresponding to the quantum energy potential Q. This equation would represent the dynamical behavior of a corpuscle with its momentum depending on the action S, related to the wave phase, hence the connection to de Broglie's guidage. The corpuscle's acceleration would then also depend on the quantum potential, which on Bohm's own account would mean "... that the forces acting on it [the corpuscle] are not only the classical force $-\nabla V$, but also the quantum force $-\nabla Q$" (Bohm and Hiley 1995, p. 30).

This same quantum potential concept, once applied to the n-body case, would be shown as early as 1974, by David Bohm and Basil Hiley (1975) to imply non-locality, in the sense that a "quantum interconnectedness of distant systems" would be valid leading to a "... notion of unbroken wholeness which denies the classical idea of analyzability of the world into separately and independently existent parts" (Bohm and Hiley 1975, p. 96).

As the authors put it, this interconnectedness feature of the world was due to the following main aspects of the quantum potential:

(a) The quantum potential $Q(X_1, \ldots, X_n)$ does not in general produce a vanishing interaction between two particles i and j as $|X_i - X_j| \to \infty$. In other words, distant systems may still have a strong and direct interconnection. This is, of course, contrary to the general requirement, implicit in classical physics, that when two particles are sufficiently far apart, they will behave independently. Such behavior is evidently necessary if the notion of analysis of a system into separately and independently existent constituent parts, which can conceptually be put together again to explain the whole is to have any real meaning.

(b) What is even more strikingly novel is that the quantum potential cannot be expressed as a universally determined function of all the coordinates $X_1, \ldots, X_n$. Rather, it depends on $\Psi(X_1, \ldots, X_n)$ and therefore on the "quantum state" of the system as a whole. In other words, even apart from the point made in (a), we now find that the relationships between any two particles depend on something going beyond what can be described in terms of these particles alone. Indeed, more generally, this relationship may depend on the quantum states of even larger systems, within which the system in question is contained, ultimately going on to the universe as a whole. (Bohm and Hiley 1975, p. 99)

This, of course, implies a substantive difference between de Broglie and Bohm. While the former would consider quantum waves to exist as real four-dimensional spacetime physical entities, the later would not. In Bohm's ontology the quantum wave is only a law-like entity and the guidage effect of the wave over the corpuscle is to be though as a pure idealistic and non-material process, converting implicate order into explicate order. This can cause conceptual tension with the idea of a "quantum force" caused by the quantum potential but, interestingly, is something that already resembles some kind of information transfer happening in Nature. A process that, on the other hand, still dismisses the possibility for a four-dimensional physical structure where nomological information can be encoded. Another of Bohm's insight is non-locality, equated with the concept of wholeness. This can be understood as a law-like (nomological) feature of the universe, once one accepts the possibility that "interaction ruling" between several systems might also be encoded in a physical underlying structure. Non-locality would follow from the necessity that each and every corpuscle, involved in an interaction, must be "informed" about all its possible behaviors in relation to all remaining possible behaviors of all other corpuscles. I will address again this topic in Sect. 12.3.

Portuguese physicist José Croca (born 1944), started from de Broglie's so called causal tradition in Quantum Mechanics developing mainly from 2003 on, yet another view about Pilot-wave theory with several original and insightful contributions. He started by reasoning that the prohibition posited by the Heisenberg uncertainty relations to well-defined trajectories for quantum corpuscles was, in fact, due to the adoption of what he called a Fourier ontology in our modes of thought about quantum phenomena. In so much that the quantum formalism uses the composition of infinite plane waves to form a corpuscle wave package, which in turn leads unavoidably to Heisenberg's uncertainty relations and to Bohr's Complementarity Principle, grounding the Copenhagen School interpretation. The Fourier treatment approach is, of course, only an ideal mathematical one, since in Nature one does not expect to have infinite waves of any sort. Croca hence proposed that a much more reasonable mathematical tool to apply to quantum mechanics would be wavelet local analysis, using Morlet solitons (Croca 2003). These are mathematically equivalent to plane waves modulated by amplitude gaussian functions. Croca's approach would, in fact, introduce a third parameter to describe quantum waves, besides wavenumber and frequency, namely the spatial spreading of the wavelet soliton, represented by the Greek letter σ (sigma) in the gaussian modulation. A quantum Morlet soliton would immediately exhibit a most favorable property, that of having well-defined wavelength and frequency, while being spatially localized. This somehow surprising feature would, consequently, avoid the problem of having to build a wave package from infinite plane waves in order to get a localized corpuscle, while allowing to generalize Heisenberg's uncertainty relations, as it will be seen in Sect. 12.4. Next, Croca went on to derive a one body wavefield non-linear differential equation, having as its natural solution precisely the Morlet soliton. He did this by starting from the two classical mechanic equations already considered by Bohm, the continuity equation and the Hamilton-Jacobi equation, although, unlike Bohm, without introducing an *ad hoc* quantum potential in the Hamilton-Jacobi equation. Croca hence obtained what he called a non-linear Master Schrödinger equation, with the nonlinear quantum potential term appearing formally in it, as a formal consequence from using the two classical equations at the start (Croca 2003, p. 72). Croca thought this to be a signature of a possible unification between the classical and quantum worlds. Significantly, one can also invert his derivation, giving thus strength to the interpretation that those two fundamental equations of classical mechanics will, in turn, result as formal consequence of a quantum Master equation, to be considered epistemologically more general. Croca also noted, as it is clear from (12.3) and (12.4) below, that if the Morlet soliton's intensity is constant, corresponding to the special case where its spatial spreading σ is exceptionally large, one gets both the linear Schrodinger equation and its usual plane wave natural solution. This strongly suggests that wavelet local analysis can be a more general approach to quantum mechanics, with the Fourier situation appearing as a particular limiting case of the former treatment. Although I have recently come to the idea, to be developed in a future work, that both dynamics, the one given by the Schrodinger linear equation, and the one given by the non-linear master equation are to be though as complementary in quantum phenomena description. That is, both are necessary to completely describe quantum events and while the later relates directly to the four-dimensional physical wavefield and the corpuscle kinematic behavior, the former describes the dynamics of the nomological information encoded in the four-dimensional wavefield.

José Croca's own thoughts about the two kinds of equations and natural solutions are the ones that follows:

$$\begin{aligned} & -\frac{\hbar^2}{2m}\nabla^2\theta + \frac{\hbar^2}{2m}\frac{\nabla^2(\theta\theta^*)^{1/2}}{(\theta\theta^*)^{1/2}}\theta + V\theta = i\hbar\theta_t \\ & \text{If} \quad (\theta\theta^*)^{1/2} = const. \quad \text{then:} \\ & -\frac{\hbar^2}{2m}\nabla^2\theta + V\theta = i\hbar\theta_t \end{aligned} \tag{12.3}$$

$$\theta = Ae^{-\frac{(x-vt)^2}{2\sigma_x^2}+i(kx-\omega t)} \underset{\sigma_x\to\infty}{\to} Ae^{i(kx-\omega t)}$$
$$\text{Implying} \quad (\theta\theta^*)^{1/2} \underset{\sigma_x\to\infty}{\to} A\ (const.) \tag{12.4}$$

Using Morlet wavelet solitons as quantum waves, Croca went on to make a wave package composition, following the same procedure as Bohr did when he used plane waves to get to the Heisenberg uncertainty relations (Bohr 1985). Croca was then able to derive the set of generalized uncertainty relations (12.5a) and (12.5b) below (Croca 2003, p. 94), containing in a consistent manner with (12.3) and (12.4) the Heisenberg's ones as a particular limiting case. As I will refer later in Sect. 12.4, one can also reason that Croca's relations also contain the Newtonian case (no correlation between position and momentum) as the other limiting case (Castro et al. 2017). The uncertainty brought about by the correlation between position and momentum (or similarly between time and energy) would now also depend on the wavelet's spatial spreading σ, as can be seen directly from (12.5a) and (12.5b).

$$\Delta p_x = \frac{\hbar}{\Delta x}\sqrt{1 - \Delta x^2/\sigma_x^2} \tag{12.5a}$$

$$\Delta E = \frac{\hbar}{\Delta t}\sqrt{1 - \Delta t^2/\sigma_t^2} \tag{12.5b}$$

Some years later, in 2015, Croca developed his quantum mechanics ontology. He introduced what he called philosophical Principle of Eurythmy (Croca 2015) as a foundational basis for the extremal principles and as a foundation principle for the pilot wave guidage effect over the corpuscle. In a certain sense, one might say that he balanced the phenomenal roles of both entities, wave and corpuscle, highlighting the ontological role of the corpuscle, as a complex active entity of its own. In his words:

> The Principle of Eurhythmy comes from the Greek word, ευριτμια (euritmia) which is the composition of the root *eu* plus *rhythmy*. Here, *eu* stands for the right, the good, the adequate; and *rhythmy* for the way, the path, the harmonic motion. The composite word means: the adequate path, the good path, the good way, the right way, the golden path, and so on (Croca 2015, p. 15) (...) The Principle of Eurhythmy states that complex systems in nature, in order to continue to exist, must "choose" and follow optimal dynamic pathways. If a given system does not – on average – follow the optimal, most efficient pathway of development, it will not survive for long. All systems, which are able to exist long enough to be observable, even at a very small scale of observation, necessarily follow the Principle of Eurhythmy. (Croca 2015, p. 38)

According to the Principle of Eurhythmy, in order to keep its structural integrity, the corpuscle would move into the regions where the intensity of the quantum wave was higher, hence explaining the origin of Born's probability rule. On the other hand, the Fermat and Maupertuis Principles would be instantiations of the Principle of Eurhythmy.

Using his Principle and following a stochastic reasoning, Croca was finally able to derive an additional second guidage formula, different from the one used by de Broglie's and Bohm, reading:

$$p = \hbar \frac{\nabla (\theta\theta^*)}{(\theta\theta^*)} \tag{12.6}$$

It is a remarkable thing that using (12.6) one can write the following heuristic relations, where in (12.7a) the second order differentiation term represents the kinetic energy associated to the quantum wave phase, while in (12.7b) the quantum potential can now be interpretated as the guidage kinetic energy of the corpuscle:

$$p_\phi = \hbar \nabla \phi \qquad \rightarrow \qquad E_\phi = \frac{\hbar^2}{2m} \nabla^2 \phi \tag{12.7a}$$

$$p_g = \hbar \frac{\nabla (\theta\theta^*)}{(\theta\theta^*)} \qquad \rightarrow \qquad E_g = \frac{\hbar^2}{2m} \frac{\nabla^2 (\theta\theta^*)^{1/2}}{(\theta\theta^*)^{1/2}} \tag{12.7b}$$

In both expressions presented in (12.7b) the term operated by the gradient in the numerator is also present in the denominator, thus providing a kind of nonlinear normalization of that term. This means that only the topological curvature of the quantum field, and not its intensity, is relevant for the pilot-wave effect. I find this to be a signature for the presence of a sort of nomological information encoded in the field, being informationally read by the corpuscle as I will propose later on.

The kinetic energy of the corpuscle will be the one associated to the wave phase in the particular case where the wave's intensity is constant, corresponding to the situation where the wave's extension is exceptionally large, compared to the corpuscle's dimension. In this case, the corpuscle moves randomly within the wave, inheriting its average kinematic characteristics from the wave's phase. In this sense, de Broglie's and Bohm's original phase guidage can be thought of as a kind of inertial momentum formula, mainly attributable to the four-dimensional physical wave's dynamic evolution. When, on the other hand, the wave's topology is asymmetric, then the kinetic energy of the corpuscle will be a composition of both energies depicted in (12.7a) and (12.7b). In this case, the corpuscle will still move within the wave but will also be highly dependent on the local topology of the wave, that is, on the topology at the particular position where the corpuscle happens to be, in a given instant, within the wave.

There is now an important remark to make. The nonlinear interaction between the quantum guiding wave and the guided corpuscle was always a motive of concern, even for de Broglie himself. A quantum real wave after traversing a beam splitter, in the realistic approach, splits in two branches: one carrying the corpuscle, and the other corresponding the so-called empty wave. For the majority of experimental cases, where the quantum wave is very large, compared to the corpuscle dimension, one can assume according to (12.7a) that the wave's energy is practically equal the energy of the corpuscle. This poses a striking objection to the existence of an empty wave in the other branch, since this empty wave seems to be a physical entity devoid of energy, as all energy seems to be in the occupied branch. In fact, it renders rather strange the pilot wave hypothesis entirely, because then one would have in general a non-energetic entity, a quantum wave, piloting a comparably high energetic entity, the corpuscle.

I wish to suggest that such a conundrum can be removed if one thinks in terms of information transfer, rather than in terms of energy balance checks. Although this is a possible way of reasoning out the of the problem, it must nevertheless be emphasized that a quantum empty wave can still have an energy. Croca has estimated that the energy of the wave would be of about 10^{54} orders less than the energy of the corpuscle (Croca 2004). This is about the same as saying that it has no energy, and of course means that whatever guiding interaction is happening between the wave and the corpuscle it must be of a nonlinear kind, in the sense that a natural amplification process must be at work. This amplification process must be able to gather energy from the vacuum or, otherwise, steer the comparably formidable energy of the corpuscle, in order to preserve energy conservation.

In a special issue about Pilot wave theory (Drezet 2023), an article by Bloch and Cohen was published relating mean quantum potential with the Heisenberg uncertainty relations, via the Fisher information (Bloch and Cohen 2022). The significant aspect I wish to note here is that the authors have shown that the mean quantum potential is proportional to the Fisher information applied to the corpuscle position measurement. In that particular context, the Fisher information represents the amount of empirical information that the measurement of the corpuscle's position provides, conditioned to the possible positions it can occupy. The original definition, as the authors have referred to, comes from Marcel Reginato, as stated in 1998:

> Consider the problem of estimating a parameter X in the presence of unknown added noise x. A measurement Y of the parameter will be related to X and x by:
>
> $$Y = X + x \tag{12.8}$$
>
> For example, Y might be a measurement of the position of a corpuscle, while X is the actual position of the corpuscle. The probability distribution $P(x)$ for the noise x will be related to $P(Y|X)$ by
>
> $$P\left(Y|X\right) = P\left(Y - X\right) = P(x), \tag{12.9}$$
>
> (...) In this case, the Fisher information I, is given by
>
> $$I \equiv \int \frac{1}{P\left(Y|X\right)}\left(\frac{\partial P\left(Y|X\right)}{\partial X}\right)^2 dX = \int \frac{1}{P(x)}\left(\frac{\partial P(x)}{\partial x}\right)^2 dx \tag{12.10}$$
>
> Reginatto (1998, p. 1775)

Bloch and Cohen made the following remark about the formulation and identities assumed by Reginato:

> (...) it is clear that for the identification of Fisher information with the quantum potential, one must have,
>
> $$\left(\frac{\partial P(x)}{\partial X}\right)^2 = \left(\frac{\partial P(x)}{\partial x}\right)^2. \tag{12.11}$$
>
> While the previous assumptions are partially justified in the literature, this assumption goes unnoticed. This statement is equivalent to,
>
> $$\left(\frac{\partial x(X)}{\partial X}\right)^2 = 1, \tag{12.12}$$
>
> which means that an increment of the "hidden" position results in an equal increment in the noise. In other words, the noise is proportional to the actual position of the corpuscle. What is the justification for such an assumption?" (Bloch and Cohen 2022, p. 117)

Providing a possible answer to this problem, I will also suggest that what is called the corpuscle position "noise" is, in fact, a kind of nomological information about a diversity of possible positions available to the corpuscle. In other words, the mathematical formalism, used to treat the uncertainty about the corpuscle position in a measurement, masks the existence of two different representations of the corpuscle position: its actual position X, the one where the corpuscle in fact is, and the overall information about all other possible positions x, that the corpuscle can potentially have. This information is encoded in the wave, being accessed by the corpuscle, via the wave's topology curvature. That is, via the encoded value of momentum that the corpuscle can assume to reach a new position from its actual one. The proportionality between the noise and the corpuscle actual position, only means that wherever the corpuscle is, there will also be the nomological information about other possible positions to be reached.

Following Bloch and Cohen about the relation between the mean quantum potential and the Fisher information, for one spatial dimension, and with $\rho(x) = (\theta\theta^*)^{1/2}$, one can thus write:

$$\overline{Q} = \frac{\hbar^2}{2m}\int \rho(x)\frac{\partial^2\rho(x)/\partial x^2}{\rho(x)}dx \tag{12.13a}$$

$$I(x) = \int \frac{1}{\rho(x)}\left(\frac{\partial\rho(x)}{\partial x}\right)^2 dx \tag{12.13b}$$

Where (12.13a) is the mean quantum potential and (12.13b) is the Fisher information for the measurement of a corpuscle's position (Reginato 1998). From these expressions we get, after some calculation:

$$\overline{Q} \propto \frac{\hbar^2}{8m} I \tag{12.14}$$

This identity is a remarkable one. Using (12.7b) one can now propose that the average guidage kinetic energy is, in fact, closely related to the overall positional nomological information encoded in the quantum wave. Surprisingly, it seems to correlate (in an informational way) energy with a set of possible locations in space, invoking the idea that such positions demand energy and therefore momentum to be reached. This gives strong bearing to the idea that the guidage effect is an information transfer phenomenon, where the corpuscle is using a sort of encoded information to acquire its state of movement, that is, momentum. It also gives credit to the possibility that the guidage effect cannot be strictly thought as an energy transfer process, since information is the actual fundamental concept to be attributed to the wave. The guidage effect seems to be more adequately described as a transduction phenomenon between information and energy. A process that in purely energetic terms will present itself as a nonlinear one. I must emphasize that at this stage it is not at all clear to me if the Landauer type of reasoning about information and energy (Landauer 1993, 1996, 1999) necessarily applies here. Anyhow, the information concept with which I am trying to cope with must be understood as a fundamental and primary phenomenal kind of information in Nature. One that is closer to what can metaphorically be envisaged a "rule or law of Nature" encoded in a quantum wave four-dimensional structure.

Before proceeding to the next section, I want to signal the theoretical work of Lucien Hardy (Hardy 2001), which also seems pertinent to the present proposals. Although the literature has developed since then, I find it quite significant that Hardy was able to show that standard quantum mechanics is a probability theory that only differs from the classical one by a simple axiom. In Hardy's own words and from his paper:

> (…) it is shown that quantum theory can be derived from five very reasonable axioms. The first four of these axioms are obviously consistent with both quantum theory and classical probability theory. Axiom 5 (which requires that there exist continuous reversible transformations between pure states) rules out classical probability theory. If Axiom 5 (or even just the word "continuous" from Axiom 5) is dropped, then we obtain classical probability theory instead. (Hardy 2001, p. 1)

Again, this seems to give strength to the hypothesis that Bohrian quantum formalism could be primarily thought as a theory about information in a very general sense, and not about the phenomena to which such information refers to. It must, nevertheless, be noted that this overall picture, where quantum waves encode information, posits considerable pressure upon quantum realists such as me, since we will have the onus of showing that empty (informational) quantum waves do exist. Quantum wave experimental detection, which also has its own history, therefore becomes another key element in the memory-wave interpretation.

To that matter a set of quantum wave detection experiments have been devised and published here (Croca et al. 2023).

12.3 Strong Nomological Physicalism

In what follows I will try to give some philosophical motivation for the possibility of having nomological encoded information in the world. This will of course be a brief account on the topic, since my main concern here is to suggest an alternative way of thinking about quantum phenomena, rather than to solve old mysteries with a few proposals, something that definitely no one should expect here.

The first sub-section will deal with the core ideas that I've gathered about a possible nomological approach to pilot wave phenomena. The second sub-section will introduce what I have called the ontology of the wave memory interpretation of quantum mechanics. Both themes define in a very initial approach what I have named "Strong nomological physicalism", the idea that one can have in Nature physical structures that can encode information about the physical behaviors of other structures.

12.3.1 Nomological Constructivism and Probabilistic Necessitarianism in Natural Laws

One of the major and deepest mysteries of Nature is its innate order and stability. To this, one might say, David Bohm answered with his implicate order concept, while José Croca forwarded the Principle of Eurythmy (Croca 2015), which in a broad sense establishes the relation between existence or being, on one hand, and order, harmony, and persistence, on the other. Around the same subject, I wish to contribute to a related topic, which can naively be formulated by asking where are the so-called "laws of Nature" written. That is, where does such natural order and harmony comes from and not so much why does it exist, which is the problem Bohm and Croca tried to address. In the prior section I gave experimental and some formal motivation to the idea that pilot-wave fields can encode information about corpuscles kinematic behaviors. Now I will try to make sense of it from a more philosophical point of view in route to what I will introduce as an ontology grounding the wave memory interpretation of quantum mechanics.

Let us start by the tension between indeterminism and determinism that divides micro and macro reality in present day Physics. I suggest that it is a dismissible tension, should one assume that the universe has a history. That is, an evolution of regularities and pattern forming that has been slowly and progressively been fixated in natural nomological memory structures during the evolution of the universe. Whatever physical entities have emerged and developed, by whatever physical processes, a probabilistic overall process has occurred, making behaviors and interactions to be describable using probability distributions, and consequently informationally encodable in nomological structures. Of course, in this universal plot there should be a few grounding and primeval "laws", ruling nomological information storage and retrieval. However, even if the existence of such rules again presupposes an *a priori* regularity in Nature, as something embedded in existence itself, this evolutionary nomological picture significantly deflates the ontological weight of the metaphysical origin of regularity. It so happens that the majority of the law-like behaviors we actually find in the world can be in principle retraced to a nomological memory, which has been constructed over time and encoded in the proper physical structures. One can even presuppose that a possible overall candidate for nomological memory storing is space itself, given its ubiquity and given that it can be viewed as the set of all possible behaviors that have occurred and that can possibly still occur in Nature by moving corpuscular entities. In this tridimensional realist view of physical phenomena, space would hence be seen as the memory of the world, while time would become the signature of all nomological information storage and retrieval processes. In what can be dubbed as a "nomological constructivist" view of Nature, nomological information retrieval (and storing) would always be a process of memory altering, rather than simply one of information flux. This means that nomological memory is always dynamic, with different timescale regularity stabilities. The main thesis of this article is that quantum waves are the physical carriers of nomological information, that is, of information about the behaviors of corpuscles. Adding to this, one can also suggest that space can be seen as an overall nomological memory structure, therefore implying that quantum waves are a sort of primeval elements that confer spatial and temporal organization to corpuscles behaviors. The corpuscles in turn, retrieving and storing nomological information in quantum waves, through their actualized behaviors and by way of the guidage effect, thus instantiating a pilot wave physical phenomenon. Since in the nomological evolution process there is a kind of *a priori* indeterminism for the majority of behaviors, one can therefore reason that determinism is but an end case of probabilistic behavior that due to habit repetition and nomological memory fixing, has stabilized, becoming a highly probable type of physical behavior at the macroscopic scale. This almost certain repeatable behavior, nomologically encoded as such, can be recognized to be deterministic, for all practical purposes, hence resolving the tension between indeterminism and determinism. This last assertion allows me to pass on to the next topic I wish to consider.

The necessitarian approach to natural regularity proposes that the laws of Nature are more than just a descriptive summary of past and future events, occurring accordingly to a regulatory scheme or upholding empirically recognized patterns (Cover et al. 2012, p. 813). Therefore, the universality and necessity of natural laws in this view are posed as real objective features of the world. I can very much relate to that belief, in so much that I think that there is a nomic necessitation upon behaviors coming from the nomological memory structure, that is, from the quantum waves upon their corpuscles. In other words, since there is no *a priori* necessity for the majority of behaviors in the world, there is a "nomological necessitarianism" at play, which results from the fact that there is nomological encoding. It is, in fact, something of a double ontological effect. On one hand, the initial encoded information could have been about a few of many possible behaviors, thus being contingent. On the other hand, once the encoding is done, the propensity for such behavior will now depend on its encoded probability of occurrence. This points to a kind of "probabilistic necessitarianism" that gains or losses its ontological necessitarian strength upon the occurrence of behaviors, depending on the dynamics of the nomological information carrier structure, that alters the encoded probability distributions of those behaviors.

Regarding the causal relations that can be hypothesized to hold between events, one should justly remember Hume when he noted that "The mind can never find the effect in the supposed cause by the most accurate scrutiny and examination. For the effect is totally different from the cause, and consequently can never be discovered in it" (Hume 1999, p. 111). This, I believe, could be due to the fact that a causal relation can now be understood as being of a nomological probability nature, which is, of course, nomologically encoded in the quantum waves and, although affecting the occurrence of behaviors in a necessitarian probabilistic way, being of an *a priori* contingence nature.

Before moving on to the wave memory ontology, there is something important to be said about interaction schemes that are to be considered, in general, possible between corpuscles or different entities. Once one assumes that strong nomological physicalism is true, thus entailing a form of necessitarianism, one must also accept that a kind of nomological non-locality is also true. The reason for this is that, besides all nomological information about the corpuscles individual behaviors, the nomological information concerning all possible interactions between the corpuscles must also be encoded. This relational encoding being a sort of second order nomological information level, coming next in the encoding hierarchy of the nomological information about each corpuscle possible behaviors. As such, whenever a corpuscle stands in some relation towards other corpuscles, it must be that all relations in which that corpuscle could stand with all other, must already be encoded or can be encoded in the nomological memory. This must surely be true for all corpuscles and, in particular, for those corpuscles that act as signal carriers in spacetime background theories, such as special relativity. This finally means that each and every interacting corpuscle, once it has assumed a certain *relata role* within a certain *relation*, must be able to retrieve the nomological information about that particular relation concerning its and all other corpuscles *relata roles*. This strongly suggests that, at least, in a nomological sense, all corpuscles are informationally connected, which points to a sort of nomologic informational holism and therefore, to a sort of nomological non-locality. It also seems highly resonant with the non-locality aspect of the quantum potential, as identified by David Bohm in his wholeness conceptualization of reality. It is equally consistent with the fact, presented in Sect. 12.2, that the quantum potential is related with the Fisher information about the corpuscle possible positions, giving strength to some kind of encoding process in a quantum wave. It should be emphasized that the fact that the nomological information is retrieved locally by a corpuscle from a pilot-wave does not mean that such information cannot be of a holistic nature. In fact, if one takes seriously the necessitarian imposition, that is, the nomic necessitation, as can be called, of such information upon the corpuscles possible interaction schemas, then non-locality follows in a reasonable straightforward manner.

Another important problem concerns the so-called configuration space representation in standard quantum mechanics. The Pilot wave realistic approach to quantum mechanics, that is, one accepting the existence of a four-dimensional spacetime grounded quantum field, could tentatively be ruled out, given the higher dimensionality of the configuration space. It can indeed be argued that to have a complete description of the n-body physical case in quantum mechanics, one must write down the appropriate expressions in a $3n + 1$ dimensional space. This describes a dynamic that cannot possibly be happening in a four-dimensional space, hence meaning that no four-dimensional quantum field exists for the n-body scenario in quantum mechanics. Much the same could be said about state superposition for one particle, but I will be only referring the configuration space objection. I suggest that this conundrum can be overcome if we admit that the $3n+1$ dynamics that we are indeed describing concerns, not the actual quantum configuration states of the corpuscles involved, but the informational representation of those states that is encoded in a four-dimensional field. In other words, Bohrian quantum mechanics, the one used in the Copenhagen School interpretation formalism, describes the dynamics of the encoded nomological information about physical states and not the physical states directly. It is thus quite natural that such a dynamic should involve all possible states and configurations of the interacting corpuscles. In fact, the required completeness, characterizing an informational description of all possible states is the reason why such description brings about the so-called superposition of states and configuration space descriptions that dimensionally outnumber usual four-dimensional space descriptions. The former would then entail a kind of apposition of distinct nomological representations corresponding to different states, mutually exclusive or otherwise, which for reasons of informational completeness should be exhaustively enumerated in the description and that standard quantum mechanics takes for state superposition.

From what I have been reasoning, in the wave memory interpretation, one can defend a sort of epistemological completeness, saying that, while Pilot-wave theory describes the dynamic evolution of the nomological information carrier, which is a four-dimensional structure dynamic, standard quantum mechanics formalism describes the dynamics of the nomological information encoded is the wave carrier structures. It is therefore necessary to relate both dynamics in a way that the two theories become one, deriving the four-dimensional quantum field from the nomological information dynamics. Adding to these two dynamics one must also consider the dynamics of the corpuscle itself, interacting with the memory carrier wave, reading and storing information, as it actualizes its possible behaviors in four-dimensional spacetime. This last kinematic description of the corpuscle can be included in the pilot-wave theory description, since this last involves both the corpuscle and the quantum wave, as both physical entities exchanging information. In what follows I introduce an ontology trying to convey the ideas developed so far.

12.3.2 The Ontology of the Wave Memory Interpretation of Quantum Mechanics

In this sub-section I wish to present a first approach to an ontology that can guide further theoretical developments within the interpretation that I am proposing. It is not a short summary of postulates as the one found in Bohrian quantum mechanics. Such a synthetic approach can come later, when one is more assured of all implicit assumptions that, I hope, will serve as building conceptual blocks for a mature theory of this sort. What I am presenting here is more an attempt to cover as much assumptions territory as possible, trying to give solid ground to future theoretical contractions, which will make the description more elegant. I have made some effort to be as complete and thorough as I could, positing in a rather intuitive way the propositions about what I hypothesize things to be. The numerical hierarchical order of the posited axioms reflects, again in a very intuitive way, the relative semantic importance of the propositions, from my own point of view. What is submitted here is intended to serve as a possible roadmap to new ideas, and not as a definitive conceptual chart.

The ontology of the wave memory interpretation of quantum mechanics.

1. Elementary particle axiom.

An elementary particle is a system composed by an elementary oscillator, called a corpuscle, and an extended ordered structure in space and time, called a quantum wave, which is finite. The wave is generated and reinforced by the corpuscle, its elementary oscillator.

1.1. Objectivity axiom.

The corpuscle, its measurable properties, and the wave coexist objectively, at all times, independently of any act of measurement made by a conscious observer.

1.2. Quantum field axiom.

A set of particles gives origin to an extended quantum field, which is a composition of quantum waves.

1.3. Identity between spacetime regions and quantum fields.

An extended quantum field represents the most fundamental state of spacetime metrical order, involving corpuscles in space and time. As such it can be identified as a spacetime region populated by its corpuscles.

1.4. Measurement axiom.

A corpuscle behavior, state or property is recognized trough what is informally said to be a measurement. The behavior of a quantum wave can be inferred from the behavior of its corpuscle. The behavior of an extended quantum field can be inferred from the behaviors of the corpuscles populating that field.

2. Nomological encoding axiom.

A quantum wave is a physical structure that encodes the probabilistic distributions of all possible fundamental behaviors in space and time of the corpuscle. The quantum wave thus acts as a nomological memory (nomology, from the Greek: *nomos*, law + *logos*, reason, order). The probabilistic distributions are called the nomological information of the corpuscle behaviors and we say informally that these behaviors are encoded in the quantum wave.

2.1. Pilot-wave axiom.

When a corpuscle expresses a given behavior, it does it so in a random and probabilistic way, accordingly with the nomological information encoded in the quantum wave. In this sense, we say informally that the corpuscle extracts nomological information from the quantum wave. Such a procedure is called the guiding effect, or the pilot-wave effect, exerted by the quantum wave on the corpuscle.

2.2. Fundamental behaviors.

There are four fundamental behaviors attributable to a corpuscle by a given observer, which are its position x, its momentum p, its time t and its energy E.

2.2.1. Properties x and p are said to be nomologically independent, that is, *a priori*, for each value x in a given x-set, any value of p in a given p-set can occur (t and E are also nomologically independent).
2.2.2. Properties x and t are nomologically dependent, that is, for each value x in a given x-set, only one value of t can occur in a given t-subset (p and E are also nomologically dependent).
2.3. Nomological information extremum conditions.
2.3.1. Position nomological information extremum principle.

A corpuscle will tend to be in regions of its pilot wave, where the nomological information encoded about position is higher.

2.3.2. Momentum nomological information extremum principle.

A corpuscle will tend to assume the states of movement within its pilot wave, where the nomological information encoded about momentum is higher.

2.4. Interactions encoding axiom.

An extended quantum field encodes the probabilistic distributions of all possible interactions in space and time between its populating corpuscles. We say informally that the corpuscles interactions are encoded in the extended quantum field.

2.5. Spacetime metric axiom.

An extended quantum field encodes, in the sense taken in 2, 2.1, 2.2, 2.3 and 2.4, all kinematic behaviors of the corpuscles populating the field. Consequently, the extended quantum field may be said to encode the spacetime metric of the region identified with that field.

3. Epistemological completeness axiom.

Quantum Mechanics is a theory that uses, in a mutually non-exclusive way, two distinct descriptions that complement themselves and are consistently related, towards a complete description of quantum reality:

(i) Pilot-wave description.

A description that uses the Pilot wave theoretical approach, dealing with the dynamics of the quantum waves, of the corpuscles, and of the interactions between the former and the later in four-dimensional spacetime. As such it is a theory about actual states, properties, and behaviors of quantum waves and corpuscles, and about how nomological information is exchanged between corpuscles and waves.

(ii) Bohrian Quantum Mechanics formalism.

A description about the dynamics of the nomological information that is encoded in the quantum waves. As such it is a theory, not about the actual states, properties, or behaviors of corpuscles, but about their encoded representation in four-dimensional wavelike physical structures, the quantum waves, also called nomological structures or nomological wave carriers. The description about nomological information uses the algebra of standard quantum mechanics, with operators and matrix calculus.

3.1. Conformity axiom.

All physical phenomena, including the dynamics of the nomological information, occur in four-dimensional spacetime and the three kinds of dynamics: of the nomological information, and of the corpuscles and quantum waves in four-dimensional spacetime, are consistently related.

4. Necessitarianism axiom or Necessitarian principle.

Necessitarianism of the nomological information encoded in a quantum wave, or encoded in a quantum field, is the metaphysical principle according to which:

(i) All possible behaviors of a corpuscle within its wave are the ones that either are already encoded or else can be encoded in a quantum wave,

And to which:

(ii) All possible behaviors and interactions of all the corpuscles populating an extended quantum field are the ones that either are already encoded or else can be encoded in a quantum extended field.

4.1. Nomological configuration.

A nomological configuration is a set of measurable values of a property which can be attributed to a given set of corpuscles. For each corpuscle there is a particular value, different from the corresponding values of all remaining corpuscles. A nomological configuration represents a possible interaction between two or more corpuscles.

4.2. Configurational nomological information.

The configurational nomological information about a given property is the probability distribution of a set of possible nomological configurations. The configurational nomological information is encoded in an extended quantum field.

4.3. Pilot-wave field axiom for the configurational nomological information.

Whenever a set of values is measured for a property in a given nomological configuration, we say informally that the corpuscles have extracted configurational nomological information from their populated quantum field. Such a procedure is called the guiding effect, or the pilot-wave effect, exerted by the quantum field on its populating corpuscles.

4.4. Nomological nonlocality axiom.

Nomological nonlocality is the metaphysical principle by which, given a nomological configuration and its related property, all configurational nomological information for that property is always available to all corpuscles in all spatial locations, within the extended quantum field populated by those corpuscles.

4.4.1. Nomological nonlocality axiom for the values in a nomological configuration.

Given that nomological nonlocality holds, nomological nonlocality for the values in a nomological configuration is the metaphysical assertion positing that, given a nomological configuration and its related property, all possible measurable values for that property can be expressed by any corpuscle in any spatial locations within the extended quantum field, populated by the corpuscles.

4.5. Necessitarian enforcing from the configurational nomological information.

Necessitarian enforcing is the metaphysical assertion positing that, given a nomological configuration and its related property, given that necessitarianism holds and that nomological nonlocality holds, then the fact that a corpuscle assumes one possible measurable value for that property, necessarily implies that each of the remaining corpuscles will assume one and only one of the remaining values for the same property in the nomological configuration.

4.6. Nonlocal and local physical phenomena contention.

Aprioristically there is nomological nonlocality, in the sense that all configurational nomological information must be completely available to all corpuscles, anywhere and at all times within an extended quantum field, in order to obtain physical phenomena. However, the way by which necessitarianism enforcing applies can be said to occur in two possible modes:

(i) Physical nonlocality mode: the value of a given property assumed by a corpuscle is independent of any travelling signal going into the vicinity where that corpuscle is, and coming from any of the vicinities where the remaining corpuscles are,
(ii) Physical locality mode: the value of a given property assumed by a corpuscle depends on at least one travelling signal going into the vicinity where that corpuscle is and coming from at least the vicinity of one of the remaining corpuscles. Each signal can also be considered a corpuscle expressing the signal suitable behavior that must be nomologically encoded in the extended quantum field.

4.6.1. Nonlocality/locality experimental discernibility: in principle, it is possible to verify experimentally if an interaction either complies to Physical nonlocality or to Physical locality.

12.4 Nomological Encoding Formalism in Quantum Mechanics

In this section I will try to provide motivation for the way nomological information can be encoded in a quantum wave. I will assume that the world initially was or in some conditions can still be a chaotic primeval place where all possible behaviors, either nomologically encoded or encodable in a quantum wave, can occur. Possibly, with the exception of a minimum *a priori* nomological order ruling quantum wave dynamics and the guidage interaction. Although this minimum nomological order requisite corresponds, once more, to the imperative imposition of a necessitarian order, apparently coming out of nowhere, the metaphysical weight of such imposition is significantly lightened by what I am proposing here. In fact, it may even happen that a sort of nomological feedback process may exist between the corpuscle and the quantum wave, providing the quantum wave with the overall nomological information for its own behavior. A process that somehow would generalize guidage as a nomological informational closed loop circuit. This is a difficult problem that I will postpone for now.

In the first section of this work, I have tried to offer motivation for the memory interpretation of Quantum Mechanics, coming from the corpuscular aspect of quantum behavior, and thus conforming to the pilot-wave approach. In this section, I will try to do the same, starting from the mathematical Bohrian formal side, now conforming to the usual wave packaging in standard quantum mechanics, as it will become clear ahead.

Let us start the argumentation by considering a very simple description involving two measurable properties in the world, let's say, P_1 and P_2. Let us also accept that the two properties are not correlated *a priori*, which means that they can, *a priori*, assume simultaneously any possible value from each of their respective domains D_{P_1} and D_{P_2}. The properties are hence said to be nomologically independent, according to the ontology introduced in the last section (axiom 2.2.1). Let us also posit that such conjoint, although independent expression of pairs of values for (P_1, P_2), has been such over a period of time that it can be described by a joint probability function f:

$$f(P_1, P_2) = \frac{\partial^2 F}{\partial P_1 \partial P_2} \tag{12.15}$$

With the normalization condition for the joint domain $D = D_{P_1} \times D_{P_2}$, such that:

$$\iint_D f(P_1, P_2)\, dP_1 dP_2 = 1 \tag{12.16}$$

The marginal probability density functions resulting from (12.16) will then be:

$$\Psi(P_1) = \int_{D_{P_2}} f(P_1, P_2)\, dP_2 \tag{12.17a}$$

$$\varphi(P_2) = \int_{D_{P_1}} f(P_1, P_2)\, dP_1 \tag{12.17b}$$

The question now is how can we encode these marginal probability distributions in waves, using for the two properties in question, $P_1 \equiv x$ for position and $P_2 \equiv k$ for the wave number (related with momentum trough $p = \hbar k$).

One possible way would be to use two distinct types of encoding waves $e(x, k)$ and $e'(x, k)$ as nomological information carriers, modulated in amplitude by each one of the marginal probabilities: $h(x, \Delta x)$ for x, and $g(k, \Delta k)$ for k. This can be formally represented using the following encoding scheme:

$$\Psi(x) = \frac{C(\Delta k)}{\sqrt{2\pi}} \int_{-\infty}^{\infty} g(k, \Delta k)\, e(x, k)\, dk \tag{12.18a}$$

$$\varphi(k) = \frac{C'(\Delta x)}{\sqrt{2\pi}} \int_{-\infty}^{\infty} h(x, \Delta x)\, e'(x, k)\, dx \tag{12.18b}$$

Where $C(\Delta k)$ and $C'(\Delta k)$ are scaling functions that depend on the spreading parameters Δx and Δk of the marginal distributions for position x and for the wave number k.

I will call (12.18a) and (12.18b) the encoding transformations for position and wave number (or momentum, using de Broglie's relation). It must be said that in (12.18a) for each position x, the marginal distribution $g(k, \Delta k)$ controls the relative weight of the particular wave, carrying the nomological information about a possible value for k. That is, carrying the probability density value, according to which such value of k (or momentum) will occur for a corpuscle at that same position x.

In an analogous manner, in (12.18b), for each possible value of k (that is momentum, using de Broglie's relation), the marginal distribution $h(x, \Delta x)$ controls the relative weight of the particular wave with that value of k, carrying the nomological information about a particular value for x. That is, carrying the probability density value, according to which such value of x will be the one corresponding to the position that a corpuscle will occupy, while exhibiting the same given value of k (or of momentum).

In all, one expects to have a large number of carrier quantum waves encoding the marginal probabilities for the occurrence of position and momentum. There will be the need for two types of nomological wave carriers, related according to the encoding transformations schema above. I must emphasize that although such a schema for one corpuscle can be imagined to be happening in four-dimensional space, the n-body case will demand a more sophisticated treatment, which I will not attempt here. Again, my main purpose in this work is only to suggest an alternative way of thinking about quantum mechanics.

The first encoding transformation (12.18a), furthermore, offers a correlational view (embedded in the sum provided by the integral) between a given position value and the momentum values nomologically available to that same position value. While the second transformation (12.18b) offers the correlational view between a given momentum value and the position values nomologically available to that same momentum value. By the expression "nomologically available" I mean that the quantum waves do encode what is possible to happen in the world (as stated in axiom 4, Sect. 12.3). It should also be mentioned that the heuristic strength of the encoding schema above is that it uses carrier quantum waves as the encoding building blocks. From (12.18a) one can in principle obtain the spatial geometry of the extended quantum field, since it is a position dependent function resulting from a sum of wave carriers, at each position.

We can now use the prior schema to encode gaussian distributions for the position x and for the wave number k (momentum). The formers are to be understood as the natural distributions for position values and for "state of movement" (or momentum) values that are *a priori* independent and nomologically available. That is, that can be expressed as actual behaviors by the corpuscles. Of course, one should consider that a position type behavior must equate to the possibility of existence of a corpuscle at some location within the quantum wave. While a state of movement behavior must be equated to the possibility of a corpuscle having a certain momentum, at some location within the quantum wave. In this state of affairs, we are not considering a state of movement as a ratio between space and time. It is to be considered as a fundamental property that has the same "ontological dignity", so to speak, as position itself. State of movement, that I am equating to momentum, would then be an intrinsic existential property of a physical entity with the same grounding importance as the entity's location or position. In other words, in the present framework a corpuscle's existence implies that it has necessarily a position and a state of movement (or momentum), aprioristically and independently of any other condition and, in fact, of each other. That is the true and most profound meaning of axiom 2.2.1 in the ontology offered in Sect. 12.3.

Let us then write the marginal distributions for x and k with a gaussian form, according to:

$$h\,(x, \Delta x) = Ae^{-\frac{x^2}{2\Delta x^2}} \tag{12.19a}$$

$$g\,(k, \Delta k) = Ae^{-\frac{k^2}{2\Delta k^2}} \tag{12.19b}$$

The encoding carrier quantum waves will be:

$$e\,(x, k) = e^{ikx} \tag{12.20a}$$

$$e'\,(x, k) = e^{-ikx} \tag{12.20b}$$

These are plane waves, infinitely spread throughout all space, in the same spirit held in Bohrian quantum mechanics formalism.

The scaling functions being:

$$C\,(\Delta k) = 1/\Delta k \tag{12.21a}$$

$$C\,(\Delta x) = 1/\Delta x \tag{12.21b}$$

To this we must now add supplementary encoding conditions, thus formally completing the encoding schema, which are given by:

$$\Psi(x) = h\,(x, \Delta x) \tag{12.22a}$$

$$\varphi(k) = g\,(k, \Delta k) \tag{12.22b}$$

More generally, to obtain (12.22a) and (12.22b) one would have to take the modulus of $\Psi(x)$ and $\varphi(k)$, since their sums are waves and not real valued distributions. For the present purpose, however, what has been written will suffice.

After the proper substitutions we finally get:

$$\Psi(x) = \frac{1}{\Delta k\sqrt{2\pi}} \int_{-\infty}^{\infty} \varphi(k) e^{ikx} dk \tag{12.23a}$$

$$\varphi(k) = \frac{1}{\Delta x\sqrt{2\pi}} \int_{-\infty}^{\infty} \Psi(x) e^{-ikx} dx \tag{12.23b}$$

These are recognizably the Fourier transformations usually applied in quantum mechanics that inevitably lead to the Heisenberg uncertainty relations.

At this point, I should emphasize that the use of plane waves as nomological carriers in the proposed scheme is only a direct consequence of the fact that these are the natural solutions of the Schrödinger linear differential equation. If we were to consider a nonlinear differential equation, such as the one proposed by Croca in 2003 (Croca 2003, p. 72) (Eq. (12.3), Sect. 12.2), then a different type of nomological wave carriers would be formally feasible for the same encoding schema. We could, in fact, and alternatively use Morlet wavelet solitons as nomological information carriers, since these are the natural solution of Croca's nolinear master equation.

Illustrating this possibility, let us now use wavelets of the form depicted in Eq. (12.4) with null average velocity and written at instant zero, contending that such simplifications will not stress the main ideas suggested. Still considering the same gaussian distributions (12.19a) and (12.19b), for x and k, we will now use as carrier quantum waves:

$$e\,(x,k) = e^{-\frac{x^2}{2\sigma_x{}^2}+ikx} \tag{12.24a}$$

$$e'\,(x,k) = e^{-\frac{k^2}{2\sigma_k{}^2}-ikx} \tag{12.24b}$$

Applying the same scaling functions (12.21a) and (12.21b) and the same encoding conditions as provided by (12.22a) and (12.22b), we will get, after the proper substitutions:

$$\Psi(x) = \frac{1}{\Delta k\sqrt{2\pi}} \int_{-\infty}^{\infty} \varphi(k) e^{-\frac{x^2}{2\sigma_x{}^2}+ikx} dk \tag{12.25a}$$

$$\varphi(k) = \frac{1}{\Delta x\sqrt{2\pi}} \int_{-\infty}^{\infty} \Psi(x) e^{-\frac{k^2}{2\sigma_k{}^2}-ikx} dx \tag{12.25b}$$

It must be realized that the carrier quantum wavelets will have a gaussian shape controlled by the parameters σ_x and σ_k. That is, they are not idealized and (might I add, strictly metaphysical), plane waves. On the contrary, these quantum wavelet solitons represent finite physical four-dimensional entities, and their finitude is a realistic and natural attribute very much common to any kind of wave phenomena. Of course, we also must expect to find quantum wavelet solitons with several values of σ_x and σ_k, meaning that the encoding scheme is more complex than is being proposed here. However, for the purpose of this analysis, I will confine myself to the simplest case, where average constant values for σ_x and σ_k will be taken.

In the same spirit of the nomological interpretation of quantum mechanics that I have been advocating thus far, I will now make a formal connection between the encoding transformations (12.25a) and (12.25b), using mother wavelet solitons, and the position-momentum generalized uncertainty relation (12.5a) suggested by Croca (Croca 2003, p. 104). I will only consider the position-momentum uncertainty relation, presupposing that one can draw similar conclusions for time and energy, something that however I will not attempt here but that can also be found in (Croca 2003, p. 104). What follows is a somehow different version of the original derivation formulated by Croca (Croca 2003, p. 94).

Using the distributions (12.19a) and (12.19b), the encoding conditions (12.22a) and (12.22b), and the encoding transformation (12.25a), we can thus write:

$$Ae^{-\frac{x^2}{2\Delta x^2}} = \frac{1}{\Delta k\sqrt{2\pi}} \int_{-\infty}^{\infty} Ae^{-\frac{k^2}{2\Delta k^2}} e^{-\frac{x^2}{2\sigma_x{}^2}+ikx} dk \tag{12.26}$$

After integrating this gives:

$$e^{-\frac{x^2}{2\Delta x^2}} = e^{-\frac{x^2}{2/\left(\Delta k^2+\frac{1}{\sigma_x{}^2}\right)}} \tag{12.26a}$$

Which leads to:

$$\Delta x^2 = \frac{1}{\Delta k^2 + \frac{1}{\sigma_x{}^2}} \tag{12.26b}$$

Solving for Δk, we then obtain:

$$\Delta k = \frac{1}{\Delta x}\sqrt{1 - \Delta x^2/\sigma_x^2} \tag{12.26c}$$

From which, using $\Delta p = \hbar \Delta k$, we finally get Croca's position-momentum generalized uncertainty relation:

$$\Delta p = \frac{\hbar}{\Delta x}\sqrt{1 - \Delta x^2/\sigma_x^2} \tag{12.27}$$

If we use the encoding transformation (12.25b), following an analogous procedure we get:

$$\Delta x = \frac{\hbar}{\Delta p}\sqrt{1 - \Delta x^2/\sigma_k^2} \tag{12.28}$$

This last expression is not the same as (12.27), since the mother wavelet parameters σ_x and σ_k, in general, should differ. To deal with that difference, and seeking consistency, we must introduce yet another restriction that I will call the encoding density condition, reading:

$$\frac{\Delta x}{\sigma_x} = \frac{\Delta k}{\sigma_k} = \rho \tag{12.29}$$

This condition will make (12.27) and (12.28) equivalent.

In expression (12.29), Δx and Δk are the intervals of all possible values for x and k, in a given physical situation for which position and momentum values can occur simultaneously. While σ_x and $\sigma_k$represent the diversity ranges of the carrier wavelets for all possible values of x and k, that can be encoded in the wavelets. Of course, formally, one always has $\Delta x < \sigma_x$ and $\Delta k < \sigma_k$, making $\rho < 1$. We can therefore see ρ as a kind of encoding density or memory storage capacity for encoding nomological information in the wavelets, as it will become clear ahead.

As $\rho \to 1$, that is, as $\Delta x \to \sigma_x$ and $\Delta k \to \sigma_k$ we can assume that the nomological memory gets saturated, and thus that a higher ordered "law-like" regime almost deterministic, will hold. On the other extreme, as $\rho \to 0$, that is, as $\Delta x \ll \sigma_x$ and $\Delta k \ll \sigma_k$, one can assume that the nomological memory is poorly populated, corresponding to a random like regime, with high indeterminacy for the possible pairs position-momentum assumed by the corpuscles in their behaviors. We will shortly return to these interpretations. The present understanding of the position-momentum generalized uncertainty relation presupposes that there is always a dynamic encoding-decoding process, happening between the corpuscles and the quantum carrier wavelets. One that alters the amount of nomological information encoded in the quantum wave carriers.

In what now follows I will use the position-momentum uncertainty relation to show that the encoding transformations (12.25a) and (12.25b), along with the encoding density condition (12.29), imply what can be called a nomological regime spectrum. This having pure chaos and utter determinism as two idealized extreme behavior situations for the corpuscles.

To make the exposition clearer, let us start by a discrete situation idealization, applying the intended interpretation to this situation as an example.

Consider two measurable properties P_1 and P_2 and their corresponding sets S_1 and $S_2$of possible values v_1 and v_2. The first set will have, by hypothesis, a number $\#S_1$ of possible values, while the second set will have $\#S_2$ possible values. If $\#S_1 \# S_2 = N_1 N_2 = M$ is the total number of possible values for P_1 and P_2 that can occur simultaneously, then for any two subsets $S'_1 \subseteq S_1$ and $S'_2 \subseteq S_2$, for the pairs of possible values $(v_1, v_2) \in S'_1 \times S'_2$ one will always have:

$$1 \leq \#S'_1 \#S'_2 \leq M \tag{12.30}$$

Now, in a more "orderly" situation, given any set $S_{12} = S'_1 \times S'_2$, the number of possible pairs of values (v_1, v_2) should be less than M, since given a value v_1 in S'_1, not all values v_2 in S_2 will be available. In other words, given any subset S'_1, only the values of a subset S'_2 of S_2 will be available, where $\#S'_2 < \# S_2$. There will be "law-like" restrictions imposing that some pairs of values are possible, while others are not. Furthermore, the nomological strength of such restrictions will have an impact on the number $\#S'_1 \# S'_2$ of possible pairs of values, lowering that number in general, for any given set $S_{12} = S'_1 \times S'_2$ of possible pairs (v_1, v_2).

If there is a one-to-one, biunivocal relation between the values in S_1 and S_2, then for any subset $S'_{1(min)}$ with just one value v_1 (the minimum number of possible values for v_1), there will be only a subset S'_2 with only one value v_2 for the only possible pair of values (v_1, v_2). And since each subset $S'_{1(\min)}$ and S'_2 will have only one element, we will have in this situation:

$$\#S'_{1(\min)} \#S'_2 = 1 \tag{12.31a}$$

One can also write the complementary situation, for which, given a one element subset $S'_{2(min)}$, we will have:

$$\#S'_1 \#S'_{2(\min)} = 1 \tag{12.31b}$$

In fact, (12.31a) and (12.31b) express a situation of maximum nomological order or correlation between the values in available pairs of values (v_1, v_2). A situation that one could call deterministic, given that there is a one-to-one biunivocal relation between v_1 and v_2.

On the other extreme end of possible nomological scenarios, we can consider that given any subset $S'_{1(min)}$ with just one value v_1 (the minimum number of possible values for v_1), there will be several possible subsets S'_2 of values for v_2, such that we can have for the largest subset (coinciding with S_2): $\#S'_{2(max)} = N_2$,. In this case, one writes:

$$\#S'_{1(\min)} \#S'_{2(\max)} = N_2 \tag{12.32a}$$

And of course, for the complementary situation, given a one element subset $S'_{2(min)}$, we can have:

$$\#S'_{1(\max)} \#S'_{2(\min)} = N_1 \tag{12.32b}$$

Both expressions (12.32a) and (12.32b) mean that for any value of v_1 in S_1, all values for v_2 in S_2 are nomologically available, and vice-versa. In other words, all pairs of values (v_1, v_2) are possible to occur, meaning that there are no "law-like" restrictions imposing that some pairs of values are possible, while others are not. In this case, one should expect to observe high indeterminacy in the behaviors of physical entities, described by the measurable properties P_1 and P_2. This would therefore correspond to the extreme indeterministic situation.

If $N_1 = N_2 = N$, one can condense the situations in (12.31a), (12.31b), (12.32a) and (12.32b) and write:

$$1 \leq \#S\prime_1 \#S\prime_2 \leq N, \quad \text{if } S_1' = \#S_{1\text{min}}' = 1 \text{ or } S_2' = \#S_{2\text{min}}' = 1 \tag{12.33}$$

This expression would only be valid for any subsets S'_1 and S'_2 if some permanent nomological restriction would be at play, fine tuning the values of $\#S'_1$and $\#S'_2$ so that (12.30) would always get. That is:

$$\left(\forall S_1' \subseteq S_1,\ \exists S_2' \subseteq S_2\right) or \ \left(\forall S_2' \subseteq S_2, \exists S_1' \subseteq S_1\right):\ 1 \leq \#S\prime_1 \#S\prime_2 \leq N \tag{12.34}$$

This not being the case, the general expression describing all situations for S'_1 and S'_2 would be (12.30).

We will now consider the continuous case, that is, intervals of possible of values for the two measurable properties P_1 and P_2. Not surprisingly, we will also consider P_1 and P_2 to be position and momentum. At this point, it is quite straightforward to interpret the position-momentum generalized uncertainty relation (12.27) as an expression similar to the nomological case as expressed by (12.34). We can formulate the similar continuous situation, writing first:

For Δx, Δp intervals of possible values for x and p that are nomologically independent:

$$\begin{aligned} &Either\ \left(\forall \Delta x \subseteq \mathbb{R}^+, \exists \Delta p \subseteq \mathbb{R}^+\right) \\ &or\ \left(\forall \Delta p \subseteq \mathbb{R}^+, \exists \Delta x \subseteq \mathbb{R}^+\right)\ such\ that\colon\ 0 \leq \Delta x \Delta p \leq \hbar \end{aligned} \tag{12.35}$$

Using (12.27) to make the proper substitutions, we write secondly:

For Δx, Δp intervals of possible values for x and p that are nomologically independent:

$$\begin{aligned} &Either\ \left(\forall \Delta x \subseteq \mathbb{R}^+,\ \forall \rho \in \mathbb{R}_0^+,\ \exists \Delta p \subseteq \mathbb{R}^+\right) \\ &or\ \left(\forall \Delta p \subseteq \mathbb{R}^+,\ \forall \rho \in \mathbb{R}_0^+,\ \exists \Delta x \subseteq \mathbb{R}^+\right)\ such\ that\colon \end{aligned} \tag{12.36}$$

$$\Delta x \Delta p = \hbar\sqrt{1-\rho^2}, \rho = \frac{\Delta x}{\sigma_x} = \frac{\Delta k}{\sigma_k}, 0 \leq \rho \leq 1\ and\ 0 \leq \hbar\sqrt{1-\rho^2} \leq \hbar$$

From (12.36) it is now possible to derive two extreme nomological situations. The first one, as before, in the discrete case, is expressed by (12.31a) and (12.31b), identified as the deterministic situation. This extreme nomological scenario corresponds to a very high encoding density ρ, and therefore to a scenario of high behavioral order. One that can ideally be equated with a physical situation of determinism. Still using (12.36), a possible formulation would be the following.

The deterministic nomological situation or Newton's nomological situation.

For Δx, Δp intervals of possible values for x and p that are nomologically independent:

$$\begin{aligned} &Either\ \left(\forall \Delta x \subseteq \mathbb{R}^{+},\ \forall \rho \in \mathbb{R}_{0}^{+},\ \exists \Delta p \subseteq \mathbb{R}^{+}\right) \\ &or\ \left(\forall \Delta p \subseteq \mathbb{R}^{+},\ \forall \rho \in \mathbb{R}_{0}^{+},\ \exists \Delta x \subseteq \mathbb{R}^{+}\right)\ such\ that: \end{aligned} \tag{12.36a}$$

$$\Delta x \Delta p = \hbar\sqrt{1-\rho^{2}},\ \rho = \frac{\Delta x}{\sigma_x} = \frac{\Delta k}{\sigma_k},\ \rho \approx 1\ and\ \Delta x \Delta p \approx 0$$

Considering position, this proposition can be interpretated saying that whatever may be the interval of possible values for position (that can occur simultaneously with values of p) and given any available nomological wave carrier (characterized by the parameters σ_x or σ_k), the encoding density ρ will always be almost at its maximum (that is, near 1).

We can see σ_x and σ_k as nomological memory capacity parameters for, respectively, position and momentum. Those parameters, however, must be finite and not too large, when compared to the possible intervals Δx and Δp of *a priori* nomological independent values. If σ_x and σ_k happen to be too large, that would mean that in the given physical situation the majority of the nomological information will remain unencoded. Geometrically speaking, we can see from (12.4), Sect. 12.2, that when σ_x (or similarly σ_k) is very large, we obtain plane waves for the memory wave carriers. That is, the diversity of values encodable in the carrier shape becomes almost null, since we will no longer have a gaussian outline for the wave's amplitude, this being, in fact, constant. Extended plane waves therefore correspond to low efficient nomological information wave carriers. In such a situation, the behavioral order will indeed be very low, giving way to indeterminacy. Alternatively, a situation of high behavioral order would mean that the available nomological memory has already reached its maximum, and hence that only a few values of position and momentum remain to be encoded. In other words, in that particular physical scenario, the occurrence of pairs of values position-momentum is under strong nomological restriction, carrying a lot of nomological weight upon the corpuscles behaviors, so to speak. The fact that $\rho \approx 1$ then implies that for any pair of intervals, that were nomologically independent, one must have $\Delta x \Delta p \approx 0$, which in turn implies $\Delta x \approx 0$ and $\Delta p \approx 0$. These formulas express the fact that for any pair of nomologically independent intervals, both intervals must be exceedingly small. We can understand this by noting that in a situation of high nomological order, a few values for x will only pair with a few values of p, and vice versa. And, furthermore, that in the extreme possible case one would have a one-to-one biunivocal relation between position and momentum, corresponding to the idealized deterministic situation advocated in Newtonian mechanics.

The second possible extreme nomological situation corresponds to the indeterministic or Heisenberg case, where there are only a few nomological restrictions ordering the corpuscles behaviors, the nomological strength of such restrictions being exceptionally low. This will be the situation similar to the one, in the discrete case above, expressed by (12.32a) and (12.32b). From (12.36) we see again that the meaningful parameter influencing the nomological scenario is the encoding density ρ. So that if such density is exceptionally low, the nomological memory will be poorly populated, implying a low amount of encoded nomological information and, therefore, allowing for the occurrence of indeterministic behavior. We can formulate this situation, using a particular version of (12.36) as follows.

The indeterministic nomological situation or Heisenberg's nomological situation.

For Δx, Δp intervals of possible values for x and p that are nomologically independent:

$$\begin{aligned} &\textit{Either}\ \left(\forall \Delta x \subseteq \mathbb{R}^+,\ \forall \rho \in \mathbb{R}_0^+,\ \exists \Delta p \subseteq \mathbb{R}^+\right) \\ &\textit{or}\ \left(\forall \Delta p \subseteq \mathbb{R}^+,\ \forall \rho \in \mathbb{R}_0^+,\ \exists \Delta x \subseteq \mathbb{R}^+\right)\ \textit{such that}: \end{aligned} \tag{12.36b}$$

$$\Delta x \Delta p = \hbar\sqrt{1-\rho^2},\ \rho = \frac{\Delta x}{\sigma_x} = \frac{\Delta k}{\sigma_k},\ \rho \approx 0 \ \text{ and } \ \Delta x \Delta p \approx \hbar$$

This proposition can be interpreted in the following way. Taking once more position as the property to reason with, whatever may be the interval of possible values that we consider for position x and given the available nomological memory (characterized by the parameters σ_x and σ_k), the encoding density ρ will always be low. This will be so, because the average length Δx for the intervals of possible values, occurring for position, is still very small, once compared with the memory capacity of the nomological wave carrier (again characterized by σ_x or σ_k). That is, the nomological situation is such that a low amount of nomological information has been or is being encoded, in that particular physical situation. In such a case we have for the encoding density $\rho \approx 0$, implying that $\Delta x \Delta p \approx \hbar$. From this we see that if Δx is extremely small, the interval Δp will be extremely large. This may be understood saying that whatever the value for position we may consider, this value can occur simultaneously with any value for momentum within an extremely large set of values for momentum. In other words, for a given value of position, any value of momentum goes. This nomological scenario corresponds to a situation of high indeterminacy. On the other hand, if Δx is extremely large, the interval Δp will be exceedingly small. This again can be similarly interpreted saying that, for any given value of momentum, any value of position can occur. This, again, translates into a situation of exceptionally low nomological order and high indeterminacy, as the one preconized in Bohrian quantum mechanics.

It should finally be reemphasized that when $\rho \approx 0$, we get from (12.5a) (or else from (12.27)) and from (12.5b)) the standard Heisenberg uncertainty relations. In the present nomological interpretation, these last correspond to a situation where the nomological memory carrier waves are plane waves, according to what was indicated in (12.4), Sect. 12.2. Once more, in the given framework, this means that when only nearly plane waves are available as memory quantum wave carriers, the diversity of the encodable nomological information is very low. That is, to have high order correlation between position and momentum, there must be high nomological memory capacity, which corresponds to a nonhomogeneous amplitude outline shape of the carriers, as depicted in (12.24a) and (12.24b), and which depends on the parameters σ_x and σ_k. It should finally be mentioned that the usual writing of the Heisenberg position-momentum uncertainty relation is given by:

$$\Delta x \Delta p \geq \hbar \tag{12.37}$$

In the context of the forwarded interpretation, this can be understood as the statement that in the Heisenberg case, the minimum measurement error of either position or momentum complies respectively either to:

$$\Delta x \approx \frac{\hbar}{\Delta p} \tag{12.37a}$$

Or to:

$$\Delta p \approx \frac{\hbar}{\Delta x} \tag{12.37b}$$

Only being larger due to experimental noise error. More to the point, for null memory storage density, one gets position-momentum uncertainty to be strictly due to lack of nomological information. Position-momentum uncertainty due to both lack of nomological information and other sources of measurement error will then provide the usual uncertainty relation (12.37).

In other words, we can also use the nomological interpretation applied to the introduced generalized uncertainty relations to distinguish between an idealized pure informational situation, involving quantum wave carriers, and the situation where one should also consider a noise error condition adding to the measurement.

12.5 Conclusion

In this article I have tried to provide motivation for a nomological interpretation of quantum mechanics, what I have called the "Wave memory interpretation of quantum mechanics". A proposal holding that corpuscles and quantum waves do exist, simultaneously, at all times, formulated in the context of a possible improved Pilot-wave theory, where quantum waves are thought to be real physical four-dimensional structures, serving as encoding carriers for a fundamental type of information in Nature, that I have called nomological information (from the Greek: *nomos*, law + *logos*, reason, order).

To give a brief account of the main ideas involved, four-dimensional quantum wave carriers encode the probabilistic distributions of all possible fundamental behaviors of the corpuscles populating those waves. The particles then behave accordingly to the probabilistic distributions encoded in the waves, exchanging nomological information with them, trough what I have equated as a pilot-wave (guidage) effect. While the waves pilot the corpuscles, these generate and reenforce the former, hence acting as moving oscillators. Another foundational idea of this interpretation is that the complete description of quantum phenomena implies the use of two dynamics that are complementary in a on exclusive way and consistently related. One is the dynamic describing both the behavior of four-dimensional quantum carrier waves in spacetime, the movement of their populating corpuscles, as well as the interaction between waves and corpuscles. This is done by means of a Pilot-wave description model, where there is a guidage effect and where one uses a non-linear Schrodinger equation to predict the evolution of the quantum wave. The other is a dynamic describing the evolution of the nomological information encoded in the four-dimensional quantum waves, where one uses the algebra of standard quantum mechanics, applying operators functional calculus, matrix calculus, and the Schrodinger linear equation.

Getting more in detail about what has been devised, in the first section I have offered motivation for the wave memory interpretation from a realistic point of view, adopting the pilot-wave framework. As already stated, this theoretical approach implies accepting the existence of quantum waves along with their corpuscles, at all times, without the need for a consciousness collapse mechanism, bringing the corpuscles into existence. It also implies understanding the guidage or piloting effect exerted by the quantum wave upon the corpuscle, as a kind of nomological information transfer process, where the corpuscle acquires nomological information from the wave, about possible positions and possible momentum values. This information is stored at the location where the corpuscle happens to be.

In Sect. 12.3, I introduced what I have called a strong nomological constructivism, associated with probabilistic necessitarianism in natural laws. This implies upholding a picture of the world where behaviors initially occur at random, progressively informing the available nomological memory and being, in turn, enforced by it. As the nomological memory builds up, it then exerts a probabilistic necessitarian imposition upon the corpuscles behaviors.

In the same section, I have also defended what I have called an epistemological completeness owing to quantum mechanics. The idea that a complete theory about quantum phenomena should envisage, in a mutually non-exclusive way, two complementary descriptions. A description about the dynamics of the memory quantum wave carrier and a description about the dynamics of the nomological information stored in the wave carrier. The first description would also include the dynamics of the corpuscle, since this depends on the nomological information exchange between the wave and the corpuscle occurring in four-dimensional spacetime. The second description would concern the dynamics of the nomological information encoded in quantum waves. Such description, of a purely mathematical formal nature, would involve all possible physical states informationally encoded in the wave, demanding higher order dimensional spaces to accomplish informational representation completeness. Both descriptions should be thought as mutually non-exclusive and consistently related. In principle, it should be possible to derive the four-dimensional pilot-wave field from the Bohrian dynamics in a way that would be consistent either with state superposition description, or with the $3n + 1$ description for the n body case in standard quantum algebra.

Also in Sect. 12.3, I have introduced a tentative ontology for the wave memory interpretation, trying to capture all ideas involved and, in particular, a feature that I have called "nomological non-locality". The idea that a kind of necessitarian enforcement coming from the memory wave carriers, should be universally acknowledged by all corpuscles, independently of their spacetime location.

In Sect. 12.4, I have tried to give an account of a possible encoding schema for nomological information in quantum wave carriers. To that purpose I have suggested that the usual Fourier treatment applied to plane waves, in standard quantum mechanics, can be used to define such a schema. From there, I have applied the same encoding schema to Morlet solitons as quantum wave carriers, instead of using plane waves. This allowed me to derive Croca's position-momentum generalized uncertainty relation (12.29), having Heisenberg's uncertainty relation as a particular case. At the end of Sect. 12.4, I have reasoned in favor of the nomological interpretation of the generalized position-momentum uncertainty relation. The relation can be viewed as a measure of encoded nomological order, characterizing the nomological correlation strength between position and momentum, depending on the amount of encoded probabilistic information in a particular physical situation.

12.6 Future Work

To conclude this article, I wish to suggest future research topics that, in my view, can contribute to the development of the nomological interpretation of quantum mechanics. First, there is the problem of actually detecting a quantum wave, the so-called empty wave. That is, a quantum four-dimensional wave without any corpuscle. This, I believe, would be a crucial experimental result paving the way to new scientific findings and philosophical thinking about quantum phenomena. A set of experiments covering several predictable conundrums, performing such task, have been proposed in a recent paper, thus providing several experimental setups for a four-dimensional quantum wave detection (Croca et al. 2023).

A second major problem solution in favor of the present interpretation would be to determine what is the proper formal representation of how probabilistic nomological information is encoded in four-dimensional quantum waves. Critically, one has to formulate the proper representation of how nomological information concerning mutually exclusive physical states is encoded. That is, and putting it in a simpler form, it is necessary to determine how the so-called superposition of states, which consequently can now be understood as a sort of logical conjunction of information about such states, can be encoded in the pilot-wave carrier. In fact, and more generally, one has to determine how can any representation in configuration space be encoded in four-dimensional quantum wave carriers.

Finally, another very important task, bringing strength to the interpretation, implies determining the proper formal relation between the dynamic describing the pilot-wave and the corpuscle behaviors, and the dynamic describing the evolution of the probabilistic encoded information in the pilot-wave.

With the present work I hope to have provided some contribution, however minor, to the understanding of quantum phenomena and the possible formulation of an overall theory that can unify and improve the available formalisms describing quantum behavior.

Acknowledgements I would like to recognize, in a most grateful way, the precious teaching and insights about quantum mechanics, generously provided to me over the years by Professor José Croca, Professor Rui Moreira and Professor Mário Gatta at the Centre for Philosophy of Sciences of the University of Lisbon (CFCUL).

This work was supported by national funds through the FCT - Fundação para a Ciência e a Tecnologia, I.P., within the scope of the R&D Unit Centro de Filosofia das Ciências da Universidade de Lisboa (CFCUL), the strategic project with the reference FCT I.P. and DOI: UIDB/00678/2020, DOI 10.54499/UIDB/00678/2020 [https://doi.org/10.54499/UIDB/00678/2020].

References

Bacciagaluppi, G., and A. Valentini. 2009. *Quantum theory at the crossroads: Reconsidering the 1927 Solvay conference*. 1st ed. Cambridge University Press.

Bloch, Y., and E. Cohen. 2022. The identification of mean quantum potential with fisher information leads to a strong uncertainty relation. *Foundations of Physics* 52: 117.

Bohm, D. 1952a. A suggested interpretation of the quantum theory in terms of "hidden" variables. I. *Physical Review* 85 (2): 166–179. American Physical Society (APS).

———. 1952b. A suggested interpretation of the quantum theory in terms of "hidden" variables. II. *Physical Review* 85 (2): 180–193. American Physical Society (APS).

Bohm, D.J., and B.J. Hiley. 1975. On the intuitive understanding of nonlocality as implied by quantum theory. *Foundations of Physics* 5 (1): 93–109. Springer Science and Business Media LLC.

Bohm, D., and B.J. Hiley. 1995. *The undivided universe: An ontological interpretation of quantum theory*. 1st ed. Routledge.

Bohr, N. 1985. *Como lectures (1928), collected works*. Vol. 6. Amsterdam: North-Holland.

Bush, J.W.M., and A.U. Oza. 2021. *Hydrodynamic quantum analogs. Reports on progress in physics*. IOP Publishing.

Castro, P., J. Croca, M. Gatta, and R. Moreira. 2017. Generalized uncertainty relations in quantum mechanics and the principles of completeness in physics. *Physical Science International Journal* 16 (4): 1–9. Sciencedomain International.

Couder, Y., and E. Fort. 2006. Single particle diffraction and interference at a macroscopic scale. *Physical Review Letters* 97: 154101.

Couder, Y., S. Protière, E. Fort, and A. Boudaoud. 2005. Walking and orbiting droplets. *Nature* 437: 208.

Cover, J.A., M. Curd, and C. Pincock. 2012. *Philosophy of science: The central issues*, Second International Student Edition. W. W. Norton & Company.

Cristea-Platon, T., P.J. Sáenz, and J.W.M. Bush. 2018. Walking droplets in a circular corral: Quantisation and chaos. *Chaos* 28: 096116.

Croca, J.R. 2003. *Towards a nonlinear quantum physics*. London: World Scientific.

———. 2004. De Broglie tired light model and the reality of the quantum waves. *Foundations of Physics* 34: 1929–1954.

———. 2015. *Eurhythmic physics or hyperphysics*. LAP Lambert Academic Publishing.

Croca, J.R., P. Castro, M. Gatta, et al. 2023. Proposed experiments to clarify the real nature of the quantum waves. *Foundations of Physics* 53: 14.

de Broglie, L. 1960. *Non-linear wave mechanics: A causal interpretation*. Elsevier.

———. 1969. *The current interpretation of wave mechanics: A critical study*. Amsterdam: Elsevier.

———. 2021. *Research on the theory of quanta*. Minkowski Institute.

Drezet, Aurelien, ed. 2023. Pilot-wave and beyond: Louis de Broglie and David Bohm's quest for a quantum ontology, Special issue. *Foundations of Physics*.

Eddi, A., E. Fort, F. Moisy, and Y. Couder. 2009. Unpredictable tunneling of a classical wave–particle association. *Physical Review Letters* 102: 240401.

Eddi, A., E. Sultan, J. Moukhtar, E. Fort, M. Rossi, and Y. Couder. 2011. Information stored in Faraday waves: The origin of a path memory. *Journal of Fluid Mechanics* 674: 433–463.

Eddi, A., J. Moukhtar, S. Perrard, E. Fort, and Y. Couder. 2012. Level splitting at a macroscopic scale. *Physical Review Fluids* 108: 264503.

Ellegaard, C., and M.T. Levinsen. 2020. Interaction of wave-driven particles with slit structures. *Physical Review E* 102: 023115.

Fort, E., A. Eddi, A. Boudaoud, J. Moukhtar, and Y. Couder. 2010. Path-memory induced quantization of classical orbits. *Proceedings of the National Academy of Sciences of the United States of America* 107: 17515–17520.

Hardy, L. 2001. *Quantum theory from five reasonable axioms (Version 4)*. arXiv. https://doi.org/10.48550/ARXIV.QUANT-PH/0101012

Harris, D.M., J. Moukhtar, E. Fort, Y. Couder, and J.W.M. Bush. 2013. Wavelike statistics from pilot-wave dynamics in a circular corral. *Physical Review E* 88: 011001.

Hume, D. 1999. *An enquiry concerning human understanding*. Oxford University Press.

Kurianski, K.M., A.U. Oza, and J.W.M. Bush. 2017. Simulations of pilot-wave dynamics in a simple harmonic potential. *Physical Review Fluids* 2: 113602.

Labousse, M., S. Perrard, Y. Couder, and E. Fort. 2016. Self attraction into spinning eigenstates of a mobile wave source by its emission back-reaction. *Physical Review E* 94: 042224.

Landauer, R. 1993. Information is physical. *Physics Today* 44 (5): 23.

———. 1996. The physical nature of information. *Physics Letters A* 217 (4–5): 188–193.

Landauer. 1999. Information is a physical entity. *Physica A: Statistical Mechanics and its Applications* 263 (1–4): 63–67.

Oza, A.U., D.M. Harris, R.R. Rosales, and J.W.M. Bush. 2014. Pilot-wave dynamics in a rotating frame: On the emergence of orbital quantization. *Journal of Fluid Mechanics* 744: 404–429.

Oza, A.U., R.R. Rosales, and J.W.M. Bush. 2018. Hydrodynamic spin states. *Chaos* 28: 096106.

Perrard, S., and M. Labousse. 2018. Transition to chaos in wave memory dynamics in a harmonic well: Deterministic and noise-driven behavior. *Chaos* 28: 096109.

Perrard, S., E. Fort, and Y. Couder. 2016. Wave-based turing machine: Time reversal and information erasing. *Physical Review Letters* 117 (9): 094502.

Pucci, G., D.M. Harris, L.M. Faria, and J.W.M. Bush. 2018. Walking droplets interacting with single and double slits. *Journal of Fluid Mechanics* 835: 1136–1156.

Reginatto, M. 1998. Derivation of the equations of nonrelativistic quantum mechanics using the principle of minimum Fisher information. *Physical Review A* 58 (3): 1775–1778. American Physical Society (APS).

Sáenz, P.J., T. Cristea-Platon, and J.W.M. Bush. 2020. A hydrodynamic analog of Friedel oscillations. *Science Advances* 6: 20.
Tadrist, L., T. Gilet, P. Schlagheck, and J.W.M. Bush. 2020. Predictability in a hydrodynamic pilot-wave system: Resolution of walker tunneling. *Physical Review E* 102: 013104.
Valani, R.N., A.C. Slim, and T. Simula. 2018. Hong–Ou–Mandellike two-droplet correlations. *Chaos* 28: 096104.

Chapter 13
What If We Lived in the Best of All Possible (Quantum) Worlds?

Valia Allori

Abstract For scientific realists, quantum mechanics is unsatisfactory because it suffers from the measurement problem. However, there are at least three promising solutions: the pilot-wave theory, the many-worlds theory, and the theory of spontaneous collapse. In this paper I argue that the measurement problem is a false problem for the realist: it was proposed as the last resort to convince the positivists that the theory is not empirically adequate. Instead, realists should focus on preserving the reductive explanatory schema that had worked so well in physics before, which requires a theory to have a three-dimensional, microscopic ontology. Incompleteness argument to this effect have been proposed in the 1920s, but effectively ignored due to the positivistic climate, other unscientific reasons, and theorems which claimed that this project was impossible. When realists re-examined quantum mechanics in the 1950s and later, they happened to focus on the measurement problem. In this paper I speculate on what would have happened if realists instead focused on finding a three-dimensional, microscopic ontology to complete quantum theory. I show that most paradoxes, puzzles, and mysteries connected with quantum mechanics would have never emerged.

Keywords Measurement problem · Completeness problem · Scientific realism · Copenhagen Dogma

13.1 Introduction

Quantum theory is one of the greatest accomplishment of scientific inquiry as it has an outstanding predictive power, both in depth and in breath: allegedly, with a single mathematical object, the wavefunction, and a simple equation, the Schrödinger equation, the theory delivers predictions which are incredibly accurate,

---

V. Allori (✉)
University of Bergamo, Bergamo, Italy
e-mail: Valia.allori@unibg.it

P. Castro et al. (eds.), *Advances in Pilot Wave Theory*, Boston Studies in the Philosophy and History of Science 344,
https://doi.org/10.1007/978-3-031-49861-9_13

and accounts for a vast variety of physical phenomena. However, its explanatory power is left to be desired: it is not clear how the phenomena the theory predicts are explained in terms of a microscopic description. In other words, scientific realists, who think that our best theories inform us about the nature of reality, consider quantum theory incompatible with realism, because the theory does not tell us a coherent microscopic story beyond the phenomena. Nonetheless, realists also think that one can provide such a picture by solving the measurement problem. This is the problem of dealing with the unobserved macroscopic superpositions (such as an instrument displaying all possible measurement outcomes at the same time) which are predicted by quantum mechanics, when assuming that the description of a physical system is entirely provided by a Schrödinger evolving wavefunction. Solutions of the measurement problem are found by denying each assumption, or by embracing the conclusion. One could for instance deny that the wavefunction always evolves according to the Schrödinger equation. The simplest way to do this is by postulating the collapse rule introduced by von Neumann: the wavefunction evolves according to the Schrödinger equation until a measurement is performed, and then it collapses into one of the terms of the superposition. Another way of doing the same thing is provided by the spontaneous localization theory of Ghirardi, Rimini and Weber which makes precise 'when' and 'where' the wavefunction collapses without referring to measurement processes. Alternatively, one could reject the claim that the wavefunction provides a complete description, which is the route taken by the pilot-wave theory: one also needs to specify the particles' position. The last option is to accept that macroscopic superpositions exist but are not observed, as defended by the many-worlds theory: 'whatever can possibly happen, will indeed happen,' not here but in another 'world,' which is suitably 'transparent' to us. These theories are commonly understood as theories about the wavefunction, namely a field in a high dimensional space. In other words, even if the world appears to us as being three-dimensional and populated with three-dimensional objects, it is rather high dimensional, and objects are fundamentally 'made of' a high dimensional field.

In this paper I argue that, contrary to the common opinion, it is misleading for the scientific realist to consider the measurement problem. There are no compelling reasons for realists to consider the measurement problem. It is a matter of historical contingency that they did, based on a series of accidents: misunderstandings implications of experimental and theoretical results, the peculiar philosophical climate, the charisma of certain people, as well as additional other-than scientific factors. Classical physics provided a reductive understanding of the macroscopic phenomena in terms of the motion of their microscopic constituents (Sect. 13.2). However, this was not the case in quantum mechanics, and therefore some people looked at a completion of the theory, looking for a deeper explanation, as it was usually done in similar situations before. In Sect. 13.3, I pretend that everybody agreed that this was the thing to be done, and I show how, outside of that scientific and philosophical context, things could have evolved. Realists would have naturally understood quantum theory as they understood thermodynamics: as a theory in need of a deeper explanation. And as thermodynamics had statistical mechanics to provide a reductive explanation of the thermodynamic principles, quantum mechanics needed to be suitably completed to provide a reductive explanation of the so-called quantum rules. If realists focused on the project of explaining the phenomena microscopically, they would never had encountered the measurement problem, and there would have been no mention of consciousness, high dimensional wavefunction ontology, undetectable many-worlds, stochastic nonlinear modifications of the Schrödinger equation. Without being distracted from the measurement problem, would have naturally considered led to a theory in which particle deterministically evolve according to highly nonclassical laws and interact nonlocally. In Sect. 13.4 I come back to what has actually happened, and I show how the measurement problem was put forward as the last of a series of arguments to convince instrumentalists, who did not care about realism, that there was something to worry about even for them: the theory is not empirically adequate. I conclude in Sect. 13.5, arguing that solving the measurement problem is not a sufficient condition for finding a satisfactory realist theory: it merely points at a problem (empirical adequacy) and it should not be thought as a way of finding a satisfactory realist quantum theory able to explain the phenomena reductively.

13.2 The Reductive Explanatory Schema

According to classical mechanics the world is made of microscopic massive three-dimensional point-particles which evolve according to Newton's law. After Maxwell, electricity was taken to consist of charged particles, namely electrons, and electromagnetic phenomena were explained by introducing a new property, charge, and a new type of force. Newton believed that light, just as matter, was made of particles. However, in the early 1800s, experiments revealed interference and diffraction patterns for light, and Maxwell's electrodynamics described light as the propagation of electromagnetic waves, supporting the theory that light is a wave. Therefore, the consensus settled that there are fundamentally two things in the world: matter, made of three-dimensional particles with mass and charge; and light, which instead is a wave vibrating in three-dimensional space. The macroscopic behavior of matter is accounted for in terms of the reductive schema, with the specification of the ways in which matter and light interact.

Classical physics explains the behavior of both earthly and celestial objects using a reductive explanatory schema in which the properties and behavior of macroscopic objects are accounted for in terms of the dynamics of their microscopic constituents, namely point-particles. Let's call this the 'reductive explanatory schema.' The story classical physics tells us about the macroscopic world is a 'geometrical' story: a table is just a table-shaped cluster of microscopic particles. Once the particles and the way in which they evolve are specified, everything else follows: the solidity of a table, the localization of a comet, the transparency of a pair of glasses, the liquidity of the water in this bottle, the compressibility of the air in this room, and so on. This project culminated in the work of Boltzmann, who developed statistical mechanics, which suitably reproduces the laws of thermodynamics in terms of the motion of microscopic point-particles. Similarly, the laws of geometrical optics, in terms of light rays, can be accounted for in terms of the vibrations of the electromagnetic waves.

To use a locution introduced by Einstein, classical physics is a constructive theory.[1] This is defined as a theory which involves the dynamical reduction of macroscopic objects in terms of the motion and interactions of their microscopic constituents. Thus, we can call this type of theories reductive theories. This is opposed to a principle theory, which is formulated in terms principles used as constraints on physically possible processes. Thermodynamics is an example of a principle theory (it has principles such as, e.g., 'energy is conserved'), while statistical mechanics is the corresponding constructive theory, which explains why such principles hold. Before the end of the nineteenth century, the goal of physics was taken to propose constructive theories, because of the reductive explanatory schema that you have explained a phenomenon when you have understood how it can be derived dynamically by the motion of its microscopic constituents. That is, this explanatory schema is reductive: microscopic entities construct macroscopic objects, and reductively explain the phenomena.

Often quantum theory is advertised as forcing us to abandon the idea that phenomena can be reductively explained in the sense I specified. We are told that we need to recognize that our language is hopelessly incomplete, that we will never be able to understand what lies beyond the phenomena, if there is such a thing. Bohr taught us that there is a classical world and a quantum world, which are complementary with one another: we need them both to get a complete picture of reality. The quantum world cannot be described in classical terms: when the wave description is adequate, the particle one is not and the other way around. The quantum world is inevitably hidden beyond our abilities. Experiments we perform on the quantum world are bound to change (or even create!) the reality beyond the phenomena. This distinctively nonclassical behavior needs to be reckoned and dealt with by recognizing that there is a paradigm shift. The reductive understanding used within classical physics was one in which we imagined the world as composed of microscopic entities moving around like little balls or sticking together to form macroscopic bodies. The quantum revolution is the acknowledgement that this type of understanding was impossible. Scientists need to swipe away the old picture and paint the canvas with new colors. We need to realize that we can only understand what is directly observable and we should give up all the desires to explain the quantum world like we did before, according to the reductive schema because we simply cannot do it.

This is what everyone has been told, this is how students have been indoctrinated for years. However, this is what Beller calls the Copenhagen Dogma: the 'rhetoric of inevitability' championed by Bohr.[2] It turns out that there is nothing inevitable about it: most of it was just very effective propaganda. This story has already been told,

[1] Einstein (1919).

[2] Beller (1999).

but now I wish to go further, and flip the inevitability claim on its head: what should have been inevitable is to notice that there is one natural extension of the reductive schema, namely the pilot-wave theory, and that all the other realist solutions of the measurement problem, as well as the idea that the wavefunction represents the ontology of the world, have received an unusual type of attention.

13.3 In the Best of All Possible Quantum Worlds

Let us imagine, starting from the experiments that prompted the development of the first quantum rules, how one could have expanded the reductive explanatory schema to the quantum domain, had not the Copenhagen hegemony won without deserving it. I will simply list the arguments that were proposed, and not the ways in which they have been misunderstood, misconstrued, or ignored, and I will follow where this naturally leads.

13.3.1 Quantum Theory as a Principle Theory

Quantum theory first begun, at the beginning of the twentieth century, as a slight modification of the classical picture to accommodate experimental outcomes (including the blackbody radiation problem, the various atomic spectra, the stability of the atom, and so on) that suggested that certain quantities are not continuous but rather are discrete or 'quantized.' The project at the time was to provide a principle theory to determine the right quantization rules for the various quantities, using the so-called Bohr–Sommerfeld quantization conditions, within the classical understating of matter as particles and light as a wave. In fact, it was difficult to come up with a constructive understanding because experiment challenged the wave nature of light (Compton and photoelectric effects) and other the particle nature of matter (particle interference). Nonetheless, some indications that one needs both particles and waves started to arise. For instance, to explain the photoelectric effect Einstein in 1905 proposed that light was made of particles, the photons, which could also account for the Compton effect,[3] and this was interpreted by de Broglie in 1924 as associating the photon to a light wave (rather than identifying light with it), so that the light wave 'guides' the motion of the photons through space.[4] Similarly, he proposed the existence of 'matter waves' associated with particles. However, de Broglie did not at the time have a viable proposal for the dynamical equation for the wave.

[3] Einstein (1905).

[4] de Broglie (1924).

13.3.2 *Wave Mechanics as a First Attempt to Maintain the Reductive Schema*

In a series of papers written in 1926–1927, Schrödinger started to develop a wave-only constructive theory, called wave mechanics, to reductively explain where the quantization rules came from. It is a theory according to which the fundamental object is a wave (along the lines of de Broglie's hypothesis of matter waves but with no particles) evolving according to the equation which now carries his name. Schrödinger's wave is now called the wavefunction.[5] Wave mechanics wished to provide a clear picture of what was going on inside an atom, along the lines of the reductive schema which has been used so far. In particular, Schrödinger could show that the hydrogen spectrum can be reproduced in terms of nodes of this wave. Particles are emergent: they are narrow wave packets, which remain isolated in a small region of space, and do not spread, giving the impression 'from a distance' of being particles.

De Broglie, Lorentz and Einstein were quick to realize that the theory was not satisfactory. First, wave packets quickly spread out, and thus the observed stable 'particle' trajectories cannot be accounted for. Moreover, the wavefunction of a system with two or more 'particles' is not vibrating in three-dimensional space, but rather it is an object in configuration space. This is the space of the (three-dimensional) configurations of all 'particles,' which is therefore high dimensional. This does not make sense in the perspective of the reductive schema: how can you explain the phenomena by introducing an object living in a high-dimensional space? If the ontology of the theory is not something in three-dimensional space, then reductionism fails. In fact, one cannot derive the macroscopic phenomena in terms of the microscopic ones, because strictly speaking there's no microscopic reality in three dimensions. Since the wavefunction lives in a high-dimensional space, we need to find a new three-dimensional ontology to save the reductive explanatory schema.[6]

---

[5] Schrödinger (1926).

[6] At the 1927 Sovlay conference de Broglie justified his introduction of particles because "it seems a little paradoxical to construct a configuration space with the coordinates of points that do not exist." (Bacciagaluppi and Valentini 2009, p. 346). Lorentz writes to Schrödinger: "If I had to choose now between your wave mechanics and the matrix mechanics, I would give the preference to the former, because of its greater intuitive clarity, so long as one only has to deal with the three coordinates x, y, z. If, however, there are more degrees of freedom, then I cannot interpret the waves and vibrations physically, and I must therefore decide in favor of matrix mechanics." Here are similar concerns expressed by Einstein in a letter to Lorentz dated May 1st, 1926: "Schrödinger's conception of the quantum rules makes a great impression on me; it seems to me to be a bit of reality, however unclear the sense of waves in n-dimensional q-space remains." (Lorentz in Przibram 1967). Similarly, here is an excerpt from a June 18th, 1926 letter that Einstein sent to Paul Ehrenfest: "Schrödinger's works are wonderful – but even so one nevertheless hardly comes closer to a real understanding. The field in a many-dimensional coordinate space does not smell like something real." Both these quotes from Einstein are taken from Howard (1990).

Schrödinger was aware of the wave packet spreading but could not find a solution. Moreover, he granted that a field in a high dimensional space could not be interpreted physically[7] and to address this problem he proposed that the square module of the wave function (a three-dimensional field) represents a charge density.[8] However, like all waves, Schrödinger's charged density waves superimpose, and their superpositions propagate to the macroscopic domain where are not observed. So, the theory is not empirically adequate when completed this way.[9]

13.3.3 The de Broglie-Bohm Theory as a Reductive Quantum Theory

In 1952 Bohm came to the rescue and showed how, by using instead a particle ontology, and relegating the wavefunction to describe the motion of particles, rather than representing a physical field, one could avoid these problems: particles are particles (so they don't' spread); and superpositions of the wavefunction have no physical meaning.[10]

Bohm wrote a second order equation for the motion of the particles. This equation was similar to the one of Newton with the addition of a distinctively quantum term, which contains a Schrödinger evolving wavefunction. The wavefunction does not represent matter, but rather has a role of a quantum potential, similar to the gravitational or electromagnetic potential, as it affects the trajectories of the particles.

The theory can be written in simpler form in terms of a (first order) guidance equation which specifies the velocity of the particles in terms of the Schrodinger evolving wavefunction.[11] This explicitly shows the nonlocality of the theory, which came to be known as de Broglie-Bohm theory, as discussed later.

[7] In fact, in 1926 he wrote: "The direct interpretation of this wave function of six variables in three-dimensional space meets, at any rate initially, with difficulties of an abstract nature." Also: "Of course this use of the q-space is to be seen only as a mathematical tool, as it is often applied also in the old mechanics; ultimately [...] the process to be described is one in space and time." (Bacciagaluppi and Valentini 2009, p. 447).

[8] He chose charge rather than mass, for instance, because he wanted to connect his theory with electromagnetism.

[9] Schrödinger (1926).

[10] Bohm (1952).

[11] Dürr et al. (1992).

13.3.4 Consequences: The Collapse of the Wavefunction

In this theory the wavefunction which evolves according to the Schrodinger equation is the wavefunction of the universe, while in general the wavefunction of systems smaller than the universe does not evolve according to this equation. In suitable situations one can however talk about the Schrodinger evolving 'effective' wavefunction of a system and discuss the interaction among different systems as if they were isolated. Thus, the system's wavefunction may evolve into a superposition, given the linearity of the Schrödinger equation. However, when the system interacts with another system, say an apparatus, their interaction may be strong enough that the wavefunction ceases to evolve according to the Schrödinger evolution. That is, their wavefunction effectively 'collapses' into one of the terms of the superposition. From that moment on, the motion of the particles in that system is governed by the 'collapsed' wave function, for all practical purposes. While it is determined in which term of the superposition the wavefunction will collapse, we cannot predict it, because it depends on the initial position of the particles, to which we have no access because of 'quantum equilibrium,' as we are going to see in the next section.[12]

13.3.5 Consequences: The Uncertainty Principle

What we can know (or: the information we can have access to) has to do with how the system we are investigating correlates with another system (the measurement apparatus) that we can access directly (because macroscopic). For instance, if we put an object on a (properly calibrated) scale, it interacts with the scale and as result the pointer moves to indicate a location labelled '1 Kg.' Because of this we acquire the information that the object on the scale weights 1 Kg. If instead a system is independent of the apparatus, then they do not interact, so there can be no correlation, and thus no knowledge of the system can be acquired by investigating the apparatus. We have independent systems when we have equilibrium, understood as in statistical mechanics, where nothing changes. So, to have information about a system we need non-equilibrium.

If the world is governed by the de Broglie-Bohm theory, the configurations of the particles are in 'quantum' equilibrium with respect to the wavefunction, which instead is not in equilibrium. So, as in statistical mechanics I cannot extract anything from a system in thermodynamic equilibrium, in a system in quantum equilibrium all we can know about it is given by the system's wavefunction (because that is the one which is not in equilibrium). That is, the maximal information we can have of where the particles are is given by their wavefunction, which is the reason why it can be called the state of the system. This limitation of what we can know about a

[12] Dürr et al. (1992).

system was first discovered by Heisenberg in 1927, who called it the 'uncertainty principle:' if we happen to know the particle position with high precision, we will be more uncertain about the particle momentum.[13]

13.3.6 Consequences: The Born Rule

The de Broglie-Bohm theory can correctly account for all the experimental result. They are statistical in character, as the theory provides only the probability distribution of the results rather than a unique result like a deterministic theory would. This is because we cannot know where the particles were, as just discussed.

It turns out that, for a typical universe,[14] if the particles of a system are initially randomly arranged according to a distribution given by the square module of the initial wavefunction of the system $|\psi_{t=0}(x)|^2$, then they will continue to be distributed that way: $|\psi_t(x)|^2$ (this property is called 'equivariance').[15] That means that, when performing an experiment to determine where a particle is, the probability P of finding it in a given place x is provided the square module of the wavefunction evaluated at that point: $P(x) = |\psi(x)|^2$. This is the so-called Born rule, who discovered it in 1926.[16]

The theory received its first empirical confirmation in 2011, when the highly non-classical character of the trajectories of the particles was experimentally observed.[17]

---

[13] Heisenberg (1927). He took his principle in a much stronger sense, as discussed in Sect. 13.4.1.

[14] Namely for the vast majority of universes, measured in terms of a typicality measure given by the square module of the universal wave function.

[15] Dürr et al. (1992).

[16] Born (1926).

[17] The theoretical trajectories have been computed in Philippidis et al. (1979), while the experimental results are presented in Kocsis et al. (2011). One will wonder how it is possible to have a measurement for the trajectories, given the uncertainty principle. This relies on the notion of weak measurement (Aharonov et al. 1988), and an ingenious experimental procedure. A weak measurement is an experiment to measure a quantity which does not disturb the system too much but, due to the uncertainty principle, gives the result with very high imprecision. Assume one weakly measures the position x_1 of a particle at time t_1. This gives an imprecise value for x_1. Repeating this many times for the same position, one obtains a statistical distribution of x_1. Taking the average, one gets a good estimate of x_1 at t_1. At a later time t_2, perform a strong (i.e. precise) measurement of x_2, the position of the particle later on. One then can compute the velocity of the particle at x_1 by dividing the two positions x_1 and x_2 (one measurement weakly, the other strongly) at the two times t_1 and t_2, by the time interval. Repeating this procedure for different positions, one obtains a velocity field, and the trajectories are calculated as the tangents to the field.

13.3.7 Consequences: Matrix Mechanics

As Bell have noticed, all experiments give results that ultimately describe where something is located: for instance, an experiment to determine the velocity of a particle, is a measurement of how far the particle has travelled in a given time. Also, a measurement of a body's weight is given by the scale pointer pointing at a given label, and so on. In general, thus, all measurements are position measurements.[18] Nonetheless, it is possible to effectively summarize experimental results in terms of matrices or linear operators, as discovered by Heisenberg, Born and Jordan in 1926.[19] To each experiment one can associate a suitable self-adjoint operator, A, and the possible experimental results are given by their eigenvalues ai: A $\psi_i = a_i \psi_i$. If one writes the wavefunction as a linear combination of the corresponding eignestates ψ_i, $\psi = c_1 \psi_1 + \ldots + c_i \psi_i + \ldots$, the probability P of obtaining one of the possible results, call it a_i , is given by the square of the coefficient c_i with which the eigenstate corresponding to a_i is weighted in the linear combination of the eigenstates of A: $P(a_i) = |c_i|^2$. This is the generalized Born rule that gives the probability of finding a given experimental result.

13.3.8 Consequences: Experiments Often Are Not Measurements

However, as first proven by von Neumann in 1932,[20] in general experiments do not reveal much about the system, because they may end up perturbing it too much.[21] Experiments which do not change the system excessively can be called 'genuine measurements' because they can be thought as revealing us a property of the system as it was before the experiment. Instead experiments which disturb the system so much to change it significantly, cannot be properly called measurements. In classical physics it is safe to assume that the system-apparatus interaction can be neglected.[22]

---

[18] Bell (1982).

[19] Born et al. (1926).

[20] Von Neumann (1932), pp. 305–324 of the 1955 English translation. See also Jauch and Piron (1963), Gleason (1957), Kochen and Specker (1967) for similar proofs. See also Pusey et al. (2012).

[21] This is not what he took himself to have proven. For that, see Sect. 13.4.4.

[22] For instance, when you are measuring the temperature of an object using a thermometer, the temperature you are actually measuring the equilibrium temperature between the object and the thermometer. However, given the physical features of the thermometer, this equilibrium temperature is sufficiently closer to the initial temperature that we can think of the temperature one reads on the thermometer as the temperature of the object before it touched the instrument.

However, as also emphasized by Bohr, this is no longer the case if the size of the instrument is about the same size of the system under investigation. In fact, for instance, when we wish to determine the position of an unmoving particle is, we have to see it first. That is, we perturb it with photons, which scatter the particle. And the more precisely we wish to know the position, the more we will have to perturb its velocity, so the less we can get a precise estimate of it, as the uncertainty principle states.

These 'no genuine measurement' proofs (or impossibility proofs against genuine measurements) sometimes are also called 'contextuality' proofs because they show that, in general, whether an experiment is destructive or not depend on the context, and when they are destructive they (obviously) reveal noting about the system. For instance, assume we want to make an experiment to determine the velocity of a particle inside of a box which, unknown to us, is not moving. To do so, we need to remove the box to see where the particle goes in a given amount of time. But at that point the particle will begin to move (because opening the box changes its wavefunction), so what we record at the end of the experiment is not the particle's initial velocity.[23] In the classical domain it is easy to perform non-destructive experiments whose results we can be safely interpret as revealing the properties of the system before the experiment.[24] Instead, in the quantum domain many experiments are destructive: their results should be taken as telling us about the interaction, rather than the system itself.

As we saw, the statistics of an experiment is suitably captured by an operator, and a series of destructive experiments (which therefore cannot be thought as measuring anything) has operators which do not commute: that is, the value of their product depends on their order. This makes a lot of sense, because if a bunch of experiments are destructive, then the order in which they are performed matters: if you read a message and then you burn it, it is different than if you burn it and then you read it.

---

[23] Notice that Einstein (1953) objected that the de Broglie-Bohm theory seems contradictory because he did not take seriously enough the fact that experiments are interactions. If we take the wavefunction of the particle in the box and solve the guidance equation, we predict that the particle does not move; while if we calculate the probability distribution of observing the particle with a given velocity as given by the Born rule, we get a non-zero distribution. Nonetheless, the contradiction disappears if one remembers that to determine the particle's velocity, we have to open the box, and thereby change the wavefunction so that the particle now moves.

[24] Even if one can construct destructive experiments which may look like true measurements but are really not. Think about taking a picture of an object with an old-fashioned film. A non-destructive experiment on the film, or a true measurement of the objects depicted in the film, would be to develop the film in a darkroom: this will tell us where the object is. Instead, a destructive experiment is looking at the film in daylight. While the former 'fixes' the objects' position on the film, the latter makes it disappear.

13.3.9 Consequences: Quantum Nonlocality

As anticipated above, the most troubling aspect of the de Broglie-Bohm theory, however, is that it is explicitly nonlocal. That is, since the wavefunction is defined on configuration space, by acting on a particle here, one affects a particle over there instantaneous, regardless of how far away the two particles are. This is in contrast with relativity, where there is an upper limit to the velocity of propagation of interactions.

So, in 1964 Bell tried to see whether one could provide a different type of reductive understanding of the quantum phenomena, assuming locality to be true. He proposed a theory in which there is some generic 'hidden' ontology, whose dynamics is governed by a Schrödinger evolving wavefunction, and calculated its empirical predictions. These are different from the predictions of the de Broglie-Bohm theory, and so one could perform a crucial test.[25] Freedman and Clauser (1972) carried out the first actual test, using an inequality, called CHSH (Clauser, Holt, Shimony, Horn), derived from the original one.[26] When the results were later confirmed in 1982 to be in favor of the de Broglie-Bohm theory,[27] it had to be accepted that locality is no longer an option, and that the de Broglie-Bohm theory is the best constructive theory which explains the quantum phenomena according to the reductive schema. Then people started to think about what this means for relativity, as it seems that the very spirit of relativity is locality, due to the existence of an upper limit in velocities.

This is where we are right now. Or better, this is where we would be, if things had gone differently.

So far, we have seen no high dimensional field ontology, no macroscopic superpositions, no stochastic modifications of the Schrödinger equation, no undetectable worlds, no contextual properties, no consciousness. Everything got shut down right in the moment people understood that one has to have a three-dimensional ontology to continue to understand things along the lines of the reductive explanatory schema which has been so successful in the classical domain. There was no reason they did not try to find a theory like that, and when they did find it, this led them to a clear and simple picture of the world, according to which particles move around in three dimensions following highly nonclassical laws, and in which the main discovery of modern science, namely nonlocality, is explicit.

But this is not what had happened. History is very different from this.

---

[25] Bell (1964).

[26] Clauser et al. (1969).

[27] Aspect et al. (1982a, b).

13.4 Resistance Is Futile

Let us briefly summarized what had actually happened, which is not as linear as the story we just presented. It is filled with people misunderstanding one another, talking past each other, ignoring or misconstruing arguments, and silencing unwanted objections. Here is a short overview.[28]

13.4.1 Matrix and Wave Mechanics

As we saw, quantum theory begun with the quantization rules, and with people's struggles trying to figure out where they came from. In 1925 Heisenberg, with the help of Born and Jordan, developed the so-called 'matrix mechanics,' a unified model to systematize experimental results using matrices. This was a formal result, in which there was no dynamics, no spacetime, no objects, but it could nonetheless reproduce the quantization rules in terms of suitable matrices, rather than reductively in terms of a microscopic dynamics. Initially, the proponents of matrix mechanics were influenced by the emerging positivistic philosophy, according to which unobservable entities should be eliminated because meaningless. They got convinced that the reason it was difficult to reductively explain the experimental results was that they were trying to fit the data in the wrong explanatory schema. The right way of explaining things is not in terms of the motion of unobservable entities but rather to account for them in terms of a mathematically elegant unified formal picture referring only to what can be observed.

To their dismay, in 1926 Schrödinger proved them wrong by proposing wave mechanics, which aimed at providing a reductive explanation of the quantum rules with a wave ontology. He also proved that his theory and matrix mechanics are mathematically equivalent, but wave mechanics was more explanatory because it is more visualizable. Heisenberg defended matrix mechanics from the lack of visualizability charge by dropping the positivist attitude and by endorsing a particle ontology. In 1927 he discovered the uncertainty principle, and claimed that there are particles, however without determinate trajectories because they cannot have positions and velocity at the same time.[29] In any case, Bohr violently quarreled with Heisenberg about his particle ontology because he wanted to keep both particles and waves (waves were also more compatible with his atomic model). In 1927 in the Como lectures Bohr anticipated his wave-particle complementarity view as a

---

[28] In this review I have followed van der Waerden (1968), Jammer (1974), Fine (1986), Cushing (1994), Howard (1990, 2004, 2007, 2013), Beller (1999), Bricmont (2016), Norsen (2017), Becker (2018).

[29] This is what he claimed his uncertainty principle forced him to accept, but this is not true, as one could have said that they have both, but we cannot know what they are with arbitrary precision.

sort of compromise.[30] Born initially endorsed Schrödinger's picture to arrive to his statistical interpretation of the wavefunction, according to which the wavefunction tells us the probability of fining a particle in a measurement, but later embraced Heisenberg's attitude. Be that as it may, the members of the Copenhagen school, guided by Bohr, hid their internal disagreements and formed a compact front against Schrödinger's wave model. He was accused not being able to explain quantum jumps, even if he did: he wrote papers accounting for the Compton and photoelectric effects in terms of wave vibrations. In 1927 at the Solvay conference, de Broglie's preliminary proposal for a complete quantum theory in terms of particles and waves was effectively shut down by a harsh criticism by Pauli, so that de Broglie abandoned his project for a long time.[31]

13.4.2 Unphysical Field Arguments Against Quantum Theory

Everyone ended up using Schrödinger's methods, undeniably more effective than the matrix model, but he was left alone in thinking about the meaning of his theory until he finally gave up (like de Broglie before him, and Bohm afterwards). This happened in great part because of Bohr's power, influence, and charisma (as well as some others' lack of these features), and the rampant positivistic philosophical climate. Some, most notably Einstein and Schrödinger, resisted to the Copenhagen idea that we should learn to live with the fact that the quantum world is hopelessly incomprehensible. However, they have been either ignored or misunderstood. As anticipated, they proposed arguments ultimately for the conclusion that as thermodynamics can be more profoundly understood in terms of statistical mechanics, so quantum theory can be better understood in terms of a deeper, microscopic theory still to be found. The first argument of this type can be summarized as follows: quantum mechanics is incomplete otherwise reductionism is false (because there's nothing moving in three-dimensional space). We already mentioned in the previous section that this argument was proposed by de Broglie, Lorentz and Einstein against Schrödinger's picture. Interestingly, this argument can be tracked down historically also to Heisenberg, who did not think it made sense to have a theory with no fundamental three-dimensional fields and with no fundamental three-dimensional physical space.[32]

As we have also seen, Schrödinger proposed his three-dimensional charge density field ontology in response. Even if the theory was still inadequate (due to the macroscopic superpositions and the spreading of the wave packet) Schrödinger never endorsed de Broglie's particles, which would have solved at least the first of these problems.

---

[30] Bohr (1928).

[31] See Bacciagaluppi and Valentini (2009).

[32] He has been reported to have said, very vividly, referring to Schrödinger's work: "Nonsense, [...] space is blue and birds fly through it" (Bloch 1976).

13.4.3 Nonlocality Arguments Against Quantum Theory

Heisenberg took the argument above to be a reinforcement of his instrumentalism. Instead Einstein looked for something in three dimensions which could represent physical entities, over and above the wavefunction. In 1927 he proposed a deterministic particle theory, which however, he later retracted because it was nonlocal.[33] Accordingly, he did not take it as a serious option, as he wanted to keep both reductionism and locality.

Instead of actively working on building a microscopic theory which would reductively account for the quantum rules, he started proposing arguments that quantum theory had to be incomplete, otherwise it would be nonlocal. There are several variations of this argument, proposed between 1927 and 1935. In 1927 at the Solvay conference, Einstein considered a particle traveling towards a narrow slit, behind which there is a curved detection screen. After the slit, because of diffraction, the wavefunction of the particle would propagate as a spherical wave, and thus hit the screen everywhere. Since instead experimentally we observe only one spot on the screen, there has to be some nonlocal interaction. Given that is unacceptable, Einstein concluded, quantum theory cannot be complete.[34] This argument was misunderstood by Bohr, who thought it had to do with the uncertainty principle,[35] but not by Heisenberg. In fact, in his 1929 Chicago lectures he conceded to Einstein that quantum theory is nonlocal.[36] However, he also claimed that it does not violate relativity when understood instrumentally as a theory of signals (in fact, such nonlocality cannot be used to send information).

In the 1930, at another Solvay conference, Einstein proposed a variation of the previous thought experiment to Bohr. Take a photon in a box, then open it and let the photon travel far away. The photon is in a superposition of energy states. However, by measuring the box's weight, one collapses the photon's energy state into one of them thereby violating locality, given that the photon is far away from the box. Bohr

---

[33] For a system with two independent components the theory would not provide independent velocities. See Belousek (1996), Holland (2005).

[34] See Bacciagaluppi and Valentini (2009), p. 440. Einstein anticipated this argument already in 1909 at a meeting in Salzburg (Bacciagaluppi and Valentini 2009, p. 198). Moreover, Schrödinger begins his 1935 cat paper presenting an example very similar to this one. Take a radioactive source which could emit particles, one at a time, in all directions. If the description provided by the wavefunction is complete, and the wavefunction evolves according to the Schrödinger equation, then the particle will be detected in all the directions at the same time, which is not what happens. So, again, either the theory is incomplete, or it is nonlocal.

[35] See Bohr (1949) for a recollection of Einstein's argument. Bohr's misunderstanding was probably due to the way the argument was phrased by Einstein, as he emphasized that when the particle passes through the slit its position is determined but its momentum is completely spread out. Presumably, the argument could have been less prone to misunderstanding had Einstein not mentioned the momentum at all, as I did in my reconstruction.

[36] He considered an experiment in which a wave packet is split by a half-silvered mirror into a reflected and transmitted packet which travel to far-separated regions. If the reflected packet is detected then the other has to immediately disappear, violating locality (Heisenberg 1949, p. 39).

however again misunderstood the argument as having to do with the uncertainty relation,[37] and the legend says that the day after he triumphantly provided a reply to Einstein in which he used Einstein's own theory of relativity against him, inflicting him with a burning and embarrassing defeat. In reality, however, they were talking past one another, as Einstein's point had not much to do with the uncertainty relations but was about the nonlocality of the theory.[38]

In a 1935 letter to Schrödinger before the cat paper, Einstein considers another experiment in which there is a box filled with a wavefunction. By inserting a partition, the box is divided into two halves, thereby splitting also the wavefunction inside. These boxes are separated in opposite directions and one of them is opened when they are far away from one another. By opening one, the content of the other object is instantly affected, again violating locality.[39]

Furthermore, the 1935 Einstein, Podolsky and Rosen (EPR) paper was designed as an incompleteness argument.[40] EPR argued that if quantum theory were complete, then an experiment on one component of a pair of entangled particles travelling in opposite directions would instantaneously determine the state of the other component of the pair, regardless of their mutual distance. This is because before the experiment systems have no properties but only acquire them afterwards. However, this again violates locality, so the only option is to admit that the theory is incomplete: there have to be 'hidden' variable, unspecified by quantum theory, which represent the pre-existing (i.e. existing before the experiment) properties that the particles possess which are revealed, rather than created, by the experiment. However, as in the case of the previous attempts, the point of the argument was completely misunderstood by Bohr,[41] and Einstein kept being painted by the growing Copenhagen school as the stubborn, old scientist who could not accept the revolutionary character of the quantum, as part of the old guard who could not let go of his ancient ways of thinking.

13.4.4 *Proofs Against the Possibility of a Quantum Reductive Schema*

There were also more legitimate reasons than these misunderstandings to discard the project of completing quantum theory in terms of hidden variables. Theorems were proposed to prove that this program cannot be carried out. Most notably in 1932

---

[37] Again, this has to do to the fact that Einstein worded his thought experiment in term of the time-energy uncertainty relation: you could either measure the energy or the time with arbitrary precision.

[38] This is how it had been reconstructed by Howard (2007). See also Nikolić (2012).

[39] Einstein to Schrödinger, June 19, 1935, in Fine (1986), p. 35.

[40] Einstein et al. (1935).

[41] The EPR argument was formulated in terms of measurements of position and momentum.

von Neumann claimed to have shown that these hidden variables are impossible. He assumed that the preexisting properties mentioned by EPR exist, and that they are revealed by experiments in terms of eigenvalues of operators. Then he showed that, under what he thought reasonable assumptions, these properties have to stand into mathematically contradictory relations. Therefore, he concluded that we cannot think of quantum theory as providing the type of reductive understanding we had in classical physics.

This proof was not sound, because some of the assumptions were not physical, and the theorems that came later on which did not use that assumption were instead relying on the idea that making an experiment on a system does not significantly change it. Instead, in general the interaction will be sufficient to disrupt the system so much that the experimental results are not the values of the properties before the experiment. In other words, these theories do not prove that hidden variables are impossible, but rather that experiments may often measure nothing of the system. However, this did not become clear until later, when Bell exposed the mistake (however, Grete Hermann quickly pointed out the problem but was ignored[42]). Also Einstein in 1938 dismissed the proof as unphysical but never wrote anything about it.[43]

So, von Neumann's impossibility proof gave the final blow to the proposals to complete quantum theory, especially coming from a living legend like von Neumann. The logic of inevitability seemed to have finally found a conclusion: we cannot do better than what the Copenhagen school is offering us. Von Neumann's proof forces us to give up our dream of reductive quantum explanation.

13.4.5 The Measurement Problem Against Quantum Theory

The arguments against the completeness of quantum theory became weaker and weaker. First, the argument for a three-dimensional ontology to preserve the reductive schema was ineffective, due to the positivist commitments of the Copenhagen school that led them to consider this type of explanation misguided because based on unobservable entities. Then, nonlocality arguments to show that at least the theory needs something more, otherwise locality is lost. But again, these arguments were impotent, because a nonlocal quantum theory does not violate relativity when understood according to the Copenhagen instrumentalism as a theory of signals. So, the last resort was to focus on what the instrumentalists seemed to care, namely experimental results, to show that quantum theory is not empirically adequate. This is what the 1935 Schrödinger cat paper is about. Schrödinger imagined a cat in a box, where a vial of poison is connected with a radioactive nucleus. If the nucleus decays, the vial is broken and the cat dies, otherwise nothing happens. If the nucleus

42 Hermann (1933, 1935). See also Bacciagaluppi and Crull (2016).

43 Shimony (1993).

is in a superposition state of 'decayed' and 'not decayed,' then, due to the linear evolution of the wavefunction describing every physical system, the superposition propagates macroscopically into a 'living' and 'dead' cat superposition. However, when we open the box, the cat is either alive or dead. We never observe such macroscopic superpositions.[44] This argument therefore pointed out that the theory is not empirically adequate. This problem is also called the measurement problem: a Schrödinger evolving wavefunction which completely describe physical systems will predict all experimental outcomes to suitably realize at the same time, which is not the case.

However, this argument failed too, as von Neumann in 1932 provided an answer to it: he postulated that the wavefunction evolves according to the Schrödinger equation unless a measurement is performed. When that happens, the wavefunction spontaneously and randomly collapses into one of the terms of the superpositions.[45] This collapse is nonlocal, but so be it: as Heisenberg pointed out, this nonlocality cannot be used to send signals. So, Copenhagen won once again, and these problems were substantially ignored until the 1950s, when people started to question that the theory did not have a unified dynamics (see Sect. 13.4.9).

In any case, the war came along, and it helped in forgetting these problems: it put on hold many collaborations, conferences, and discussions, and pushed for technological results over conceptual ones. So, everyone ended up following along the mainstream Copenhagen vision without too many questions.

13.4.6 Bohm's Rediscovery of de Broglie's Beginning of a Reductive Theory

Slowly, things got back to normal in the 1950s. After having written a textbook in quantum theory, Bohm (1952) started to think harder about the theory and rediscovered de Broglie's work, expanding it: while he was at Princeton, he proposed a theory in which particles evolve according to an equation defined

[44] In the weeks that preceded the publication of the cat paper, in a letter to Schrödinger, Einstein discusses an example in which there is a substance which may spontaneously combust. If it does combust, it makes some gunpowder explode, otherwise nothing happens. The substance microscopic superposition 'combust' and 'not combust' is amplified macroscopically into a superposition of 'explosion' and 'not explosion,' which is not observed. He wrote: "The system is a substance in chemically unstable equilibrium, perhaps a charge of gunpowder that, by means of intrinsic forces, can spontaneously combust, and where the average life span of the whole setup is a year. In principle this can quite easily be represented quantum-mechanically. In the beginning the ψ-function characterizes a reasonably well-defined macroscopic state. But, according to your equation, after the course of a year this is no longer the case at all. Rather, the ψ-function then describes a sort of blend of not-yet and of already-exploded systems." In A. Fine (1986). This is remarkably similar to the argument proposed by Schrödinger few months later.

[45] Von Neumann (1932), chapters V ad VI, pp. 184–237 of the 1932 German version and pp. 347–445 of the 1955 English translation. Reprinted in Wheeler and Zurek (2003): 549–647.

in terms of a Schrödinger evolving wavefunction. He developed his theory as a microscopic and reductive understanding of quantum mechanics, not as a solution of the measurement problem. He wrote Newton's equation of motion and showed that one can reproduce the quantum predictions if one adds a quantum potential, which is written in terms of a Schrödinger evolving wavefunction.

In this theory the wavefunction never collapses but there are no macroscopic superpositions. In fact, matter is made of particles, and the only term of the wavefunction that effectively guides their motion is the one under which the particles are, so that we can forget, for all practical purposes, the other terms.

Bohm was encouraged by Einstein to pursue this theory, but effectively ostracized by everybody else because he dared to put the Copenhagen dogma in doubt. Moreover, his sympathies towards communism got him exiled in Brazil, without a passport, so he could not even travel to defend his theory from the Copenhagen attacks. After few years mostly spent in exile, he effectively stopped defending his theory.

Einstein, while initially supportive of Bohm, did not like the explicit nonlocality of the theory. Also, he did not like the fact that later on Bohm proposed it as a theory of a field in configuration space,[46] which Einstein thought did not make any physical sense. So he got convinced that quantum theory had to be completely changed rather than supplemented: he wanted to keep both locality (because of relativity) and three-dimensionality of the ontology (because of reductionism), and there was no proposal for a constructive theory which was also local.

13.4.7 Contextuality of the de Broglie-Bohm Theory

Most people dismissed the theory: there had to be something wrong with it, because von Neumann proved that what he did was impossible. Even those who accepted that the theory could get around the impossibility proofs, thought that it could do it at great cost, namely postulating contextual properties. That is, when we think of a property of a system we think of it as some sort of timeless characterization: the spin along direction x is 'up,' and that is what it is. If we set up an experiment to measure such a property, the experiment is supposed to give us its value. If we repeat it after we have measured another property, say its spin under another direction, it should give us the same value as before. Instead, people thought, in the de Broglie-Bohm theory the properties depend of the context: if we measure spin along x, then its value changes depending on whether we measure it after we have measured the spin along another direction. This contextuality of the de Broglie-Bohm theory was taken as a disadvantage of the theory, as an indication that if we want to think in terms of some microscopic picture, then we end up with weird properties.

---

[46] See Bohm (1980).

However, as Bohm himself explained in his 1952 paper but it is still today not enough appreciated, his theory does not fall prey of the impossibility proof not because it has contextual properties but because experiments change the system so that if we perform another experiment, we should not expect to find as a result the value of the property it had at the beginning.

13.4.8 Bell on the Impossibility Proofs and Bell's Theorem

After having seen Bohm "doing the impossible" (namely avoiding the impossibility poofs) Bell proved the impossibility proofs assumed mistakenly that one can measure properties without modifying the system.[47] The de Broglie-Bohm theory assumes that experimental apparatuses may change the system, so it can complete quantum theory bypassing these proofs, but it is nonlocal. So, Bell started to wonder whether one could complete quantum theory while also preserving locality, like Einstein would have wanted. So, in 1964 Bell picked up again the EPR argument: the perfect correlations observed in entangled systems can only be locally explained by the existence of preexisting values. By assuming that these values exist, Bell derived an inequality which holds for his theory but not for quantum mechanics, and that can therefore be used as a crucial test. This test was later performed confirming the quantum mechanical predictions.

From this Bell wanted to argue that any theory which matches the quantum predictions had to be nonlocal. However, most people thought he proved that hidden variables as predicted by Einstein (local or not) are impossible, and took Bell to have provided another impossibility proof, similar to the one of von Neumann. This was true, in the sense that experiments do no reveal preexisting properties, but this is not what Bell wanted to prove, for one thing because he already knew that. Rather, he wanted to prove that you have to have nonlocality to reproduce the quantum predictions. But it surely did not help to get the point across that he started from EPR, which included the assumption that experiments can reveal (preexisting) properties of the system. In fact, he knew was this assumption was false (indeed, this is what he himself said von Neumann got wrong in his impossibility proof), while others still took it seriously. Nevertheless, to remedy this confusion, later Bell reformulated his inequality without passing from EPR, therefore without assuming that there are any preexisting properties, and only assuming the locality condition.[48] This shows that any theory that matches the quantum mechanical predictions has to be nonlocal. Nonetheless, for a variety of reasons, controversies are still open about what Bell did or did not prove.

[47] Bell (1966). This paper was written however in 1964.

[48] Bell (1975).

13.4.9 The Other Solutions of the Measurement Problem

The research on the meaning of quantum theory barely continued in an organized way after Bohm was ostracized. While Bohm wished to provide a reductive understanding of the quantum phenomena, but was shut down by the impossibility proofs, starting from the late 1950s, the attention of those isolated people still interested in understating the theory slowly started to shift from the problem of suitably completing quantum theory to get a microscopic description, to solving the measurement problem. Some researchers started to question the orthodoxy, which now included von Neumann's collapse postulate, not in virtue of its non-three-dimensional ontology or its nonlocality, but from within the dogma. For instance, Everett, a student at Princeton, started from the measurement problem and argued that the von Neumann collapse was unsatisfactory: what is a measurement, after all? In contrast with Schrödinger, Einstein, de Broglie, and Bohm, he was not motivated by realism. Rather he was bothered by the imprecision of the von Neuman formulation and the inconsistency of applicability of the theory. He thus proposed, in his 1957 dissertation, an original way to make sense of macroscopic superpositions: he postulated that the various terms of the superposition express relative states of affair of an observer. One could think of the wavefunction as splitting into different worlds, one for each term of the superpositions.[49] Naively Wheeler, his advisor, enthusiastically discussed this idea with Bohr, who instead did not react well, and ultimately destroyed Everett's career, who abandoned academia for military research. Feynman, who heard about this theory from Wheeler, found the many-worlds idea preposterous. The whole idea was forgotten for years by everyone except DeWitt, a cosmologist friend of Wheeler, who in 1970 published a paper on quantum theory and reality. This soon followed by the DeWitt and Graham book on many-worlds (1973), in which they address the problem of cosmology: where is the observer if the system is the universe? In 1985 Deutsch, who had attended an Everett lecture in 1977, published a paper on quantum computing, arguing that one could make sense of the reason why quantum computers are faster than classical ones, only assuming that there is a world for each branch of the wavefunction where the calculations happen. The many-world theory, not adding anything to the standard theory, unlike Bohm, and not modifying the Schrödinger equation (like collapse theories, see below), seemed to many as the best compromise among the solutions of the measurement problem (if one forgets about the inflated metaphysics, that is). Because of this, it slowly grew in the sympathies of physicists. In particular for cosmologists, who already realized that the Copenhagen view was inadequate, and for people working on computation, after Deutsch's results on quantum computers. The many-worlds theory had the problem to show that the different worlds do not interact with one another, otherwise we would observe them. Decoherence theory, which started being developed by Zeh, provided an answer: physical systems, by

[49] Everett (1957).

interacting with their environment, lose coherence, and therefore their ability to interfere with themselves.[50]

The measurement problem was also mentioned by Wigner, who thought that the von Neumann collapse was real, and that it was triggered by consciousness,[51] and Rosenfeld, Bohr's faithful soldier, immediately attacked.

In the 1970s, the first tests of Bell's inequality started to be performed, with confirmation of the violation of Bell's inequality in 1982. Moreover, the first conferences on quantum foundations were being held, all focusing on the measurement problems (1970 in Varenna, 1973 in Erice).

In the same period, people were critic of the von Neumann collapse postulate because of the introduction of two fundamental evolutionary laws, and worked on incorporate the collapse process into the dynamical equation of the wavefunction, to provide a unified dynamics of the macroscopic and the microscopic. The spirit of their work was more similar to the one of Everett rather than the one of Einstein, Schrödinger, de Broglie, and Bohm, in the sense that they did not focus on the ontology (so they were not bothered by the high dimensional wavefunction) or on the nonlocality of the collapse. Rather they wanted to make the Copenhagen recipe precise. Unlike Everett, however, they did not want to introduce many worlds in the process. This project culminated in the dynamical collapse models of Ghirardi, Rimini and Weber.[52] In this theory, the wavefunction evolves according to a stochastic nonlinear equation so that it collapses at random times, into random places with a frequency that depends on how big the system is, so that macroscopic objects collapse almost instantly.

After their proposal, Bell started to advertise the idea that there are three ways of making realist sense of quantum theory, legitimizing the many-worlds theory and the dynamical collapse models: assuming measurements have results, "either the wavefunction, as given by the Schrödinger equation, is not everything, or it is not right" (Bell 1987).

This is the story as we know it. Most researchers who found themselves wondering about the meaning of quantum theory ended up looking at the measurement problem, and took the de Broglie-Bohm theory, the Everett many-worlds theory and the theory of dynamical collapse as equally plausible ways of making sense of quantum theory, each with their own pros and cons. For instance, the de Broglie-Bohm theory is often said to have the problem of being nonlocal, the many-worlds theory has the problem of probabilities, and the spontaneous collapse theory seems *ad hoc*.

In any case, given that all these theories have the wavefunction in it, it became natural to think of it as the ontology of the theory, embracing what was initially thought to be impossible: a high-dimensional ontology. One could argue that this is the straightforward way of interpreting these theories: they have an equation for the

[50] Zeh (1970).

[51] Wigner (1961, 1963).

[52] Ghirardi et al. (1986).

wavefunction, so the wavefunction is the ontology.[53] Otherwise one could argue that this is the only way to make the theory local again, even if in configuration space: Bell showed that the reality is nonlocal, but if we have an ontology in configuration space, we can still have that type of locality.[54] At any rate, this view has grown in popularity and now goes under the name of wavefunction realism. This growing field contributed to put even more bad light on the de Broglie-Bohm theory, because it does not fit the schema: if one sees the theory as having a wavefunction ontology living in a high dimensional space, one has to either postulate that there is also a three-dimensional space in which particles live, or that there is only one particle living in the same space of the wavefunction. In both cases, the picture of the theory provides of the world is extremely revisionary and makes it seems like the presence of the particles is unnecessary: why postulate the particles, if you can do without, like in the many-worlds theory?[55] The point, however, is that you cannot do without particles, if you care about having a reductive explanation of the phenomena. But it's easy to miss this point if you focus on solving the measurement problem.

13.5 The Measurement Problem Is a Red Herring

Which argument should the realist have looked at? I have argued that the natural thing to look at would have been the unphysical field problem, because it is the one which allows to see what the problem is if one wishes to preserve the reductive schema. In this section, let's see how the various arguments for incompleteness are distinct in what they are asking. Three of such arguments were proposed:

1. Unphysical field arguments (Sect. 13.4.2): Quantum mechanics is incomplete otherwise reductionism is false (because there's nothing moving in three-dimensional space);
2. Nonlocality arguments (Sect. 13.4.3): Quantum mechanics is incomplete otherwise relativity is false (because quantum theory is nonlocal);
3. Macroscopic superpositions arguments (Sect. 13.4.5): Quantum mechanics is incomplete otherwise it is empirically inadequate (because it predicts unobserved macroscopic superpositions).

These arguments against the Copenhagen dogma were systematically ignored (like the physical field problem) or misunderstood (like the nonlocality problems). Ironically, as we have seen, the only argument that received attention later, namely the measurement problem, was only proposed as a last resort to convince the positivists that their theory was empirically inadequate.

---

[53] Albert (1996).

[54] Ney (2021).

[55] Brown and Wallace (2005).

The unphysical field problem is the strongest of these arguments. It was proposed to show that one needs to complete the theory with a three-dimensional, microscopic ontology if we wish to preserve the reductive schema of explanation of the phenomena. And why should we abandon such schema, if we do not need to? So, this argument states that the wavefunction cannot be interpreted as a physical field vibrating in space. The nonlocality arguments instead focus on the observation that if we do not complete quantum theory one way or the other then we lose locality. While the first argument tells us what to do (add a three-dimensional, microscopic ontology), the locality argument does not tell us how to complete the theory: it merely tells us that it needs to be done. Presumably, this drift is due to the fact that the members of Copenhagen school cared less about the microscopic description and more about the results. However, as mentioned already, Heisenberg did not care about quantum theory being nonlocal. So, the last attempt was in 1935, when Schrödinger formulated the incompleteness argument in macroscopic terms, the only terms positivists would care about: if quantum mechanics were complete, then there would be unobserved macroscopic superpositions. The measurement problem is the problem of empirical adequacy of the theory: we need to suppress these superpositions, regardless of whether the theory is local or not, regardless whether we complete it with a three-dimensional, microscopic ontology or not.

So, among all three, the measurement problem is the weakest of the problems a realist needs to solve. Indeed, it allows for *ad hoc*, nonlocal, imprecise solutions like the von Neumann collapse rule. Why should realists be bothered by such a problem? Rather, realists should care about having a three-dimensional, microscopic ontology from which they can build a reductive schema, and then the problem becomes to find the correct ontology. The natural choice would be between particles and fields, like people had done until then. Take a field ontology, like Schrödinger's charge density proposal. Even assuming we can control the field packet's spreading through decoherence, the theory shows unobserved macroscopic (three-dimensional) superpositions. It is as if we postulate that matter is a wave, but then we do not see interference. What should we conclude from this? One option is to accept that we were wrong, and matter is not a field. So, the best way to solve it is by adding particles, which is exactly what the de Broglie Bohm pilot-wave theory does. The second option is being stubborn: insist that matter is a field and adopt a many-worlds picture to account for the fact that we do not see macroscopic superpositions, even if they exist. That is, each term of the superpositions exists but they are in different worlds which do not interact. However, this second option seems completely *ad hoc* and unnecessary, as the first option (particles) is much simpler, more sensible, and less revisionary. Before the advent of the quantum weirdness, no one would have dared to propose anything like that. Moreover, in contrast with the second option, the first option allows for the reductive schema. The goal of Schrödinger and Einstein was to find the corresponding constructive theory to explain quantum theory, seen as a principle theory. They were looking for the corresponding 'statistical mechanics' for 'thermodynamics.' This is what the first option provides. Instead many-worlds provides a more complex principle theory, in which the principles apply to unobserved and unobservable worlds. What is the advantage of doing that?

Are there other alternatives? Not really; certainly not if we care about the reductive schema. The other option proposed to solve the measurement problem is to modify the Schrödinger equation (stochastically and nonlinearly). This provides a more elegant, and unified principle theory, but not a constructive theory. If one looks for one in this context, one would have to add a three-dimensional ontology to the spontaneous localization dynamics. But what would be the point of complicating the wavefunction evolution, if one can obtain the same result with the simpler Schrödinger evolution?

Needless to say, if someone thinks the collapse of the wavefunction is not real (like in Bohm and Everett) or it is dynamical (like in collapse models), there is no reason to think physics needs consciousness to provide a complete description of reality.

So, I argue, by focusing on the measurement problem, namely the problem of suppressing unobserved macroscopic superpositions, one is giving too much credit to ideas that would have never been considered, had history been different. Realists should have focused on preserving the reductive schema, to the extent that it was possible. Thus, for this reason in the 1920s people argued that quantum theory needed a three-dimensional ontology. In this framework, the idea of a wavefunction ontology would have been dead since then, and together with it the many-worlds theory and the spontaneous collapse model, with the pilot-wave theory as the only surviving proposal.

13.6 Nonlocality and Reductive Explanation

The locality argument discussed above asks us to complete the theory in a local way. However, Bell showed that we cannot have that: nature is nonlocal, and the pilot-wave theory embraces this nonlocality in the most straightforward way.[56] So, we can keep reductionism and the classical explanatory schema in terms of three-dimensional entities moving around and composing macroscopic bodies. However, there are nonlocal correlations between these entities. This creates a true conflict with the theory of relativity, understood as a theory about reality rather than signals.[57] This is the true quantum revolution: not the paradigm shift in terms of the type of understating we can achieve, but a new feature of the world.

So, as we have seen, in the pilot-wave schema the reductive explanation schema is preserved, but we have nonlocality. Now, one may naturally ask what would have happened if we dropped the reductive explanation. In other terms, could we somehow preserve locality if we ditch reductive explanation? Some wavefunction

[56] See e.g. Bricmont (2016), Norsen (2017).

[57] See Maudlin (2011).

realists have argued that we can: they ask us to entertain the possibility that the true ontology is the wavefunction in its high dimensional space, and they argue that this is the price to pay to maintain locality.[58] However, the locality they preserve is locality in the space of the wavefunction, namely high dimensional configurations pace. So, it is unclear whether this is indeed the locality that people like Einstein cared about.

Be that as it may, we have finally arrived at the present state of affairs: we have to deal with the fact that reality is different from what we think it is, as it is nonlocal.[59] And we would have arrived immediately to this conclusion had history been different and had the pilot-wave theory been accepted when it was proposed. We could have devoted our research time to try to reconcile relativity with quantum theory, instead of trying to solve the measurement problem.

References

Aharonov, Yakir, David Z. Albert, and Lev Vaidman. 1988. How the result of a measurement of a component of the spin of a spin-1/2 particle can turn out to be 100. *Physical Review Letters* 60: 1351.

Albert, David Z. 1996. Elementary quantum metaphysics. In *Bohmian mechanics and quantum theory: An appraisal*, ed. J. Cushing, A. Fine, and S. Goldstein, 277–284. Kluwer.

Aspect, Alain, Jean Dalibard, and Gérard Roger. 1982a. Experimental test of Bell's inequalities using time-varying analyzers. *Physical Review Letters* 49: 1804–1807.

Aspect, Alain, Philippe Grangier, and Gérard Roger. 1982b. Experimental realization of Einstein-Podolsky-Rosen-Bohm Gedanken experiment: A new violation of Bell's inequalities. *Physical Review Letters* 49 (2): 91–94.

Bacciagaluppi, Guido, and Elise Crull. 2016. *Grete Hermann – Between physics and philosophy*. Springer.

Bacciagaluppi, Guido, and Antony Valentini. 2009. *Quantum theory at the crossroads: Reconsidering the 1927 Solvay conference*. Cambridge: Cambridge University Press.

Becker, Adam. 2018. *What is real? The unfinished quest for the meaning of quantum physics*. New York: Basic Books.

Bell, John S. 1964. On the Einstein Podolsky Rosen Paradox. *Physics* 1: 195–200. Reprinted as Ch. 2 in *Speakable and unspeakable in quantum mechanics*. Cambridge: Cambridge University Press (1987). (written in 64 published in 66, reprinted as ch 1 "On the problem of HV in QM" in the Speakable Unspeakable book).

———. 1966. On the problem of hidden variables in quantum theory. *Review of Modern Physics* 38: 447–452. Reprinted as Ch. 1 in *Speakable and unspeakable in quantum mechanics*. Cambridge: Cambridge University Press (1987).

[58] Ney (2021).

[59] Some claim that there is another assumption in Bell's theorem, namely statistical independence, which we could reject, thereby saving locality. There are two ways of rejecting statistical independence, one is superdetermnism, and the other is retrocausality. Roughly put, in superdeterminism not only the future is determined by the past and the laws, but also, it could not have been otherwise: random selections of samples to detect correlations is impossible, thereby it is questionable whether it is worth assuming this instead of nonlocality. A similar reasoning seems to hold for retrocausality, and it is unclear how different that is from nonlocality. See Bricmont (2016) for a discussion.

———. 1975. The theory of local beables. TH-2053-CERN, 1975 July 28. Reprinted in *Epistemological Letters* 9(17): 11–24 (1976), in *Dialectica* 39: 86 (1980) and as Ch. 7 in *Speakable and unspeakable in quantum mechanics*. Cambridge: Cambridge University Press (1987).

———. 1982. On the impossible pilot wave. *Foundations of Physics* 12: 98–99. Reprinted as Ch. 17 of *Speakable and unspeakable in quantum mechanics*. Cambridge: Cambridge University Press (1987).

———. 1987. Are there quantum jumps? In *Schrödinger. Centenary celebration of a polymath*, ed. C.W. Kilmister, 41–52. Cambridge: Cambridge University Press. Reprinted as Ch. 22 in *Speakable and unspeakable in quantum mechanics*. Cambridge: Cambridge University Press (1987).

Beller, Mara. 1999. *Quantum dialogue: The making of a revolution*. Chicago: University of Chicago Press.

Belousek, Daniel. 1996. Einstein's 1927 unpublished hidden-variable theory: Its background, context and significance. *Studies in History and Philosophy of Modern Physics* 27 (4): 437–461.

Bloch, Felix. 1976. Heisenberg and the early days of quantum mechanics. *Physics Today* 29 (12): 23–27.

Bohm, David. 1952. A suggested interpretation of the quantum theory in terms of "hidden" variables I and II. *Physical Review* 85: 166–193. Reprinted in: Wheeler, J.A., and W.H. Zurek, eds. 1983. *Quantum theory and measurement*, 356–396. Princeton: Princeton University Press.

———. 1980. *Wholeness and the implicate order*. Routledge.

Bohr, Niels. 1928. The quantum postulate and the recent development of atomic theory. *Nature* 121(suppl.): 580–590. Reprinted in: Wheeler, J.A., and W.H. Zurek, eds. 1983. *Quantum theory and measurement*, 87–126. Princeton: Princeton University Press.

———. 1949. Discussion with Einstein on epistemological problems in atomic physics. In *Albert Einstein: Philosopher-scientist, vol. 1*, ed. P.A. Schilpp

Born, Max. 1926. Zur Quantenmechanik der Stoßvorgänge. *Zeitschrift für Physik* 37(12): 863–867. English translation, On the quantum mechanics of collision processes. In Wheeler, J.A., and W.H. Zurek, eds. 1983. *Quantum theory and measurement*, 52–61. Princeton: Princeton University Press.

Born, Max, Werner Heisenberg, and Pascual Jordan. 1926. Zur Quantenmechanik, II. *Zeitschrift für Physik* 35: 557–615. English translation, On quantum mechanics II. In *Sources of quantum mechanics*, ed. B.L. van der Waerden, 321–386. Amsterdam: North-Holland Publishing (1968).

Bricmont, Jean. 2016. *Making sense of quantum mechanics*. Springer.

Brown, Harvey R., and David M. Wallace. 2005. Solving the measurement problem: de Broglie–Bohm loses out to Everett. *Foundations of Physics* 35: 517–540.

Clauser, John F., Michael A. Horne, Abner Shimony, and Richard Holt. 1969. Proposed experiment to test local hidden-variable theories. *Physical Review Letters* 23 (15): 880–884.

Cushing, James T. 1994. *Quantum mechanics: Historical contingency and the Copenhagen hegemony*. Chicago: University of Chicago Press.

de Broglie, Louis. 1924. *Recherches sur la théorie des quanta*. Ph.D. thesis, Université de Paris. *Annals de Physique* 3(10): 22–128 (1925). English translation, de Broglie, L. 2004. On the theory of quanta. *Foundation of Louis de Broglie* ed.

Deutsch, David. 1985. Quantum theory, the Church-turing principle, and the universal quantum computer. *Proceedings of the Royal Society of London A* 400: 97–117.

DeWitt, Bryce S. 1970. Quantum mechanics and reality. *Physics Today* 23 (9): 30–35.

DeWitt, Bryce S., and Neil Graham. 1973. *The many-worlds interpretation of quantum mechanics*. Princeton: Princeton University Press.

Dürr, Detlef, Sheldon Goldstein, and Nino Zanghí. 1992. Quantum equilibrium and the origin of absolute uncertainty. *Journal of Statistical Physics* 67: 843–907.

Einstein, Albert. 1905. Über einen die Erzeugung und Verwandlung des Lichtes betreffenden heuristischen Gesichtspunkt (On a Heuristic Point of View about the Creation and Conversion of Light). *Annalen der Physik* 17(6): 132–148. English translation, in: Ter Haas, D., ed. 1967. *The old quantum theory*, 91–107. Pergamon Press.

———. 1919. What is the theory of relativity? *The London Times*. Reprinted in Einstein, A. 1982. *Ideas and opinions*, 227–232. New York: Crown Publishers, Inc.

———. 1953. Elementare überlegungen zur Interpretation der Grundlagen der Quanten-Mechanik. In *Scientific papers presented to Max Born on his retirement from the Tait Chair on Natural Philosophy in the University of Edinburgh*, 33–40. Edinburgh: Oliver and Boyd. English translation, "Elementary Considerations on the Interpretation of the Foundations of Quantum Mechanics," https://arxiv.org/abs/1107.3701.

Einstein, Albert, Boris Podolsky, and Nathan Rosen. 1935. Can quantum-mechanical description of physical reality be considered complete? *Physical Review* 47: 777–780. Reprinted in: Wheeler, J.A., and W.H. Zurek, eds. 1983. *Quantum theory and measurement*, 138–143. Princeton: Princeton University Press.

Everett, Hugh III. 1957. Relative state formulation of quantum mechanics. *Review of Modern Physics* 29: 454–462. Reprinted in: Wheeler, J.A., and W.H. Zurek, eds. 1983) *Quantum theory and measurement*, 315–323. Princeton: Princeton University Press.

Fine, Arthur. 1986. *The shaky game: Einstein, realism and the quantum theory*. Chicago: University of Chicago Press.

Freedman, Stuart J., and John F. Clauser. 1972. Experimental test of local hidden-variable theories. *Physical Review Letters* 28 (938): 938–941.

Ghirardi, GianCarlo, Alberto Rimini, and Tullio Weber. 1986. Unified dynamics for microscopic and macroscopic systems. *Physical Review D* 34: 470.

Gleason, Andrew M. 1957. Measures on the closed subspaces of a Hilbert space. *Indiana University Mathematics Journal* 6 (4): 885–893.

Heisenberg, Werner. 1927. Ueber den anschaulichen Inhalt der quantentheoretischen Kinematik and Mechanik. *Zeitschrift für Physik* 43: 172–198. English translation, The physical content of quantum kinematics and mechanics. In Wheeler, J.A., and W.H. Zurek, eds. 1983. *Quantum theory and measurement*, 62–84. Princeton: Princeton University Press.

———. 1949. *The physical principles of the quantum theory*. New York: Dover.

Hermann, Grete. 1933. *Determinismus und Quantenmechanik*. manuscript. Dirac Archives DRAC 3/11.

———. 1935. Die naturphilosophischen Grundlagen der Quantenmechanik. *Abhandlugen der Fries'schen Schule* 6 (2): 75–152.

Holland, Peter. 2005. What's wrong with Einstein's 1927 hid-den-variable interpretation of quantum mechanics? *Foundations of Physics* 35 (2): 177–196.

Howard, Don. 1990. 'Nicht sein kann was nicht sein darf,' Or the prehistory of EPR, 1909–1935. In *Sixty-two years of uncertainty*, ed. A.I. Miller, 61–111. New York: Plenum.

———. 2004. Who invented the 'Copenhagen interpretation'? A study in mythology. *Philosophy of Science* 71: 669–682.

———. 2007. Revisiting the Einstein-Bohr Dialogue. *Iyyun: The Jerusalem Philosophical Quaterly 56. Special Issue Dedicated to the Memory of Mara Beller*, 57–90.

———. 2013. Quantum mechanics in context: Pascual Jordan's 1936 Anschauliche Quantentheorie, chapter 11. In *Research and pedagogy: A history of quantum physics through its textbooks*, ed. M. Badino and J. Navarro. Berlin: Edition Open Access.

Jammer, Max. 1974. *The philosophy of quantum mechanics*. New York: Wiley.

Jauch, Josef M., and Constantin Piron. 1963. Can hidden variables be excluded in quantum mechanics? *Helvetica Physica Acta* 36: 827.

Kochen, Simon, and Ernst Specker. 1967. The problem of hidden variables in quantum mechanics. *Journal of Mathematics and Mechanics* 17: 59–87.

Kocsis, Sacha, Boris Braverman, Sylvain Ravets, Martin J. Stevens, Richard P. Mirin, L. Krister Shalm, and Aephraim M. Steinberg. 2011. Observing the average trajectories of single photons in a two-slit interferometer. *Science* 332: 1170–1173.

Maudlin, Tim. 2011. *Quantum non-locality and relativity*. 3rd ed. Blackwell.

Ney, Alyssa. 2021. *Finding the world in the wave function*. New York: Oxford University Press.

Nikolić, Hrvoje. 2012. EPR before EPR: A 1930 Einstein-Bohr thought experiment revisited. *European Journal of Physics* 33: 1089–1097.

Norsen, Travis. 2017. *Foundations of quantum mechanics: An exploration of the physical meaning of quantum theory*. Springer.

Philippidis, Chris, Christopher Dewdney, and Basil J. Hiley. 1979. Quantum interference and the quantum potential II. *Nuovo Cimento B* 52: 15–28.

Pusey, Matthew F., Jonathan Barrett, and Terry Rudolph. 2012. On the reality of the quantum state. *Nature Physics* 8: 475–478.

Przibram, Karl. 1967. *Letters on wave mechanics: Schrödinger, Planck, Einstein, Lorentz*. New York: Philosophical Library.

Schrödinger, Erwin. 1926. Quantisierung as Eigenwertproblem (Quantization as an Eigenvalue problem) *Annalen der Physik* 4(79): 361–376; 489–527; 80, S.: 437–490; 81, S: 109–140. Reprinted in Schrödinger, E. 1927. *Abhandlungen zur Wellenmechanik*, Johann Ambrosius Barth. English translation, Schrödinger, E. 1928. *Collected papers on wave mechanics*. Glasgow.

———. 1935. Die gegenwärtige situation in der Quantenmechanik. *Naturwissenschaften* 23(48): 807–812. English translation, "The present situation in quantum mechanics," reprinted in: Wheeler, J.A., and W.H. Zurek, eds. 1983. *Quantum theory and measurement*, 152–167. Princeton: Princeton University Press.

Shimony, Abner. 1993. *The search for a naturalistic world view, vol. 1. Scientific method and epistemology, vol. 2, natural science and metaphysics*. Cambridge: Cambridge University Press.

van der Waerden, Bartel L., ed. 1968. *Sources of quantum mechanics*. Amsterdam: North-Holland Publishing.

von Neumann, John. 1932. *Mathematische Grundlagen der Quantenmechanik*. Berlin: Springer. English translation, *Mathematical foundations of quantum mechanics*. Princeton: Princeton University Press (1955).

Wigner, Eugene. 1961. Remark on the mind-body question. In *The scientist speculates*, ed. I.J. Good, 248–302. Heineman. Reprinted in: Wigner, Eugene. 1967. *Symmetries and reflections*, 171–184. Bloomington: Indiana University Press. Reprinted in: Wheeler, J.A., and W.H. Zurek, eds. 1983. *Quantum theory and measurement*, 168–181. Princeton: Princeton University Press.

———. 1963. The problem of measurement. *American Journal of Physics* 31: 6–15. Reprinted in: Wigner, Eugene. 1967. *Symmetries and reflections*, 153–170. Bloomington: Indiana University Press. Reprinted in: Wheeler, J.A., and W.H. Zurek, eds. 1983. *Quantum theory and measurement*, 234–341. Princeton: Princeton University Press.

Zeh, H. Dieter. 1970. On the Interpretation of measurement in quantum theory. *Foundations of Physics* 1: 69–76. Reprinted in: Wheeler, J.A., and W.H. Zurek, eds. 1983. *Quantum theory and measurement*, 342–352. Princeton: Princeton University Press.

Chapter 14
How the History of Thermodynamics Informs the Hidden Variables Debate in Quantum Mechanics

Adam Forrest Kay

Abstract The main purpose of this paper is to inform all those interested in quantum foundations about several very striking historical parallels between modern quantum theory and thermodynamics in the 1800s. These parallels deserve to be more widely known as they comment on the status of the hidden variable debate. Some general observations about science and some reasons to believe in deterministic hidden variables are also given.

Keywords Quantum foundations · Thermodynamics · Hidden variables · Bell's theorem · Hydrodynamic quantum analogues

14.1 The Status of Hidden Variables

In October 2022, the Nobel Prize was awarded in the field of quantum foundations. It was shared between Alain Aspect, John Clauser, and Anton Zeilinger, 'for experiments with entangled photons, establishing the violation of Bell inequalities and pioneering quantum information science'.

The citation went into further detail about why we should care about the Bell inequalities. According to the Nobel Committee (Outreach, 2022):

> In the 1960s, John Stewart Bell developed the mathematical inequality that is named after him. This states that if there are hidden variables, the correlation between the results of a large number of measurements will never exceed a certain value.

Quantum theory predicts the correlation exceeds this certain value, and the laureates were honored because their experiments vindicated the quantum prediction. In other words, hidden variables do not exist.

A. F. Kay (✉)
Department of Mathematics, Massachusetts Institute of Technology, Cambridge, MA, USA
e-mail: afkay@mit.edu

P. Castro et al. (eds.), *Advances in Pilot Wave Theory*, Boston Studies in the Philosophy and History of Science 344,
https://doi.org/10.1007/978-3-031-49861-9_14

The award was predictably followed by a blizzard of popular science explainers, both articles and videos, informing us that 'The Universe is Not Real', and asking 'Has quantum mechanics proved that reality does not exist?'

The Nobel announcement is puzzling for anyone even moderately conversant with the quantum foundations literature, but also unsurprising. It is indeed common to find inaccurate comments along these lines in the physics community. Maudlin (2014) gives the example of *Physics World*, the official magazine of the Institute of Physics, which posted a video less than 100 s long which purports to explain Einstein's qualms with quantum mechanics. The crux of the argument comes at the end, where the narrator (Martin Archer) states:

> people who have tried to make so-called hidden variables theories, which means that there is sort of a secret proper state that we don't know about, can't seem to do it. Bell's theorem actually says it's impossible. So I'm afraid, Einstein, it looks like God does play dice.

As Maudlin rightly remarks, this displays 'complete incomprehension of what Bell did, what Einstein thought, what the situation with respect to deterministic "hidden variables" theories is'.

Those who have read slightly more about the foundational issues here know that the common view is flatly wrong. This brings us to the second, more subtle error involved in discussions of Bell's theorem. It is quite common by now to make the observation above, and correct it with citations from both Bell and Einstein, regarding the issue of locality and *spukhafte Fernwirkung*, spooky action at a distance. Maudlin does this in the paper just cited, as does Bricmont in his book (Bricmont, 2016). Wiseman (2014) is aware of the choice, and frames it in a way that explains somewhat why the Nobel committee could blithely state that hidden variables do not exist:

> As Bell proved in 1964, this leaves two options for the nature of reality. The first is that reality is irreducibly random, meaning that there are no hidden variables that 'determine the results of individual measurements'. The second option is that reality is 'non-local', meaning that 'the setting of one measuring device can influence the reading of another instrument, however remote'.
>
> Most physicists are localists: they recognize the two options but choose the first, because hidden variables are, by definition, empirically inaccessible.

It is certainly wrong to claim hidden variables are 'by definition' empirically inaccessible. One can of course define hidden variables in this way, but that is to define oneself out of the discussion. Later in this essay we will prove this. But that is not the relevant mistake. The real mistake is in the identification of two core assumptions.

As the above quote makes clear, the common line is to consider the assumptions of reality and locality. Then, we deny hidden variables *tout court*, or allow them and say that Bell's inequality proves the existence of 'spooky action at a distance', but in a way that does not break relativity, since information cannot travel faster than the speed of light. This view is supported by Bohmians because David Bohm's 1952 pilot-wave theory is explicitly nonlocal. Indeed, this position was supported by Bell himself, who admired both Bohm and de Broglie. This fact is ironic, considering the opinion of most physicists that Bell 'disproved hidden variables'.

While reality and locality are certainly the two most obvious assumptions, there are others. The assumption of measurement independence (associated with the freedom of experimenters to alter measurement settings) was also known to Bell, and has been the focus of some recent work (Hossenfelder and Palmer, 2020).

But there is also a fourth assumption of *statistical compatibility*, interesting because deeply hidden and unknown to most people. Bell's inequalities are, in fact, a particularly simple case of a type of inequalities that have been studied by mathematicians for over 150 years.

> Abstracted from any physical assumption, Bell type inequalities have a long history, beginning in mid-nineteenth century. They have important applications in various branches of mathematics and mathematical physics: Combinatorial theory, probability theory, propositional logic, the theory of computational complexity, the Ising spin model and neural networks, and of course, the foundations of quantum mechanics (Pitowky, 1989).

Such inequalities were first explored by George Boole in his work on probability theory in *The Laws of Thought* (Boole, 1854). He was the first to show that, given three random variables x, y, z and their pairwise joint distributions $P(x, y)$, $P(y, z)$, $P(z, x)$, the joint probabilities must satisfy certain inequalities in order to even be possible. If the three joint probabilities violate Boole's inequalities, they violate what Boole called 'the conditions of a possible experience'. In the example he gave the Royal Society, drawn from probabilities of three symptoms of a disease occurring in pairs, Boole said 'the evidence is contradictory' (Boole, 1862).

For example, consider the joint probabilities given in Vorob'ev (1962):

$$P(x = 1 \cap y = 1) = P(x = 0 \cap y = 0) = 0.5$$
$$P(x = 1 \cap z = 1) = P(x = 0 \cap z = 0) = 0.5$$
$$P(y = 1 \cap z = 0) = P(y = 0 \cap z = 1) = 0.5$$

Examination will show that all three of these conditions cannot be simultaneously satisfied. What this means mathematically is that the probability $P(x, y, z)$ cannot be assigned a non-negative value consistent with the joint probabilities given, and thus cannot be formed. There is no way of connecting the data given in the joint probabilities into a single probability space. In this case the random variables are said to be *incompatible*.

Boole's problem has attracted the attention of mathematicians for 150 years. In 1936, the Italian mathematician Carlo Bonferroni worked on it, and managed to generalize the conditions to an arbitrary number of random variables. These are called the Bonferroni inequalities. In the 1940s the French mathematician René Fréchet wrote two books on the subject. In 1955 another French mathematician, Jean Bass, proved some necessary conditions for the existence of a probability space of three random variables (Bass, 1955). Then in 1959, five years before Bell published his epochal paper, the Russian mathematician Nikolai Vorob'ev generalized the result of Bass to any number of random variables (Vorob'ev, 1962). Many other mathematicians have contributed and work in the field continues to this day.

As has been shown, largely independently by several different authors, Hess (2005), Khrennikov (2008), and Nieuwenhuizen (2009), the Bell inequalities are closely related to the Boole inequalities, and are a special case of the theorem of Jean Bass, obtained in 1955. The true significance of the Bell inequalities can therefore be seen to be a result of classical probability theory that were first explored before Bell was even born.

According to these authors, the violation of Bell's inequality implies an inability to construct a single probability space for hidden variables that can connect the pair-wise joint probabilities of measurement angles used to formulate the theorem. The physical significance of the statistical incompatibility is attributed to contextuality in a sense more broadly construed than that generally associated with the Kochen-Specker theorem. The measurement apparatus introduce fundamentally different measurement environments, or contexts, which condition the results, generating statistical incompatibiliy (Khrennikov, 2008, 2009).

If these ideas of incompatible contexts can be more broadly developed and turn out to be sound, it would explain the violation of Bell's inequalities not as evidence that reality does not exist, or that entangled particles undergo spooky action at a distance. Rather, all dynamics are locally real but the measurement apparatus plays a fundamental role.

This view is all the more plausible, considering two deeply inter-related facts. First, it is well known that quantum mechanics 'lacks a physical description of the measurement process' (Hossenfelder and Palmer, 2020), and indeed that the two standard postulates of quantum mechanics (evolution according to Schrödinger's equation and the collapse postulate) are not even mutually consistent. This is the measurement problem. Something is up with measurement.

Second, quantum theory does not include any time dynamics on the Compton scale, which is inconceivably far from human scales and consequently very hidden. For all we know, such dynamics may exist. Indeed there are reasons for thinking they do. Already in 1925 de Broglie was talking about 'the oscillation so intimately tied to the very existence of matter' (de Broglie, 1925) and recent hydrodynamic analogs of quantum phenomena are very close to this picture. Bush and Oza, in a review of work in the field, observed that '[de Broglie's] physical picture suggested that many of the quantum mysteries may be rooted in an unresolved dynamics on the Compton scale, the scale of particle vibration' (Bush and Oza, 2021).

For example, the analog of quantum corrals in Harris et al. (2013) involves an emergent statistics identical to that given by quantum theory, but on a macroscopic time scale visible to the naked eye. The bouncing frequency in these experiments is roughly 50 Hz, whereas the Compton frequency is 1.6×10^{21} Hz. The wavelength of the probability distribution of particle position within the droplet corral is 4.75×10^{-3} m, whereas the statistical wavelength of an electron in a quantum corral is on the order of 10^{-9} m. In their report on the hydrodynamic corral, the authors conclude

> Whether or not the statistical description provided by quantum mechanics represents a complete description of physical reality was the subject of the celebrated debate between Einstein and Bohr. Whatever the case may be in quantum mechanics, the linear statistics is clearly an incomplete description of our fluid system and is underlaid by a complex, nonlinear, pilot-wave dynamics.

The existence of the droplet experiments, alongside the discrepancy of scale, makes it very reasonable to suppose that some physical, deterministic dynamics, currently unknown, are giving rise to the statistics. If this is the case, we are simply living through another period of scientific confusion, many of which have occurred before (I will describe one in detail shortly). This is the reasonable, natural conclusion, far more likely than invoking the same discrepancy of scales to support the many-worlds interpretation, which is what Spyridon Michalakis did in a recent Vox interview (Abad-Santos, 2022):

> Let's say we only perceive 100 frames per second, something like that. We can be aware of our lives and choices we make, but then the frame rate of the universe where you could be flickering between different timelines is 40 orders of magnitude above that.

If one is going to invoke an enormous discrepancy of scale, why use it to support many worlds when the same idea also can support hidden dynamics?

We conclude that the significance of the experiments done by the laureates is completely misunderstood by the very committee that awarded them the most coveted prize in the world. Maybe the experiments prove that there are no hidden variables and therefore we should be antirealists about core physical variables like position and momentum. Or, maybe they prove that the hidden variables are nonlocal, and that physical systems can affect each other instantaneously over arbitrary distances. Or, just maybe, they prove that the particle and the measurement device have some hidden, currently unimagined interaction which results in the hidden variables depending on the measurement apparatus in such a way as to make the correlations measured in Bell tests statistically incompatible.

There may yet be a classical explanation for the correlations observed in Bell experiments. I personally believe this is the case, and that the current accepted view that 'Bell disproved hidden variables' is an enormous canard. However well-deserved this Nobel prize may be, the reasons for its bestowal could well be viewed by posterity as the nadir of confusion reached in the twenty-first century regarding the true meaning of atomic theory. In other words, I am hopeful that this is as bad as it will get. Various developments in quantum foundations, including hydrodynamic quantum analogues, work on context dependence, and the unprecedented ease of finding information, make me optimistic that we will soon escape what Carver Mead called 'the dark ages of theoretical physics' (Mead, 2002).

In this essay, I would like to present another, historical reason to be optimistic. There is a deep and thoroughgoing parallel between thermodynamics in the nineteenth century and quantum physics in the twentieth century. The parallel is so striking that it is a great mystery to me why it is not more widely known.

To fully adumbrate my conclusions, we should believe that hidden variables will one day be discovered in the atomic world. This is for two reasons. First, the main formal and cultural characteristics of quantum theory in the twentieth century are basically identical to those of the theory of heat transfer in the nineteenth century. Second, hidden variables were eventually discovered in the theory of heat.

Here are the main parallels:

1. There were multiple interpretations of the nature of heat, just as there are multiple interpretations of quantum mechanics;
2. Positivists explicitly denied hidden variables would ever be found to explain the heat equation, just as they did in the twentieth century (and today) in quantum theory;
3. Both theories are statistical;
4. The heat equation and the Schrödinger equation are extremely close, differing only by a factor of i.

14.2 The Debate Over Heat

In 1804, when Fourier did his original researches into heat flow, there were two theories of the nature of heat. The dominant theory was that heat is a *type* of matter, an element alongside hydrogen and oxygen. This special matter manifested as an invisible fluid, called 'caloric'. Its properties were: it could not be created or destroyed, was self-repulsive, was very light or weightless, and was so subtle that it permeated the finest pores of matter.

The alternative theory held that heat was a *state* of matter. It was caused by the 'insensible motions' of the matter, insensible in that we could not feel the motion of the writhing iron the way we can feel a writhing snake. We can only feel that motion in the form of heat. Newton, Hooke, Boyle and Locke were all of the view that heat is motion. However, the theory was hard to develop, and none of these men could describe the motions better than Boyle's 'various, vehement, and intestine commotion of the parts among themselves' (Boyle, 2000).

Caloric was dominant because it was simply a better theory. For example it had a story 'explaining' specific heats. It was a surprise when Black discovered in the 1760s that if he took two material samples of equal volume and temperature, but different nature, there was no clear rule determining how much heat would be required to raise both samples by the same temperature. For example, contrary to both intuition and what the theory of vibration would seem to imply, it took much more heat to raise a milliliter of water from 100°F to 101°F than it did to raise the same amount of mercury from and to the same temperature. Yet mercury is over 13 times heavier than water.

The calorists explained this by saying that caloric had an 'affinity' for different types of matter. It was attracted to and clung to mercury with much greater ease than water. Thus, heat flow in mercury was accelerated. Humphrey Davy noted that this was no explanation at all, merely begging the further question of what caused this selective attraction. Still, it was better than the vibrationists could do, and the interplay of fluid particles which were self-repulsive yet variously attracted to all other matter gave the caloric theory a rich range of possibilities.

Another phenomenon the vibrational theory had trouble explaining is what we today call adiabatic cooling. A Frenchman named Gabriel Jars had noticed something in a Hungarian mine in 1758: if air compressed by a column of water over 40 metres high was suddenly allowed to escape, 'compact ice was deposited on any object placed in the way of the issuing air'. Cleghorn was able to 'explain' this phenomenon with caloric:

> because, with the air... now being rarified, the repulsion between the fire particles was diminished and so fire flowed out of the thermometer.

The caloric theory became highly articulated and complex, for example in the work of the chemist and doctor Joseph Marie Socquet. Similar to how ice, fluid, and vapor are three different states of water, Socquet thought that caloric was also highly dynamic, and could exist in a state of composition, capacity, temperature, or radiance. Each one of these had a different behavior that interfaced in a complicated way with the observed properties of heat and matter.

Caloric also served as scaffolding for attacking serious problems in mathematical physics. Laplace, probably the most famous of the calorists, used the theory to resolve an old discrepancy between the predicted and observed speed of sound that had foiled Newton, Euler, Lambert, Lagrange, both Bournoullis and d'Alembert. Laplace and Biot found a solution that stands today, and while the caloric has been removed from its account, it inspired their ideas and helped crystallize them. As Roller estimated (Roller, 1961):

> Indeed, as a conceptual scheme serving to correlate most of the facts then known about heat, and as a convenient way to think about these facts or even to predict new ones, [caloric theory] was proving itself to be far more useful than the opposing and relatively undeveloped doctrine that heat is a form of motion.

Today, of course, we know that the vibrationists were right. Heat is motion. Or, more accurately, the average of molecular kinetic energy. And we admire those who rebelled and dissented against the majority view of their time. Count Rumford, Humphrey Davy and Thomas Young come to mind.

There are two takeaways here. The wrong theory was more scientific for a long time, but more importantly, there were multiple interpretations of the nature of heat.

Fourier, after several struggles, finally managed to publish his *Théorie Analytique de la Chaleur* in 1822, and it was soon recognized as a landmark of mathematical physics. But the heat equation, since it only described the flow of heat through time, was completely incapable of deciding between the caloric and the vibrational theories. Naturally, both sides agreed on the facts of heat flow. Because Fourier's equation described these facts, both sides could have claimed it for their own. The math gave a rule that worked, but for mysterious reasons. To clear up the mystery of why the math worked, the heat equation needed an interpretation.

I need not develop the modern parallel since everyone knows that the Schrödinger equation has many interpretations. Just like in the 1800s, today we have a differential equation that works for mysterious reasons.

14.3 Positivism

Another striking parallel has to do with the importance of postivism in both theories. The term was invented around 1832 by August Comte, probably the first philosopher of science. His basic idea was that there are three types of knowledge, each with its corresponding mode of cognition. He called these types the theological, metaphysical, and positive. We would today, without much loss, call these religious, philosophical, and scientific. Theology is associated with the mode of imagination, metaphysics with disputation, and positivism with demonstration.

This relatively mundane distinction between religion, philosophy and science took on dynamic life with the further claim, central to Comte's entire program, that the positivistic type of knowledge was the most mature, important, and final one. Below the positivistic, inferior and antecedent to it, was the metaphysical, and below that again, the most primitive and worst, was the theological.

Comte called this the 'Law of Three Stages'. He believed that all fields of human endeavor, be they physics, biology, or politics, went through these stages of thought in order, and that human beings themselves went through these stages as they matured (Comte, 1896):

> Now, each of us is aware, if he looks back upon his own history that he was a theologian in his childhood, a metaphysician in his youth, and a natural philosopher in his manhood.

This is where positivism's famed, and distinctive, rejection of metaphysics comes from. Comte saw metaphysics as a mere transition from theological fantasies to scientific realities. And he saw in Fourier the ideal modern exponent of his ideals. Comte dedicated his *Cours de Philosophie Positive* to Fourier, and never missed an opportunity to admire him, mentioning him several times in the same breath as Newton.

Positivistic ideas were in the air in France during the 1820s and 30s, and had many sympathizers, but as *The Oxford Handbook of The History of Physics* points out (Buchwald and Fox, 2013), Fourier was

> an intellectual leader of the younger French physicists who staged a silent revolt against Laplacian physics, an important part of which was a positivistic indifference or hostility toward Laplacian hypothesizing about microphysical forces and structures.

Fourier could provide an antidote to Laplacian hypothesizing because Fourier restricted himself only to the mathematical expressions of phenomena, without inquiring into their 'hidden natures'. The very beginning of his book is practically a one-sentence manifesto of positivism:

> Primary causes are not at all known to us; but are subject to simple and constant laws, which we may discover by observation, and the study of which is the object of natural philosophy.

The book, in 639 pages of mathematical exposition, does not mention the word caloric a single time. To Comte, this was a very beautiful thing because the 'true nature' of heat was a metaphysical fiction, an atavistic spectre of theology. Nothing could come of discussing primary causes and hidden natures but endless dispute, as the debate between the calorists and vibrationists over the last hundred years had demonstrated. In one passage, Comte praised Newton's theory of gravitation, and then in the same breath, looking for one more outstanding example of true knowledge, Comte said:

> I will chose the beautiful series of researches by M. Fourier on the theory of heat... In this work, which is so eminently positive in its philosophical character, the most important and most precise laws of thermology find themselves unveiled, without the author inquiring one single time into the hidden nature of heat. Without him mentioning, except to indicate emptiness, the so agitated controversy between the partisans of the material caloric and those who made heat consist in vibrations of a universal ether.

Comte admired Fourier for restricting his scope to the flow of heat alone, which he called 'restraining himself in the research to an entirely positive order'. Doing otherwise, and attempting to approach the hidden nature of heat, would be to 'throw himself into inaccessible problems'. They are inaccessible not because we are not ready for them, but because the human mind can never know them. As Fourier said in his opening sentence: primary causes are not at all known to us. Attempting to approach primary causes with hypotheses just ends in 'interminable discussions', because this is an attempt to make pronouncements on 'questions necessarily insoluble for our intelligence' (Comte, 1831).

Now let us turn to the positivism of quantum mechanics. It has long been observed that the Copenhagen interpretation has something positivistic about it. The very views on atomic physics that displeased Einstein so much were, ironically, grounded in the success of special relativity. With various frequency, Bohr, Heisenberg, Bohm and Pauli all stressed the success Einstein had by following Mach's views and abandoning the material ether as a metaphysical fiction. Indeed, Heisenberg himself took this approach in his first paper on matrix mechanics, which begins with a Fourier-esque line

> The objective of this work is to lay the foundations for a theory of quantum mechanics based exclusively on relations between quantities that are in principle observable.

According to Heisenberg in *Physics and Beyond*, Einstein recanted his own early positivism and warned Heisenberg that the positivism of his new quantum mechanics would 'one day get you into hot water' (Heisenberg, 1971). Heisenberg did not listen. A couple years later, in his famous paper discussing the uncertainty relations, Heisenberg's concluding thoughts leave no doubt whatsoever. He writes (Heisenberg, 1927):

> In view of the intimate connection between the statistical character of the quantum theory and the imprecision of all perceptions, it may be suggested that behind the statistical universe of perception there lies hidden a 'real' world ruled by causality. Such speculations seem to us—and this we stress with emphasis—useless and meaningless. For physics has to confine itself to the formal descriptions of the relations among perceptions.

This view was also held by Bohr, who wrote in 1927 (Heilbron, 1988):

> In our description of nature the purpose is not to disclose the real essence of the phenomena, but only to track down, as far as it is possible, the relations between the manifold aspects of our experience.

The positivism of the Copenhagen interpretation, and consequently all of quantum physics for the past 100 years, has been repeatedly stressed by modern scholars. Arthur Fine, for example, wrote that 'where it really matters Bohr invariably lapses into positivist slogans and dogmas' (Fine, 1986). Mara Beller said that 'the Copenhagen interpretation… rests on two central pillars—positivism' being the first, and Christian de Ronde speaks of the 'Bohrian-positivist alliance' (de Ronde, 2021).

Nor was this connection obvious only in retrospect. The Copenhagen interpretation was understood even by contemporaries to be positivistic. Schrödinger wrote a letter in 1940 to Arthur Eddington, explaining that he was unimpressed by 'alleged grand philosophical revelations' created by quantum physics, calling Heisenberg and his cohort the 'enthusiastic champions of that new positivistic outlook' (Schrödinger, 1940).

Other physicists, working slightly later, also observed the same dynamic. David Bohm characterized the quantum orthodoxy as follows (Bohm, 1986a):

> somehow people were turning a method of calculation into an explanation of reality. They wanted to say that their method of calculation corresponded to reality. It was part of the current positivist approach.

This retreat from explanation into positivistic exclamations about the limitations of knowledge is what caused Einstein (2011a) to say about the Copenhagen spirit that 'this epistemology-soaked orgy should come to an end'.

14.4 Statistical Description

Finally, let us consider the statistical nature of both theories. The question Comte thought was 'insoluble to human intelligence' was solved a few short years after his death. In fact, the groundwork for this solution was laid a century before.

In his *Hydrodynamica*, published in 1738, Daniel Bernoulli explained Boyle's law as follows: the pressure of the air on the container is the sum force of the molecules hitting the container walls. Reducing the volume of a container increases the pressure on the walls because it increases the number of collisions. Increasing the volume has the opposite effect: the molecules have more space to roam and so there are fewer collisions between themselves and the walls. Rather than air particles being somehow springy, as Boyle had imagined, or hard to compress because of caloric's self repulsion, they contained a certain amount of *vis viva*, or living force, which was entirely dependent on their motion and how much they weighed. It was this living force, which is two times what we today call kinetic energy, that was pressing back on the container walls.

Bernoulli's kinetic theory of gas structure was not particularly influential for a hundred years after it came out. Most people thought that gas molecules adopted a lattice formation due to long-range repulsive forces between them, and then did not move unless disturbed. This was called the *static* theory of gases.

The German scientist Rudolf Clausius was one of the founders of thermodynamics and the creator of the term 'entropy'. By 1850 he was convinced the kinetic theory of gases was true, but he did not publish anything about it until 1857. Less than a year later, an objection was published by Buys-Ballot, a Dutch chemist and meteorologist. Paraphrasing, he asked that if these molecules are going so fast,

> How then does it happen that tobacco-smoke, in rooms, remains so long extended in immovable layers? How can clouds of smoke exist in open air? How could carbonic acid gas remain so long in an open vessel?

In short, why does it take gases of a different kind so long to mix? Clausius was able to answer this question, again with Bernoulli's idea of molecules colliding into each other all the time. They only manage to move in straight lines for a little while before they collide with another molecule and rebound. This straight-line motion was called by Clausius the *free path*, and he was able to compute a probability distribution of free paths, depending on the properties of the gas.

The diffusion properties of gases turn out to shed considerable insight into the diffusion properties of heat. If heat is the mean molecular kinetic energy of the particles making up a bar of iron, we can use Bernoulli's ideas to understand *why* heat behaves the way it does. The kinetic energy in a system will, over time and through an unimaginable number of collisions, spread evenly. If two neighboring regions of iron have different temperatures, then the probability distributions of speeds must be different. But because they are neighboring, they will interact with each other. The aggregates with a lower average speed will start going faster as they steal kinetic energy from the aggregates with a higher average speed, which of course start going slower. The system will ultimately seek a single probability distribution of speeds. If the distribution does not change from point to point in the iron bar, everywhere must have the same temperature. This is called thermal equilibrium, and no heat is flowing.

We have, in fact, just discovered why the heat equation works. We have solved Comte's 'inaccessible problem', and uncovered the hidden nature, the primary cause, of why temperature behaves as it does. This is exactly what we expect to happen as knowledge advances. A differential equation may be mysterious at first, but it must work for a reason. In time, this reason will be uncovered and lead into surprising places. As understanding of heat matured and combined with other investigations, it became thermodynamics. This resulted in much more than an understanding of heat. It yielded insights of a fundamental character, into energy, entropy, information, work, life, causality, and time.

Now, bringing the discussion back to quantum mechanics, the statistical nature of its predictions is characteristic of the theory. The debate about completeness *is* the debate about whether the quantum story will repeat the story of heat or not. From the very beginning, many authors made the comparison between the quantum statistics and those of thermodynamics. For example, when he began to consider the possibility of hidden variables, von Neumann wrote (1932):

> Explanations by means of hidden parameters have (in classical mechanics) reduced many apparently statistical relations to the causal foundations of mechanics. An example of this is the kinetic theory of gases... Whether or not an explanation of this type, by means of hidden parameters, is possible for quantum mechanics is a much discussed question.

Similar quotes are to be found the writings of Einstein, Born, de Broglie and Bohm, among others. When Einstein said 'God does not play dice', what he meant is that the statistical character of quantum mechanics is no different than that of any other statistical theory. In other words: statistical theories emerge from averaging over unknown variables within deterministic theories. When he argued with Bohr over the completeness of quantum mechanics, the argument was about whether or not future physics would remain statistical. Bohr (1949) admitted as much in his reconstruction of the long debate:

> The question at issue has been whether the renunciation of a causal mode of description of atomic processes involved in the endeavours to cope with the situation should be regarded as a temporary departure from ideals to be ultimately revived or whether we are faced with an irrevocable step...

And Einstein (2011b) on this subject is characteristically eloquent:

> It seems certain to me that the fundamentally statistical character of the theory is simply a consequence of the incompleteness of the description... It is rather rough to see that we are still in the stage of our swaddling clothes, and it is not surprising that the fellows struggle against admitting it (even to themselves).

This is the essence of the question behind hidden variables: Will the statistics of quantum mechanics turn out to be like those of heat? And we can see now that those physicists who assume hidden variables are 'by definition' or 'in principle' never observable, are simply committing the same error of circularity that has been identified in arguments for completeness since the 1960s. In Bohm's words (Bohm, 1957),

> the conclusion that there is no deeper level of causally determined motion is just a piece of circular reasoning, since it will follow only if we assume beforehand that no such level exists.

If hidden variables are never observable, then quantum mechanics will turn out to be complete. But to *assume* they are never observable is to settle the question through the way it is asked, a clear *petitio*. If hidden variables *are* uncovered eventually, those who claimed they would never be found will have made a mistake identical to Comte's, when he claimed the hidden nature of heat would never be known.

The similarities uncovered so far are perhaps suggestive, but are not particularly compelling. One could imagine these parallels arising by coincidence. But there is one more we have not mentioned, the strongest paralell of them all, which sets all the others into context and makes it seem extremely unlikely that we are dealing with coincidence.

14.5 Nearly Identical Equations

We restrict ourselves to one spatial dimension for simplicity without any loss of generality. Writing u_t for $\partial u/\partial t$ and u_{xx} for $\partial^2 u/\partial x^2$, the heat equation and the Schrödinger equation are respectively:

$$u_t = u_{xx} \qquad\qquad u_t = iu_{xx}.$$

Here u is temperature, or the wavefunction. The Schrödinger equation differs from the heat equation by a factor of $i = \sqrt{-1}$ on the right hand side.

Such mathematical similarities are very suggestive but we need only note the profound differences in the behavior of differential equations created by a mere difference in sign to understand that even the slightest change in an equation must be carefully examined. We can come to a more clear understanding by eliminating the requirement for imaginary numbers.

Writing the wavefunction $u = f + ig$ where both f and g are real functions, a simple algebraic transformation allows us to break the Schrödinger equation into a coupled partial differential equation:

$$g_t = +f_{xx}$$
$$f_t = -g_{xx}.$$

The heat equation states that the change of temperature in time is given by the spatial laplacian of the temperature (the curvature, in 1D). If a point in space is below the average of its surrounding points, its laplacian will be positive, meaning the change of temperature in time will be positive. In plain English: if a point is cooler than its surroundings, it will get hotter with time. Conversely if a point is hotter than its surroundings, it will get cooler. Concentrations of heat will tend to spread out or diffuse.

The two coupled functions f, g that form Schrödinger's equation are almost like this, but rather than depending on themselves, the way heat does, they depend on each other. Not only that, the sign is flipped in the second equation, which means that it models a *concentration* rather than a diffusion process. Thus quantum theory discusses some sort of phenomenon which, were we to imagine it as physical, involves a tug-of-war between a diffusion and a concentration process.

The closeness of the two differential equations gives us quite a strong reason to believe that hidden variables will one day be discovered in quantum mechanics, and that the many parallels between the story of heat and the story of the quantum are more than mere coincidence. To this formal parallel we can add another, pointed out by Hossenfelder. The Schrödinger equation in density matrix form is 'remarkably similar to the classical Liouville equation' (Hossenfelder and Palmer, 2020), which is known as the fundamental equation of statistical mechanics. The correspondence is very close:

$$\rho_t = \{H, \rho\} \qquad\qquad i\rho_t = [H, \rho].$$

Here ρ is the classical or quantum probability density, H the Hamiltonian, the $\{\}$ are Poisson brackets and the $[\,]$ is the commutator. Again, we have the quantum case including the factor of $i = \sqrt{-1}$. Noting that the Liouville equation is linear in ρ despite its underlying dynamics having no such requirement, Hossenfelder remarks:

> To us, this close formal similarity between the two equations strongly suggests that quantum physics, too, is only the linear probabilistic description of an underlying non-linear deterministic system.

In both the heat and Liouville equations we have extremely close mathematical relationships between classical and quantum theories. Both are statistical. We know that the classical theory has a deterministic underpinning. Whether or not the quantum theory does as well is the entire question of hidden variables.

14.6 Conclusions

It is not a surprise that positivistic sentiments arise around statistical theories. The primary investigative method of positivism is to clear the path for short-term formal developments through a rejection of deeper investigations. With the sole exception of quantum mechanics (and some toy counterexamples), statistical theories are *all* characterized by averaging over unknown quantities. Positivism has a fraught relationship with unknown quantities, being inclined to reject them as 'metaphysical' and therefore unscientific.

No positivistic approach would have a problem with moving from clear quantities into statistical statements about their ensembles. But historically, positivism has resisted the opposite direction whenever science did not know what quantities gave rise to the statistics. The primary example of this is the Maxwell-Boltzmann distribution and how it determines the heat equation. If Mach were alive today, he would admit the existence of atoms, because they can be observed. But he denied their existence when they could not be observed. The catch is that if people like Bernoulli, König, Clausius, Maxwell, Boltzmann and so on followed Mach's principle of thought economy and did not allow themselves to imagine unobservable quantities, the progress of science would have been severely restricted and atoms would have remained unobserved. Just like ignorance itself, this capacity to strengthen its own limitations is what makes positivism dangerous.

If Comte had had his way, everyone would have forgotten about the 'empty' and 'completely insoluble' debate between the caloric and vibrational theories. They would have renounced such questions and applied themselves to 'real' problems, which positivism alone could identify. And we know now that this would have been a complete disaster, freezing progress indefinitely until people rebelled against Comte's sentiment that the true nature of heat was unknowable. Happily, the disaster was averted. Many people worked on both the caloric and vibrational hypotheses. The caloric theory burned itself out just as the vibrational theory was accelerating. Eventually the combined work of many investigators proved not only that the vibrationists were right about heat, but also that Comte and Fourier were dead wrong about primary causes. Eventually, the nature of heat and the meaning of the heat equation were specified down to the smallest detail.

The positivistic disaster *did* happen in the case of quantum mechanics. At first this was not felt, since the quantum theory was a 'horn of plenty' in Bell's words, and supported decades of formal development. But eventually, because the rejection of hidden variables and the denial of primary causes became the accepted view, physics entered a *cul de sac*. The proposed solutions (Copenhagen, many worlds, etc.) are so outlandish because the obvious solution is off the table by assumption.

How would Einstein view the situation today? To him, it was 'certain' that quantum mechanics was incomplete. To a mind fixed with this certainty, it would be clear that the defining characteristic of our current theoretical climate is that of bad taste in theory construction. It would be clear that Carver Mead's diagnosis of a 'dark age of theoretical physics' is not far off the mark because serious thinkers are forced to sit at the table with those who say things like: 'I will pound the table and insist that the wave function is the sum total of reality' (Carroll, 2019).

This has happened because the physics establishment is convinced, as the Nobel citation demonstrates, that hidden variables do not exist. The conviction acts as a kind of mold, softening the hard wood of physical intuition. It led to the von Neumann-Wigner 'consciousness' induced collapse of the wave function, which introduced subjectivism and led to Bertrand Russell lamenting 'the shadow physics of our time' (Bricmont, 2016). It has led to modern physicists citing the Buddhist philosopher Nagarjuna, a practice that Bell also lamented ('It is irresponsible to suggest that technical features of contemporary theory were anticipated by the saints of ancient religions...by introspection' Bell, 1987). It has led to the Everettian multiverse, an anti-empirical theory that reads like a practical joke on the credulous and a cost paid by the field for lacking philosophical integrity.

Those who have followed the history of the Bohr-Einstein debate in its origin and as it was pursued by later exponents know just how weak the arguments for completeness always were. It is a long story of vague and questionable philosophizing, the puffing up or misreading of weak proofs, authoritarian pronouncements (what Beller called a 'rhetoric of inevitability'), dogma, hero worship, and groupthink. Causal alternatives, like de Broglie's double solution, were not disproved but simply ignored. This history is a for another time, but my experience has been that the more one learns about the actual arguments that were made for completeness, the less well-founded completeness appears.

How to interpret the quantum formalism is no great mystery. It 'seems certain', to use Einstein's parlance once again, that it will one day be understood as a kind of statistical mechanics of an underlying causal description. The real mystery is why the quantum formalism was taken to be so authoritative. Those afflicted with the Copenhagen spirit really did think it was *endgültig*, the end of the road, in Heisenberg's words. This has created enormous confusion. Why would anyone simply assume that a statistical theory had no causal underpinning? It is hard to find the words to express how entirely unnatural that is. Both Bohm and Bell shared this basic intuition. Bohm said (Bohm, 1980):

> I have never been able to discover any well-founded reasons as to why there exists so high a degree of confidence in the general principles of the current form of the quantum theory.

And Bell, speaking of the theory's supposed inescapable indeterminism, said (Bell, 1990):

> When I read history, it always amazes me that the founding fathers could come so confidently to that conclusion. Up to then the tradition in physics was that if something was unpredictable it was because it could not be fully controlled. And for me the natural assumption would be [the traditional one].

The caloric theory of heat was dominant for a hundred years before it was swept away entirely. Bernoulli had the right idea a hundred years before the kinetic theory of gases was developed. This is actually the common situation and there are many other examples: extramission, comets, phlogiston, miasma, geocentrism, d'Alembert's paradox and spontaneous generation come to mind. Confusion measured in centuries is completely normal in the history of science.

Particles are small and fast and complicated, and there is zero reason to believe that we uncovered all their secrets, much less 'the sum total of reality', with a statistical theory discovered when people were still using the telegraph. Bell's theorem is impressive, and has greatly advanced the debate, but it has not shown that we are forced to choose between unreality and nonlocality. We could, in fact, have subtle dependence on measurement, which makes all the sense in the world.

There can never be a full proof that some future knowledge will not sweep away hidden assumptions and reveal to us an atomic theory of inner logic and beauty that is as inevitable and perfect as it is classical and deterministic. As we work towards advancing the frontiers of science within a culture of quantum completeness, disrespect of the weight-bearing columns of the temple of knowledge, ignorance of the past, and anti-philosophical mathematization, we may take courage from Bell's own words (Bell, 1987):

> Long may the spirit of Louis de Broglie inspire those who believe that what is proven by impossibility proofs is lack of imagination.

References

Abad-Santos, A. 2022. *Everything Everywhere All at Once, explained by a quantum physicist.* Retrieved from https://www.vox.com/culture/23024945/everything-everywhere-all-at-once-multiverse. Accessed on December 4, 2022.

Bass, J. 1955. Sur la compatibilité des fonctions de répartition. *Comptes Rendus 240*: 839–841.

Bell, J.S. 1987. *Quantum Mechanics*. Cambridge: Cambridge University Press.
Bell, J.S. 1990. *Indeterminism and Nonlocality*. CERN: Center of Quantum Philosophy of Geneva. Retrieved from https://www.youtube.com/watch?v=3l9HtG-VZCU&t=10s
Bell, J. 1993. *The Ghost in the Atom*, ed. P.C.W. Davies and J.R. Brown. Cambridge: Cambridge University Press.
Bohm, D. 1957. *Causality and Chance in Modern Physics*. London: Routledge and Kegan Paul.
Bohm, D. 1980. *Wholeness and the Implicate Order*. London: Routledge.
Bohm, D. 1986. *Interview of David Bohm by Maurice Wilkins*. Niels Bohr Library & Archives, American Institute of Physics. Retrieved from www.aip.org/history-programs/niels-bohr-library/oral-histories/32977-4
Bohr, N. 1949. Discussion with Einstein on epistemological problems in atomic physics. In *Albert Einstein: Philosopher Scientist*, ed. P.A. Schilpp. New York: MJF Books.
Boole, G. 1854. *The Laws of Thought*. Walton & Maberly.
Boole, G. 1862. On the theory of probabilities. *Philosophical Transactions of the Royal Society B London 152*: 225–252.
Boyle, R. 2000. *The Works of Robert Boyle*, ed. M. Hunter and E.B. Davis. London: Pickering & Chatto.
Bricmont, J. 2016. *Making Sense of Quantum Mechanics*. Berlin: Springer.
Buchwald, J., and Fox, R. 2013. *The Oxford Handbook of the History of Physics*. Oxford: Oxford University Press.
Bush, J.W.M., and Oza, A.U. 2021. Hydrodynamic quantum analogues. *Reports on Progress in Physics* 84: 1–44.
Carroll, S. 2019. *Something Deeply Hidden: Quantum Worlds and the Emergence of Spacetime*. London: Penguin Press.
Comte, A. 1830. *Cours de Philosophie Positive*, vol. 1. Paris: Bachelier.
Comte, A. 1831. *Cours de Philosophie Positive*, vol. 2. Paris: Bachelier.
Comte, A. 1896. *The Positive Philosophy of August Comte*, ed. H. Martineau (Trans.). London: George Bell and Sons.
de Broglie, L. 1925. Sur la fréquence propre de l'électron. *Comptes rendus hebdomadaires des séances de l'Académie des sciences, t. 180*: 198–450.
de Ronde, C. 2021. Mythical thought in bohr's anti-realist realism (or: Lessons on how to capture and defeat smoky dragons). https://api.semanticscholar.org/CorpusID:251741583
Einstein, A. 2011a. Letter to Erwin Schrödinger: Correspondence with H. A. Lorentz, Max Planck, and Erwin Schrödinger, ed. K. Przibram. Philosophical Library/Open Road.
Einstein, A. 2011b. *Letters on Wave Mechanics*, ed. K. Przibram. New York: Open Road.
Fine, A. 1986. *The Shaky Game: Einstein, Realism, and the Quantum Theory*. Chicago: University of Chicago Press.
Fourier, J. 1822. *Théorie Analytique de la Chaleur*. Paris: Firmin Didot.
Fox, R. 1971. *The Caloric Theory of Gases: From Lavoisier to Regnault*. Oxford: Clarendon Press.
Harris, D.M., Moukhtar, J., Fort, E., Couder, Y., and Bush, J.W.M. 2013. Wavelike statistics from pilot-wave dynamics in a circular corral. *Physical Review E 88*(011001): 1–5.
Heilbron, J.L. 1988. The earliest missionaries of the Copenhagen spirit. In *Science in Reflection: The Israel Colloquium: Studies in History, Philosophy, and Sociology of Science*, ed. E. Ullmann-Margalit, pp. 201–233. Dordrecht: Springer Netherlands.
Heisenberg, W. 1925. Quantum-mechanical reinterpretation of kinematic and mechanical relations. *Zeitschrift für Physik 33*: 879–893. https://doi.org/10.1007/BF01328377
Heisenberg, W. 1927. Über den anschaulichen inhalt der quantentheoretischen kinematik und mechanik. *Zeitschrift für Physik 43*: 172–198.
Heisenberg, W. 1971. *Physics and Beyond: Encounters and Conversations*. New York: Harper & Row.
Hess, P.W.K. 2005. The bell theorem as a special case of a theorem of bass. *Foundations of Physics 35*: 1749–1767.
Hossenfelder, S., and Palmer, T. 2020. Rethinking Superdeterminism. *Frontiers in Physics 8*: 139. https://doi.org/10.3389/fphy.2020.00139

Khrennikov, A. 2008. Epr-bohm experiment and bell's inequality: quantum physics meets probability theory. *Theoretical and Mathematical Physics 157*: 1448–1460.
Khrennikov, A. 2009. *Contextual Approach to Quantum Formalism.* Berlin: Springer.
Maudlin, T. 2014. What Bell did. *Journal of Physics A: Mathematical and Theoretical 47*: 42.
Mead, C. 2002. *Collective Electrodynamics: Quantum Foundations of Electromagnetism.* Cambridge: MIT Press.
Narasimhan, T.N. 1999. Fourier's heat conduction equation: History, influence, and connections. *Reviews of Geophysics, 37*(1): 151–172.
Nieuwenhuizen, T.M. 2009. Where bell went wrong. *AIP Conference Proceedings 1101*(1):127–133. https://doi.org/10.1063/1.3109932
Outreach, N.P. 2022. *Press Release.* Retrieved from https://www.nobelprize.org/prizes/physics/2022/press-release/
Pitowky, I. 1989. From george boole to john bell - the origins of bell's inequality. In *Bell's Theorem, Quantum Theory and Conceptions of the Universe*, ed. M. Kafatos, pp. 37–49. Berlin: Springer.
Roller, D. 1961. *The Early Development of the Concepts of Temperature and Heat: The Rise and Decline of the Caloric Theory.* Cambridge: Harvard University Press.
Schrödinger, E. 1940. Letter to A. S. Eddington: Correspondence with H. A. Lorentz, Max Planck, and Einstein, Albert, ed. K. Przibram. Philosophical Library/Open Road.
von Neumann, J. 1932. *Mathematical Foundations of Quantum Mechanics.* Princeton: Princeton University Press.
Vorob'ev, N. 1962. On the theory of probabilities (N. Greenleaf, Trans.). *Theory of Probability and Its Applications* 7(2): 147–163.
Wiseman, H. 2014. Bell's theorem still reverberates. *Nature 510*: 467–469.

Chapter 15
Form, Function, and Continuity: Knowledge on the Move

Gildo Magalhães

Abstract During historical processes, transformations are collectively carried out over a certain time span, and when examined over a long period they appear to be an ongoing development, without breaks. Nature's processes likewise exhibit an apparent continuity. However, undulatory descriptions in physics have unveiled how aspects of continuity can be associated with discontinuities. "Pilot" or "empty" waves have opened up new windows to explain hitherto mysterious phenomena. The author is convinced that such approach can be extended to other domains of natural history, such as fluid mechanics, chemistry, geology, cosmology, and others. A particularly fruitful application of this approach can be found in the biological field. Although few contemporary biologists continue to support a genetically functional, i.e., causal, morphology, the evolution of form cannot be isolated from the development of function, both displaying the paradoxes between continuity and discontinuity. Emphasis on a certain teleology is not only unavoidable, but actually it is recommendable to bridge the discontinuity gap, as applicable to various sciences. This text is an effort to notice how entangled certain historical and epistemological problems can become.

Keywords Form · Function · Continuity · Evolution · Restricted teleology

15.1 Introduction: Continuity x Discontinuity

In historical processes, transformations are collectively established, generally over a long period. For example, the complex transition from feudalism to mercantile capitalism is only understood in the relatively long term, since elements of a new mode of production can be detected in the fast urbanization displayed during earlier parts of European Middle Ages, which later resulted in the foundation of

---

G. Magalhães (✉)
University of São Paulo, São Paulo, Brazil

P. Castro et al. (eds.), *Advances in Pilot Wave Theory*, Boston Studies in the Philosophy and History of Science 344,
https://doi.org/10.1007/978-3-031-49861-9_15

commercial banks, like Monte dei Paschi di Sienna (1472). This does not prevent the recognition that a greater diffusion of commercial capitalism happened only afterwards, with the help of technical innovations displayed in the great Portuguese and Spanish navigations, and other factors as well.

There are different levels in the apparent continuity of historical developments. Social and technical innovations, such as mentioned in the previous paragraph, and which may appear as suddenly irrupting in the scenario, represent themselves the materialization at a given moment of already ongoing, and sometimes much older, trends. Inventions and novelties may spread quickly or less rapidly, such as the plow with iron core, or the mills and forges that replaced a significant part of the manual work during the European Middle Age, or book printing with movable types, or the formation of political parliaments, and many others. The same controversy between continuity and discontinuity is attached to concepts usually discussed in the history of science and technology, such as "scientific revolution" or "industrial revolution". Would such changes be smooth and gradual, or on the contrary abrupt, causing jumps and sudden disruption between the old and the new? Perhaps the best alternative is not a clear-cut borderline. There is continuity, in the sense that nothing ever comes out of sheer nothing, there are always roots in the past for anything. On the other hand, there is discontinuity, because there are events with such powerful implications, that life is significantly affected by them. Consider, for example, a mixture of continuity and discontinuity as displayed by the Protestant Reformation in the 1520's, breaking away from the Roman Church, yet keeping many of its articles of faith and practices.

It is interesting to note that such problems of transition appear associated to every historical evaluation of events and, in the particular case of the history of science, it was illustrated by historian George Sarton with an image of the progress of science across time represented by a continuous ramp. However, according to this author, if we could get closer, and inspect a shorter timespan we would see that such a continuous line does not exist, but the straight ramp resembles instead a stairway with steps. Looking closer and closer, each stair rung also becomes a sequence of smaller rungs, and so on (Sarton 1956, 20–23).

On the other hand, moving on now from humanities to natural sciences, we live within a tradition in which it is generally accepted that there are no leaps in nature, that is, the development of natural processes over time can be, for the most part, represented by a continuous function. It is as if our minds seem to resist discontinuity. In several instances, an alternative procedure is called upon, i.e., maybe this impression of continuity would be nothing but a problem of scale: at the human level we do not perceive such discontinuities.

It is intriguing then to ask whether continuity is purely illusory. Is a natural process metaphorically like a straight line formed by points – and is this only a mathematical idealization, like a function limit? It might be recalled that this comparison introduces us into the mathematical challenge of the continuum problem, when this quest was in essence deepened in the second half of the nineteenth century with Georg Cantor's notion of transfinite sets. It has become common ground to think that an imaginary continuous line (without beginning and end) must encompass all real numbers. Even if this is true, how can it be reconciled with our picture of the physical universe, where there are no ideal points or lines made up of points corresponding to all real numbers? The question is even more intricate, since there is no proof of the particular type of infinity associated to such a line, although we can prove that points forming a real continuum are non-countable. We do not intend to pursue this problem farther in the present text, as it is not our concern here, but interested readers are referred to an abundant bibliography on the subject of the mathematical continuum – see for example the bibliography cited in (Magalhães 2013).

In the physical nature, one usually assumes that any material movement can be described by a function that associates a variable like the space occupied by an object to another variable as time. Intuitively we assume that these concepts of time and space are continuous, so that an object performing the movement will pass through all "points" in a given space, and all corresponding "instants" in time, but we do not know whether space and time are really continuous at all levels of description, or if that is not just another simplification, valid for practical purposes at our human scale. How are then change processes and their discontinuities to be understood in the realm of nature? The riddles posed by movement to ancient Greek philosophers are still an epistemological problem today.

One can rightly argue that mankind is part of the nature, therefore it is not expected that things should be drastically different in natural history from what was already hinted at for human sciences in the case of historical events, which can be analyzed as a mixture of continuity and discontinuity. Certainly, there may be peculiarities in terms of the decision power, or degrees of freedom, with which humans are endowed, as contrasted to non-human phenomena.

The question that we believe to be relevant at this moment is perhaps metaphysically grounded, although it bears an empirical character as well: Why would nature not proceed continuously in its transformations? In favor of the usual continuous human outlook, we should remember that continuity equations have been very useful in science, as they are linked to aspects of optimization principles that have been proposed to better understand nature (Magalhães 2012). On the other hand, discontinuities usually appear when analyzing changes in chemical, physical, biological, geological, cosmological, and other natural domains. For example, quantum theory, which was introduced starting at the end of the nineteenth century, came to modify the conceptual framework ingrained with continuity. This change came about based on theorizations about the so-called black body radiation, later extended to all radiations of electromagnetic energy, and finally to everything that is usually called matter and energy. It began thus to be clear that nature does exhibit a discontinuous, "quantum", character, at least in terms of some entities usually encountered in physics and chemistry theories.

However, the torrid disputes around how to interpret and to justify ambiguities related to the new quantum theory soon appeared on the stage, and a non-causal attitude, adopted by the predominant Niels Bohr's Copenhagen School interpretation, led to an epistemology rooted in probabilistic explanations for quantum phenomena. To counter this trend, physicist Louis de Broglie proposed that every physical phenomenon could be explained in terms of waves, which in turn admitted a concept called "pilot" or "empty" waves (also called "theta waves", in the Lisbon School terminology). Consequently, a kind of pilot wave (the continuous "mother-wave") has been proposed as the headquarter and propagator of singularities in the sub-quantum environment, and hence discontinuities. Such waves, acting collectively, realize potentialities (or trends) as discontinuities that may manifest themselves individually, but not with the same intensity and quality of their collective behavior – for a more comprehensive description of the Lisbon School of quantum physics, see (Croca and Araújo 2010).

Even the interior collectivities that make up bodies directly observable at our human scale can demonstrate the importance of the joint collective behavior in common transition phenomena, such as freezing water. Transitions are therefore not an individual issue, but a collective one, in which the whole itself transforms the phenomenon and covers it with a non-linear action – as can be seen, for example, in the well-known problem of linear indeterminacy of the gravitational attraction of three bodies, which appears with an even increased complexity for a higher number of constituents.

Let us assume at this point that discontinuities are somehow associated with the wave, rhythm nature of matter/energy in the universe. Recently, there has been increased evidence that some visible macroscopic phenomena are linked to waves, with a change of phase achieved in order to collectively adjust behaviors, so that different collectivities harmonize with each other. Two puzzling examples of this are the "quantum" activation of light emission by a large number of fireflies in forests, and the auto-synchronization of oscillating metronomes on a rigid rolling platform.[1] Another similar demonstration of rhythm is provided by the phenomenon of electromagnetic pendulum coupling, as a particular illustration of this principle of collective behavior (Doubochinski and Tennenbaum 2008). Hence, continuity and discontinuity would be just different manifestations of a single process of a higher order, evidencing the organization of matter and energy.

Elsewhere, we have endorsed and developed the argument that in the Universe there is a non-deterministic tendency to spontaneously form order and complexity, which over historical time has translated as formation of more permanent structures, such as chemical elements, molecules, inorganic and organics bodies, interstellar and stellar matter, constellations, etc. (Magalhães 2022). This is also understood as a tendency for each being to continue existing, even if it means transforming and creating new structures, overcoming the opposite tendency towards disorder. In this perspective, living beings emerged, and then organized themselves in a constant and collective manner, in such a way that, when examined for a sufficiently long duration, there seems to be a causal purpose in this organization, or a certain degree of teleology, alongside some degree of freedom to escape rigid determinism.

[1] Cf. vídeos shown at https://www.youtube.com/watch?v=Aaxw4zbULMs, and https://ncase.me/fireflies

Because of the maximization of the suggested trend towards organization displayed by life, we will now turn to this specific area of natural history, namely biology. In the processes of biological evolution, paleontological discontinuities used to be attributed to gaps in the fossil record. Later, this opinion changed, after "saltationism" evolution was proposed.[2] The implication was that evolution does not produce species differentiation in a gradual (continuous) fashion, but rather there are evolutionary "jumps", at least considering geological time scales.

It seems that this riddle of biological discontinuity has much to do with the problem of the appearance of forms favorable to specific functions, or in other words, the tension between changes in morphology as well as transformations in physiology of living beings. As a matter of fact, in our understanding, and as hinted above, the same question also arises at other observational levels, and not limited to biology. For example: returning to quantum physics, in the subquantum world, why do "particles" (or "akrons", in the Lisbon School terminology) concentrate energy, while waves are in charge of its propagation? Another example: why do crystals have the geometric structures we know? Or also: why do atoms have their structure concentrated in nuclei, but are prone to form assemblies in the form of chemical molecules through bonds outside the nuclei? And we could continue, on and on, with such questioning, noting once again that all transitions arising from these examples represent discontinuities.

We prefer, as said before, from this point on to dwell somewhat more deeply in the biological area, because it lends itself particularly well to discussing the above questions in a way that illuminates other areas, as evolution has continued to combine forms with functions in an intriguing way. For this reason, in this text, we will examine how morphogenesis was particularly examined by three twentieth-century authors who thought outside the dogmatic orthodoxy "box" that has dominated developmental biology. To enforce the endurance of the discussions, and to somehow dissipate what might be considered as a "whiggish" outlook in History, we will present these authors out of the chronological order, going from the more recent to the older works. It should also be noted, naturally, that the tension between function and form of course appeared much earlier than the few authors that we will briefly revisit here.

[2] Richard Goldschmidt defended a saltationism interpretation of evolution already in 1940. Niles Eldredge and Stephen Jay Gould proposed a variant of this idea, as an alternative to phyletic gradualism, with their evolution theory by "punctuated equilibria" in 1972. Gould's reasoning has been exposed in his famous books and articles. Particularly interesting is his application of discontinuity as a valid explanation for major geological transformations in "The godfather of disaster" (Gould 1992).

15.2 The Rebellion of Forms

The Spanish scientist Jorge Wagensberg (1948–2018) analyzed the question of evolution from a quite original point of view, considering some forms preferentially exhibited and used by Nature.[3] This biophysicist associated the function of different natural forms with a mathematical formulation of their performance in terms of evolution, arriving at an innovative result connected with the idea of progress. Starting with information theory concepts, he established a mathematical identity to express the capacity of a being to continue existing.[4] Translated into common language, this identity states that progress would then be verifiable for a being – that is, a living being –, if along its natural history, the species' capacity to survive increases, in spite of the environment, or in other words, the species evolves. According to Wagensberg, to solve this problem nature favored evolution in multiple scales through nine fundamental forms, following a typology of functions applicable to living beings, but in certain cases also to inanimate beings (Wagensberg 2002 and 2004).

One starting point for his discussion is the shape of the circle. The circular form is connected with the functions of rotation, generation, defense, and minimal energy dissipation. Associated with the flat circle, its rotation around the diameter produces the sphere, which protects and contains the maximum volume for a minimum surface, a property which is suitable for beings relatively fixed in space. Examples of application in this way are a variety of fruits, as well as organs such as the eye, and fundamental generation cells, as the ovum. Oncolites also enable the isotropic growth of microorganisms around rounded corners in successive layers, and spherical symmetries are abundant in Nature. Planets and stars are approximately spherical, as are treetops. The circumference in nature can deform into an ellipse, for example due to the superposition of an aerodynamic function, as in the cross-section of fish, or turn into cylinders or conical shapes, such as tree trunks.

A second form considered fundamental is the hexagon, which can serve the function of compactly paving a surface, that is, without leaving gaps, as for example in the bee's honeycomb, which is made up of hexagonal columns, saving wax and space. Other examples of application of this form can be noticed on the top of tortoise shells and in insect eyes. If the available space is not flat, but a sphere, the best paving is achieved with pentagons surrounded by five hexagons, as illustrated by the official soccer ball.

The third identified form, and a quite ubiquitous one, is that of the segment, whose function has been to locate or to connect elements, as for example tree branches do. The segment also enables displacements of a living being or its

[3] Wagensberg created a whole science museum ("La Caixa") in Barcelona that impressively illustrates all of his concepts regarding evolution.

[4] In its core, this idea is related to the notion of *conatus*, as used by the philosopher Baruch Spinoza in the seventeenth century.

extensions, like the limbs, and it serves as a measurement standard, as well as a resistance element, as for example in the web of spiders. Associated with the segment, we find the angle, which can be visualized as the bending of a segment around a vertex, with the functions of pointing, gathering, concentrating, penetrating. Examples of these functions are the spines of plants and trees, or the force concentration to ease penetration, as exemplified by the teeth of sharks and other predators, or also the force distribution of ribs in relation to the spine bone.

A fourth natural form is that of the spiral, which allows the function of dense packing, as exemplified by the tails of various animals, such as the chameleon and the seahorse, or the rolled snouts of insects such as the moth. We can also mention in this respect, that another associated function for the spiral is that of permanent growth, maintaining an original space-saving shape, which occurs within many seashells, but can also be identified in other natural structures.[5]

What could be imagined as a spatial transformation of the spiral combined with a translation movement is the helix, which can be considered a fundamental fifth form, due to its natural importance. This shape has the additional function of gripping, and also of compacting or intensifying the grip, as in the coiled tail of various arboreal monkeys. A similar function can also be verified in helical trajectories of seeds falling from trees, which increases the probability of being carried away by the air to new territories. Helices are abundant in various vegetal forms in dense rain forests. In addition to Wagensberg's examples of compacting, we remind that the spiral is present also in DNA packaging using the double helix.

The parabola represents a sixth form. It serves to emit, receive, concentrate on a focus, as in a parabolic mirror or an antenna. One example of this function is the bat's spatial sensing system of reflected waves in its parabola-shaped auditory organ.

The catenary is a seventh form, it functions to hold. For example, a suspended vine accommodates the shape of a linear mass density to its own weight, a property also displayed by spiders' web lines. The catenary revolution around an adequate axis generates a minimum-area surface, which is the form assumed by a water and soap film limited by two circles. Wagensberg stresses also the inverted catenary arcs extensively used, for example, by architect Gaudi, due to its resistance capacity to withstand the weight of the material from which it is constructed, without collapsing.

An eighth fundamental form is that of the wave, whose function is to displace, diffuse, communicate. For instance, the fish and reptiles move by producing transverse waves with their bodies, water mammals swim according to vertical transverse waves, and worms move according to longitudinal waves More generally, the wave is fundamental also in the inorganic world, because waves make it possible

[5] It is noteworthy that spirals conform to the concept mathematically identified as the Golden (or Divine) Section, studied and intensively used as an aesthetic function by ancient Greeks, and rediscovered in the European Middle Ages and Renaissance. The ubiquity of this proportion has been extensively shown in nature – see, for example (Huntley 1970).

to move information without displacing matter. We have already highlighted the significance of waves in forming various natural rhythms.

Finally, a ninth fundamental form is the fractal, a shape that was only more recently described in mathematical terms. The fractal is capable of filling, subdividing and colonizing a given space, with the property that the whole has parts that in turn look exactly like the very whole, and those parts subdivide in similar parts, etc. Examples of fractal elements are as varied as lightning discharges, cauliflower heads, fern leaves, the set of branches in trees and in turn their secondary branches, the circulatory system of vessels in animals and plants; and neurons also form a fractal web in the brain.

Of course, there may be more natural shapes than those fundamental ones sketched by Wagensberg. It is interesting that a deeper understanding of forms and their functions can thus be achieved by starting to answer a basic question: how did living beings succeed in face of a hostile and dangerous environment, and achieved progress in their survival? However, in this matter Wagensberg assumed the orthodox, Darwinian standpoint: for him, all this progress was achieved by hazard, and gradually incorporated into life. We also observe that Wagensberg's work, although very helpful, does not enter into the problem of continuity and discontinuity, perhaps because he also adhered to another probabilistic description, the orthodox interpretation of quantum physics.

15.3 Growth and Form

Although Wagensberg was influenced in his reflections on form and function by a classic work, that of the Scottish biologist D'Arcy Wentworth Thompson (1860–1948), this was an author who in turn did not believe in Darwinism as a valid explanation for evolution (Thompson 1992).[6] Thompson focused on symmetries, mathematical proportions and transformations of living beings, to explain morphogenesis. He highlighted the initial impact he felt when learning the idea discussed by Immanuel Kant (1724–1804), who stated that the living organism is already the *telos*, or the final cause, of its own generation and development process, thus playing the role of both an "architect" (planner) and a "builder" (executor) of a plan. I remind that this is not far away from what philosophers like Friedrich von Schelling (1775–1854) and his circle later advanced for the formulation of a philosophical movement, the so-called German *Naturphilosophie.*

Thompson agreed, in particular, with the Aristotelian concept of "entelechy" for the living being, as had been famously reaffirmed by German zoologist Hans Driesch (1867–1941), that the embryo incorporates a strongly epigenetic (vitalist)

[6] First published in 1917, Thompson had the opportunity to revise the book for its second edition in 1942.

feature of causality and intentionality (Driesch 1908).[7] This idea also shows an influence of *Naturphilosophie*, implying that vital functions and teleology are intertwined, and outside the scope of purely mechanist explanations – life cannot be understood with recourse to non-life. One should remember that German *Naturphilosophie* diverged from Enlightenment ideology in this respect. Jean Le Rond d'Alembert (1717–1783), for example, and for that matter those philosophers inclined to mechanist views in general, denied that there are "final causes" or teleological ones.[8] This position prevented from looking for any cause directly relating growth and form, for the function appears to be satisfied only by a whimsical result of blind chance. The "why" is thus abandoned in favor of "how" evolution proceeds.

We will not dwell here in the extensive and detailed mathematical description of the many animal species that Thompson carried out very aptly using mathematics, showing how function and form are so deeply entangled that the final results do work out magnificently. It is important to note, as Thompson registers, that the evolution as narrated by Charles Darwin in *The Origin of Species* is rather disappointing, because with his explanation Darwin failed to account for the branches' formation in the "tree of life" presented by him.

Thompson remarked that it looks as if there would be infinitely small continuous variations in the evolution of species. Why then are there so many gaps in continuity? This is not due to the widely acknowledged imperfection of the fossil record. The argument, he pointed out, is that a barrier must be crossed to move from family to family, without which there is no longer the possibility of direct comparisons between them. How do birds descend from reptiles, mammals from earlier quadrupeds, or in turn quadrupeds from fish, invertebrate vertebrates, echinoderms, and mollusks from coelenterates, or even one group of protozoa from another?

If one cannot go from invertebrates to vertebrates, or from a coelenterate to a worm, Thompson concluded, it could partly be because there is a "principle of discontinuity", inherent in all classification systems, whether in mathematics, physics or biology. This principle becomes evident when considering the discontinuities, as revealed, for example, by the lines of the spectrum of the chemical elements, or by the families of crystals, or by Dalton's atomic law, or by the table of chemical elements itself.

The forceful conclusion of Thompson is that this is not only due to mathematical description, but Nature itself proceeds in leaps, whether in the inorganic or organic world, without recourse to the many intermediate steps that would necessarily

---

[7] This was presented by Driesch in his Gifford Lectures of 1907–1908, which caused a great impact on natural scientists.

[8] Entry "Final Causes" – Portuguese translation (D'alembert 2015, 45–48). I have argued elsewhere, in opposition to such reductionism, that various formulations of optimization principles seek to understand nature, and its causal organization, denying that their existence must be relegated to simple accidental chance.

approach continuity. Variations and mutations are in practice discontinuous, against the mathematical idealization of continuity.

15.4 Form and Function

Slightly before Thompson wrote his best-known book, there appeared another work by his contemporary, the Scottish biologist and fish specialist Edward Stuart Russell (1887–1954). This was dedicated to a detailed historical reassessment of natural history treaties, almost all from the nineteenth century, showing how scientists explained the genesis of function (physiology) combined with genesis of form (morphology), especially when changes in form occurred that affected the way living beings work (Russell 1916). Russell was in favor of Lamarckian heritability, and followed a holistic approach, influenced by German *Naturphilosophie*, and specially by Goethe. We will not discuss the bulk of Russell's comprehensive account, which is outside the scope of this text, but will concentrate on a few main points of interest here.

In the *Metamorphosis of Plants* (1790), Johann Wolfgang von Goethe (1749–1832) remarked that we try to understand things by separating them into parts. It is possible to separate a living organism into its structural elements, but we cannot recompose it or give it life by putting those parts back together. The study of morphology can reveal the internal meaning of the parts constituting the organism, but without approaching it in a fixed way, except as a transitory phase of such constitution. The form itself is of interest only as a manifestation of the living being's internal activity. For Goethe there is a "constitutive force" that carries the "idea" of the organism. Living beings strive to manifest this idea, sometimes this looks to us as if they were conveying art-like works produced by nature. This conception is valid not only for an individual in formation, but for the whole set of living organisms (this is the so-called "unity of plan", applicable to all types of life).

The French naturalist Étienne Geoffroy Saint-Hilaire (1772–1844) believed in evolution following a kind of discontinuity (similar to what is referred as "saltationism") and maintained that historically there was only one fundamental type of animal. He postulated transformations that would reduce several animals to vertebrates (a "unity of composition", added to unity of plan), disregarding function compared to form, thus favoring exclusively a morphological perspective. The unity of plan and composition would result from a force exerted at the atomic level.

Saint-Hilaire's countryman Georges Cuvier (1769–1832), who pioneered the areas of comparative anatomy and paleontology, on the contrary, was adept of a teleological transformation in which function would take precedence over form. It is instructive to compare Saint-Hilaire and Cuvier in terms of evolution. An evolutionary theory of the type defended by Saint-Hilaire is similar to the proposal that Charles Darwin (1809–1882) would later make, starting in 1859, i.e., one that assumes that variation occurs by chance and not as a direct response to environmental changes. For Saint-Hilaire, form determines function, adapting existing material to meet new demands, but for Cuvier, it is function that determines form and creates new organs in response to emerging functional needs.

The leading English biologist Richard Owen (1804–1892) believed that organic forms result from the action of two antagonistic principles.[9] One of them would be responsible for the vegetative repetition of structures and the "unity of organization", and could produce diverse structures such as the vertebrate skeleton, the pentagonal symmetry of the echinoderms or the crystalline spicules of sponges – in the same way as the repetition of crystal forms in the inorganic world is part of the organization plan of Nature. The other principle, of a teleological nature, would dynamically respond for the functions of living beings in an adaptive way, capable of producing the diversity of organic beings while aiming at the development of the function.

The German Karl Ernst von Baer (1792–1876), when studying embryology, concluded that in the past a kind of "creative thinking" in cosmic space also brought together scattered masses, and of spherical form. The same principle happened to create life forms from the beginning, out of dust on the surface of metallic planets. For him also, evolution functioned by a kind of "saltationism", not by way of random and continuous variations, but as a result of that general principle of creative organization, which is a teleological property. According to his defense of teleology, the purpose of life is part of the purpose of the entire universe.

Opposite to Karl von Baer, the German physiologist Theodor Schwann (1810–1882), who is well-known because he extended the concept of cell to all living beings, conceived a different theory of the organism. In contrast with the teleological vision of a vital force, which would guide activities towards a purpose, Schwann had a radical materialist conception, and hence believed that there are only blind (random) forces in Nature, along the theories of Saint-Hilaire and Darwin. At the time, "atomist" materialists saw each isolated cell as a vital and independent unit: cell morphology left physiology (function) in the background.

As exposed above, such opposition between purposeful and purposeless natural actions had already arisen, Goethe, Kant, Cuvier, Owen, and von Baer favoring the teleological explanation. Cuvier, in special, defended on the contrary that the improvement of a function, i.e., an enlarged capacity to realize a goal, brings with it the improvement of the respective organization of the living being, i.e., the form. Nature had the property of increasing the complexity of beings by giving them new organs and new faculties, as time advanced. And, he concluded, in the universe known to us, man resulted the most complex being there is.

The famous theory of the French naturalist Jean-Baptiste Lamarck (1744–1829) also goes somewhat in the direction taken by Cuvier and postulates an active action

[9] This is in line with the ideas entertained by Friedrich Schelling, as a building block for *Naturphilosophie*.

of self-adaptation by the organism.[10] For Lamarck, a contemporary of Cuvier, as well as later for Owen, there were also two forces at work responsible for the variability of beings, with different intrinsic properties. One (which Lamarck called "vital force") that tends to produce an increase in complexity, and the other (called by him "circumstance", equivalent to the environment) that tends to disturb, but also to diversify this increase. The problem is to correlate the causes of changes in form along with the respective changes in function. How do the parts coordinate to form the whole and guarantee the essential condition of survival?

Lamarck did not have the explanation to account in detail for the action of the two forces, yet his ideas left a permanent influence in various other naturalists. The stage would drastically change in the second half of the nineteenth century, when teleology was increasingly considered by many leading scientists as theological and to have been definitively banned by the evolutionary theory of Charles Darwin, after he published his major contribution, *On the origin of species,* assuming that the adaptations are for the most part fortuitous. Only relatively few naturalists contradicted this Darwinist postulate, and continued to advocate a genetically functional, i.e., causal morphology, in which the evolution of form could not be isolated from the evolution of function.

Wilhelm Roux (1850–1924), the well-known German zoologist and embryologist, tried in the beginning to compromise between purely mechanistic materialism and metaphysical vitalism. Notwithstanding his earlier efforts, he turned up adopting Lamarck's idea that external stimuli can produce new structures. These, over many generations, end up becoming hereditary, even in the eventual absence of such stimuli. Roux believed that Darwinian natural selection cannot explain the origin of a functional structure and the functional harmony of the structure, as thousands of cells would have to vary orderly together in order that some real advantage could arise in the struggle for survival, which is highly improbable to happen with success through individual selection. By studying the properties of blood vessels, Roux concluded that they must be interlaced with the functioning of tissues and organs. Form is not something fixed, it is an always dynamic manifestation of functional activities, capable of responding to foreseen and unforeseen contingencies. Function creates the organic need, and only after many generations does the organ appear before the function is actually and regularly performed. The differentiation of tissues in the embryo would be due in large part to a qualitative division of the cytoplasm and not to the nucleus, according to Roux.

The English writer Samuel Butler (1835–1902) is known for his fierce opposition to Darwinian evolution in Great Britain. He proposed that living beings are active and intelligent agents, continuing what their ancestors already accomplished and suffered, through a kind of intense, but unconscious memory, which acts repeatedly until becoming an automated memory. Variations therefore are "remembered" and become hereditary, in the way of a psychological process, with the same "nature" as

[10] It is also part of the *Naturphilosophie* principles to conceive self-organization as a natural phenomenon.

this type of memory. Accordingly, we are more unconscious and have less control over, for example, our circulation and digestion processes, which are extremely ancient functions, linked even to our invertebrate ancestry. Who would be the designer of this intelligent design, asked Butler? For man it is man himself, he answered, not individually, but mankind during his entire existence and even beyond the human species, that is, from the dawn of life to the present moment.

Organisms could automatically repeat inherited functions, but for Butler they could also create novel functions in response to new needs when facing environmental adversities. Ontogeny then would not be a simple recapitulation of phylogeny, since the universe is not deterministic. It should be taken into account, however, that there is a similarity between the laws that govern evolution, so that parallel responses may occur, even without following a deterministic trackway. Evolution could therefore have started over again many times and from very different primordial cells, such that the character of discontinuity (saltationism) might prevail in evolutionary development, rather than continuity.

Butler's proposal would possibly satisfy those who subscribe to a monistic interpretation, like Leibniz', or a pan-psychic one, like Spinoza's. Under this light, all things that exist would be considered "animate", regardless of whether they are alive or not, albeit in different degrees. Thus goes the path followed by Cuvier and von Baer, and resumed by scientists even in the twentieth century, such as the already cited Hans Driesch (1867–1941), who did not hesitate to get rid of Darwinian evolutionary dogmatism, and courageously defended vitalism and teleology, in the footsteps of the classic views of Aristotle.

Russell synthesized extraordinarily well the different philosophical positions around the riddle of function, form, and their interplay. His excursion to these ideas was not only applicable in his time but has continued to be relevant. Due mostly to ideological factors, contemporary science does not seem to have paid enough attention to his achievement.

15.5 Final Considerations

From the three authors just reviewed, as well as others mentioned in their works, it is possible to draw some conclusions that look quite bold regarding what is accepted as the orthodoxy dominating evolution since the "modern synthesis", that makes up the fundamental stone of current biological thought.

There is some theoretical and practical evidence (in the fossil record) that support the idea that evolutionary transformation does not manifest itself as a continuous function. The great taxonomic divisions of macroevolution, like phyla, do not seem to have been constituted out of the continuous and uninterrupted differentiation of species, but would rather have erupted in a relatively short time, perhaps not so suddenly to be called a "revolution", yet the transitions may still be considered abrupt.

Each new taxonomic grouping (be it those higher in the classification, like "class" or "phylum", be it lower, like "genus" or even the basic "species") demonstrates appropriation of forms for the internal organs, which in turn also developed from changes in form adapted to functions, in a teleological way such that usually function comes before form, in terms of timing.

Morphogenesis might be linked to the non-linearity of evolution, which in turn discloses itself in discontinuous "leaps".[11] In other words, both form and function are manifestations of the tendency to increasing complexity and harmony, which unifies and gives a certain "continuity" sense, even if materialized by discontinuities, a statement which needs not to be paradoxical.

In the present text, I pointed out how illustrative the history of science can be, and showed how three different scientists met the challenge of understanding questions that are beneath the surface of the problem, and how useful their results can still be to our contemporaneity. The history of ideas is an interesting guide to follow the conceptual development of knowledge, and how it moves towards a better understanding of nature.

One final note is that the natural unity of the organic and the inorganic worlds, seemingly too far apart, can be appreciated by a convergence in both cases of a eurythmic tendency, initially glimpsed at in physics, more specifically in the case of quantum theory. Eurhythmy has been used in the context of the Lisbon School to express a trend to optimum use of natural collective properties. This may look surprising, because human society as part of nature is not usually imagined in this unitary context, but in fundamental social activities, such as economy, and its explanatory descriptions such as history, this trend to exist with increasing complexity and harmony might derive from an approximation at first unsuspected of the man-built environment within the natural world (Magalhães 2015).

References

Croca, José, and João Araújo, eds. 2010. *A new vision on Physis. Eurhythmy, emergence and nonlinearity*. Lisboa: FCT.

D'alembert, Jean Le Ronde. 2015. *Enciclopédia, vol. 3 (Física e Matemática)*. São Paulo: UNESP.

Doubochinski, Daniel, Jonathan Tennenbaum. 2008. On the general nature of physical objects and their interactions, as suggested by the properties of argumentally-coupled oscillating systems. arXiv:0808.1205.

Driesch, Hans. 1908. *The science and philosophy of the organism*. London: Adam and Charles Black.

Gould, Stephen Jay. 1992. *Bully for the brontosaurus*. New York: W. W. Norton.

Huntley, H.E. 1970. *The divine proportion. A study in mathematical beauty*. New York: Dover.

[11] Non-linearity and the collective behavior characterize complexity, as some biologists have begun to better understand at the end of the twentieth century – see, for example (Solé and Goodwin 2000. Other authors like Eva Jablonka have also contributed for a revival of vitalist and epigenetic approaches.

Magalhães, Gildo. 2012. Emergence: Continuity or break up? A look beyond physics and behind economics, communication at the Colloquium on *Emergence and Non-Fundamental Metaphysics*, Lisbon.

———. 2013. On a possible contribution of transfinite mathematics towards eurhythmy. In *Space, time and becoming*, ed. José Croca et al. Lisboa: Faculdade de Letras da Universidade de Lisboa.

———. 2015. Some reflections on life and physics: Negentropy and eurhythmy. *Quantum Matter* 4 (3).

———. 2022. *The web of the universe. Evolution and eurhythmy*. Chisinau: Lambert.

Russell, Edward Stuart. 1916. *Form and function*. Pinnacle Press (reprint 2009).

Sarton, George. 1956. *History of science and the new humanism*. New York: George Braziller.

Solé, Ricard, and Brian Goodwin. 2000. *Signs of life. How complexity pervades biology*. New York: Basic Books.

Thompson, D'Arcy Wentworth. 1992. *On growth and form. The complete revised edition*. New York: Dover (reprint 1942)

Wagensberg, Jorge. 2002. El progreso: un concepto acabado o emergente? In *Observatorio siglo XXI. Reflexiones sobre arte, cultura y tecnologia*, ed. José Tono Martínez. Barcelona: Paidós.

———. 2004. *La rebelión de las formas*. Barcelona: Tusquets.

Chapter 16
John Bell's Unpublished Notes About de Broglie's Pilot Wave

Augusto Garuccio and Angela Laurora

Abstract In this contribution we present and discuss an unpublished note of John Bell on the de Broglie's Pilot Wave. The note is the result of the analysis of the scientific correspondence and other original documents made available by Mrs. Mary Bell, wife and colleague of John, performed form one of us (A.L.) in a PhD project devoted to the historical reconstruction of the scientific thought of John Bell and confirms the highly original character in theoretical research. Among Bell's original documents, a manuscript entitled "Notes for a history of the pilot wave" confirms that the Irish researcher considered central the idea of the pilot wave. We will present, moreover, a comparison between the unpublished note and the short writing published in 1988 in memory of de Broglie, that suggests that Bell had in mind to write a more elaborate work on this subject. Unfortunately, its sudden and untimely disappearance has put the end word to this project.

Keywords John Bell · Pilot wave · Unpublished manuscript · Historical reconstruction · Didactical approach · Varenna conference

16.1 Introduction

In 1935 the paradox formulated by Einstein, Podolsky and Rosen was one of the main topics of the research about the foundations of Quantum Mechanics and John Bell (1928–1990) played a key role in its debate, allowing to deal with aspects that have led to new developments of conceptual relevance and, subsequently, the application relevance.

A. Garuccio (✉) · A. Laurora
Centro Interuniversitario – Seminario di Storia della Scienza, Università degli Studi di Bari Aldo Moro, Bari, Italy
e-mail: augusto.garuccio@uniba.it; angela.laurora@uniba.it

P. Castro et al. (eds.), *Advances in Pilot Wave Theory*, Boston Studies in the Philosophy and History of Science 344,
https://doi.org/10.1007/978-3-031-49861-9_16

Bell was able to stimulate research on the EPR paradox, orientating the work of all those who have dealt with these issues and the importance of his scientific work is now widely recognized by the scientific community.

Recently the possibility of using the analysis of the scientific correspondence and other original documents made available by Mrs. Mary Bell, wife and colleague of John, has allowed to provide a contribution (Laurora 2014) that integrates the historical reconstruction of the scientific thought of Bell and confirms the highly original character in theoretical research.

The debate on the issue opened in 1935 by Einstein, Podolsky and Rosen – with the publication of the article "Can the Quantum-mechanical description of physical reality be Considered Complete?" (Einstein et al. 1935) – remained, for about 30 years, limited to a philosophical level and its development, not only related to the belief of the protagonists, was made prudent by a previous book by the famous mathematician John von Neumann, *Mathematical Foundations of Quantum Mechanics*, which in 1932 had shown the impossibility to complete the Quantum Mechanics in causal and local terms proposed by Einstein (Von Neumann 1932).

In 1952, however, the physicist David Bohm published "A Suggested Interpretation of the Quantum Theory in Terms of Hidden Variables" (Bohm 1952) in which, instead, he showed to be possible that completion of Quantum Mechanics and reformulated the paradox considering a configuration very close to a possible physical and experimentally feasible situation.

It is the reading of Bohm's article that caught the John Bell's interest and increased the reasons that had led him to take an interest in the issue since he was a student.

In 1964 he published "On the Einstein-Podolsky-Rosen paradox" (Bell 1964), a pioneering article in which he proposed the famous inequality that bears his name.

The article, having received a lukewarm reaction, gave rise to a renewed interest in the research on Quantum Mechanics foundations, following the publication of Clauser, Horne, Holte and Shimony's article in 1969 on the proposal of experimental tests to verify the Bell's inequality and contributions and discussion at the Varenna Conference in 1970 (Bell 1971).

The publication of the results of the first generation of experimental tests conducted between 1972 and 1976 and those of the second generation experiments carried out between 1976 and 1982 stimulated Bell's reflections and allowed to broaden interest and participation in the debate.

It was a time when, however, the debate took place still in informal ways, as evidenced by the role played by the meetings on the question of hidden variables organized between 1973 and 1982 by the Gonseth Association and the consequent dissemination of the "Epistemological Letters", who collected the contributions to the debate and then allowed the comparison and refinement of their ideas before their publication in some appropriate journals.

Between 1982 and 1990, a period characterized by the continuation of reflection on key issues and third generation experimental tests, there was a Bell's greater involvement at conferences to which – compared to those of the Seventies – he contributed by seminars that were particularly marked by the importance of spreading the already consolidated ideas and to disseminating them even through the interviews he released between 1983 and 1990.

In March 1987, Bell collected, in the volume *Speakable and unspeakable in quantum mechanics* (Bell 1987), the scientific papers published between 1964 and 1986, including those that deal with de Broglie's theory. In the Preface to the volume, Bell suggested reading those articles that, in his opinion, allow to understand the essential aspects of the theory formulated by the French physicist. They are the articles that have allowed to know John Bell's point of view about the pilot wave model and chronologically preceding the newly discovered document.

16.2 Notes for a History of the Pilot Wave

Among John Bell's original documents, a manuscript entitled *Notes for a history of the pilot wave* on the pilot wave was found, an argument that – as indicated by the Irish physicist himself – always interested him.

He himself points out in a short written article entitled *Louis de Broglie's guiding wave* in the collection of memories and personal testimonies on Louis de Broglie published in 1988 by the Fondation Louis de Broglie to remind the French scientist about one year since its disappearance (Bell 1988).

- 1 -

NOTES FOR A HISTORY OF THE PILOT WAVE

J.S. Bell
CERN - Geneva

The magnificent discovery of de Broglie waves, in 1923, won for Louis de Broglie immortal fame. Is it not strange then that his efforts to understand the physical status of his waves commanded so little attention among his great contemporaries?

This contribution, not being a scientific publication, is virtually ignored by literature but is valuable because it clearly documents Bell's thinking about it:

> The magnificent discovery, in 1923, of de Broglie waves, won for Louis de Broglie immortal fame. Is it not strange then that his efforts to understand the physical status of those waves commanded so little attention among his great contemporaries? His 'pilot wave' picture, of 1927, has fascinated me since I first heard of it. And when I did first hear of it, in 1952 when it was rediscovered and developed by David Bohm, I was indignant that I had not been told of it before. [...] But the courage and vitality of de Broglie, in his late maturity,

> in breaking from a heavily prevailing orthodoxy, remains for me admirable and inspiring. [...] I was indignant that I had not been told sooner of the pilot wave picture. And I am indignant that even now most students are not told of it at all. For it is a powerful antidote to the complacent notion that the ambiguity and subjectivity of contemporary physical theory are dictated by experimental facts.

We note that this short note highlights his 'politics' position, so to speak, on the interpretation of Quantum Mechanics clearly expressed in the last lines of his contribution to homage to de Broglie.

Otherwise, John Bell had already dealt with the pilot wave in 1982, on the occasion of the 90th birthday of de Broglie, preparing for that celebratory event a contribution entitled *On the impossible pilot wave*, then published on the "Foundation of Physics" in October of the same year and later included in the volume *Speakable and unspeakable in quantum mechanics* (Bell 1993).

> Moreover, the essential idea was one that had been advanced already by de Broglie in 1927, in his 'pilot wave' picture. But why then had Born not told me of this 'pilot wave'? If only to point out what was wrong with it? Why did von Neumann not consider it? More extraordinarily, why did people go on producing 'impossibility' proofs, after 1952, and as recently as 1978? When even Pauli, Rosenfeld, and Heisenberg, could produce no more devastating criticism of Bohm's version than to brand it as 'metaphysical' and 'ideological'? Why is the pilot wave picture ignored in text books? Should it not be taught, not as the only way, but as an antidote to the prevailing complacency? To show that vagueness, subjectivity, and indeterminism, are not forced on us by experimental facts, but by deliberate theoretical choice? I will not attempt here to answer these questions.

The rediscovered manuscript confirms Bell's interest in this topic and is particularly significant as it addresses the issue in a historical perspective, trying to provide an organic picture of the scientists' reflections that had been questioned on the subject, organizing them chronologically and with reference to bibliographic sources.

The document consists of 12 typed pages. The first three pages are dedicated to Bell's reflections, ending with the following statement: 'Paris, 5 November, 1987'. The manuscript is then placed after March 1987 when de Broglie died and before the publication of *Speakable and unspeakable in quantum mechanics*.

Let us note that Bell writes in the introduction to first edition of this book: "the papers [on quantum philosophy], from the years 1964–1986, are presented here in the order, as far as I now can tell, in which they were written" and this information probably explains why the manuscript was not inserted in the book.

However, it is not apparent that this work was subsequently published and, in fact, it is not even present in the updated edition of the same volume published in 2004. A systematic search among secondary sources did not provide any feedback, so it is reasonable to consider the manuscript almost certainly unpublished.

On page 4 of the document, Bell reports the referenced bibliography and in appendix, consisting of 8 pages, he encloses the text of the considerations that the various scientists did in conjunction with two events that Bell considers most relevant to the subject: the Fifth Solvay Conference on *Électrons et photons* in

1927 and the publication of the collection of writings *Louis de Broglie, Physicien et Penseur* in 1952 (George 1953).

In the pages of the manuscript concerning reflections, he starts from the idea of the pilot wave proposed by Louis de Broglie in 1923 and reflects on the consideration given to this concept by the founding fathers of Quantum Mechanics[1]:

> I have myself been fascinated by his 'pilot wave' picture since I first heard of it – when it was rediscovered and developed by David Bohm in 1952. And I have been intrigued by the reception given to this idea. So I have been led to seek out especially the opinions of the great founding fathers on this matter. I have put together here, in Appendices, what seem to me to be the essential passages in the writings I have found.

The interest in this subject originated from the reading of David Bohm's work in 1952. Bell was fascinated by the publication so much that this was definitely the decisive push for his involvement as a protagonist in the debate on the Quantum Mechanics foundations:

> I have myself been fascinated by his 'pilot wave' picture since I first heard of it – when it was rediscovered and developed by David Bohm in 1952. And I have been intrigued by the reception given to this idea. So I have been led to seek out especially the opinions of the great founding fathers on this matter. I have put together here, in Appendices, what seem to me to be the essential passages in the writings I have found.

As already mentioned, in particular he evaluates the views expressed by them on the occasione of two events that, as already pointed out, Bell considers relevant to the subject: the Fifth Solvay Conference on Electrons et Photons in 1927, which saw the most prestigious physicists of the time confront the formulation of the new quantum theory, and the publication in 1952, on the occasion of the 60th birthday of de Broglie, of a collection of writings *Louis de Broglie, Physicien et Penseur*, to which contributed the protagonists who had participated in the debate until that year:

> There were two particularly suitable occasions for such opinions to be expressed. The first was in October 1927, the 5th Solvay Conference when de Broglie first presented his ideas. The second was in 1952, when a festschrift was put together for de Broglie's 60th birthday – just after, as it happened, the work of Bohm became known.

Concerning the Fifth Solvay Conference in 1927, he indicated those who, among the protagonists of the debate, responded to de Broglie's proposal:

> All the great men were at the 1927 meeting: Einstein, Born, Bohr, Schrödinger, de Broglie, Pauli, Heisenberg, Dirac – in order of age. Only Einstein, Schrödinger and Pauli responded to the Broglie. Einstein thought that "Mr. de Broglie is right to look in that direction". (Indeed, Einstein had himself had very similar ideas about the combination of electromagnetic waves and photons.) But he gave non support beyond this remark. Schrödinger said that attempts to put particles and waves together could be at best temporary expedients. He was already in love with the wave, did not want particles at all, and so it was for the rest of his life. Pauli dismissed the pilot wave picture on the grounds that, after an interaction, a system is not seen to go into one of its stationary states. The reply of de

[1] Quotations take into account handwritten notes by Bell at the margin of the typed document.

Broglie was hesitant, and the point was left for Bohm to clear up. In fact, the idea that a system must always be in one of its stationary states was a hangover from the old Bohr atom. It does not appear now in any careful formulation of the modern theory. Pauli seems not to have returned to the point.

Likewise, examining the volume *Louis de Broglie, Physicien et Penseur* in 1952, Bell specifically identifies those who provided a further commentary to de Broglie's idea. Regarding the other physicists, who did not contribute to the collection of writings, he quotes (their) some comments that he considered helpful for his remarks:[2]

In the 1952 festschrift, Einstein and Pauli commented again on the pilot wave theory, and what seems to me the essence of their remarks I include in the Appendices. Schrödinger did not return to it, and after 1927 I know of no comment by him on the pilot wave idea. Born and Heisenberg were present in 1952, but wrote about other things. So I include in the Appendix a relevant remark by Born in a letter to Einstein, and one of his notes to the collected Born-Einstein letters [3]. And I include some relevant passages from a 1955 survey [4] by Heisenberg. Bohr and Dirac were absent in 1952. I represent Bohr here by a remark from a letter to Pauli in August 1927, and by another attributed to him by Heisenberg, and by some of 1952 analysis of Rosenfeld [2], who seems to have appointed himself the champion of Bohr. And I quote some passages from a 1963 article by Dirac [5].

After examining the various contributions, Bell highlights Bohm's work in 1952 as an occasion for reopening the reflection on de Broglie's idea and the opinions of the various players in the debate:

The situation had been very much changed by the 1952 work of Bohm. Louis de Broglie had himself taken up again his old ideas. The great men had to admit, with visible reluctance it seems to me, that the pilot wave picture could not be faulted logically, nor distinguished experimentally from the orthodox theory. They cast around, rather desperately it seems to me, for other reasons to reject it. They called it 'metaphysical' and 'ideological'. They need not detain us. Pauli pointed out that a careless modification of the picture could get it into trouble with the exclusion principle. So if we modify the pilot wave picture, hoping to improve on quantum mechanics, let us be careful. And they deplored the lack of symmetry between q and p. The ordinary quantum formalism, like classical canonical mechanics, has such symmetry – until one looks at a realistic Hamiltonian

The last reference is to Paul Dirac:

It is refreshing in the end to come to the opinions of Dirac – comforting as they are for those who (like Einstein, de Broglie, Bohr and the others) worry greatly about the meaning of quantum mechanics. Dirac thought that one should not worry very much about that. He thought that the puzzles were associated with the provisional state of the theory, and that looking back from the vantage point of a more developed theory we will regret the time spent trying to interpret an underdeveloped one. He insisted too on beauty in the equations, and on this ground neither the pilot wave picture nor the more elaborate 'double solution',

[2] In the quotation bibliographic references are maintained in the same form as they appear on page 4 of the typewriter and below as well as annotated by Bell: [2] Louis de Broglie, Physicien et penseur, André George, ed. (Albin Michel, Paris 1953), [3] The Born-Einstein Letters, M. Born, ed. (MacMillan, London 1971), [4] Niels Bohr and the Development of Physics, W. Pauli, L. Rosenfeld and V. Weisskopf, eds. (Pergamon, London 1955), [5] P.A.M. Dirac, Scientific American, May 1963, p. 45

preferred by de Broglie, would have appealed to him. However, in some respects, he is definitely in the de Broglie camp – in his gentle scepticism about 'complementary' and his openness to the idea that clarity and precision might one day prevail in theoretical physics

Bell concludes his reflection by emphasizing the importance of pilot model in the teaching of Quantum Mechanics:

Most contemporary theoretical physicists, busy as they are, making important developments in the manner of Dirac, have little time for questions of meaning. If they think of these things at all, they are likely to think of themselves as following Bohr and 'complementarity'. If they have heard of the pilot wave, they are likely to think vaguely that it has somehow, probably by Pauli, been refuted. It has not. And it remains a powerful antidote to the complacent idea that the ambiguity and subjectivity of orthodox quantum mechanics is dictated by experimental facts. I think that all students of quantum mechanics should be exposed to the pilot wave picture, to broaden their minds.

16.3 Conclusions

The comparison between the unpublished and the short writing published in 1988 in memory of de Broglie would suggest that Bell had in mind to write a more elaborate work that then had to summarize – taking into account the expiration due to de Broglie's commemoration – in the one published in the collection of writings in honor of the French Nobel Prize.

However, the finding of the manuscript among Bell's unpublished documents confirms that the analysis of issues that since young Bell considered central has continued to engage him even in mature ages and suggests that he has kept that draft work in view of an accurate and more elaborate reflection on the idea of the pilot wave, starting from the detailed historical reconstruction of the debate on that theory. Unfortunately its sudden and untimely disappearance has put the end word to this project.

References

Bell, J.S. 1964. On the Einstein-Podolsky-Rosen Paradox. *Physics* 1 (3): 195–200.

———. 1971. Introduction to the hidden-variable question. In *Foundations of quantum mechanics*, ed. B. d'Espagnat, 171–181. Proceedings of the International School of Physics "Enrico Fermi", 49th Course (Villa Monastero, Varenna, 29 June–11 July 1970). New York/London: Academic.

———. 1987. *Speakable and unspeakable in quantum mechanics*. Cambridge: Cambridge University Press.

———. 1988. Louis de Broglie's guiding wave. In *Louis de Broglie que nous avons connu*, ed. Fondation Louis de Broglie, 27. Paris: Fondation Louis de Broglie., Conservatoire national des arts et métiers,.

———. 1993. *Speakable and unspeakable in quantum mechanics*. Cambridge: Cambridge University Press.

Bohm, D. 1952. A suggested interpretation of the quantum theory in terms of hidden variables. *Physical Review* 85 (2): 166–179.

Einstein, A., et al. 1935. Can quantum-mechanical description of physical reality be considered complete? *Physical Review* 47 (10): 777–780.

George, A., ed. 1953. *Louis de Broglie, Physicien et Penseur*. Paris: Albin Michel.

Laurora A. 2014. *John Bell e le ricerche sul paradosso di Einstein, Podolsky e Rosen*. PhD thesis in History of Science, Università degli Studi di Bari Aldo Moro.

Von Neumann, J. 1932. *Mathematische Grundlagen der Quantenmechanik*. Berlin: Verlag von Julius Springer.

Chapter 17
The Quest for the Ontic Nature of the Quantum Waves

J. R. Croca

Abstract The quest for the ontic nature of theoretical proposed entities has a long tradition in physics. Here we briefly present the efforts related with the demand for the nature of the quantum waves.

Keywords Ontic nature of the quantum waves or De Broglie waves · Traditional quantum physics · Nonlinear causal quantum physics · Crucial experiments

17.1 Introduction

In the history of physics, the quest for the ontic nature of some theoretically proposed entities has a long tradition. In the end, the question was and is, weather some of these proposed entities do have a real physical nature or, on the contrary, are mere conceptual helpful tools.

It is common knowledge, for those familiar with the history of science, that the more pragmatic researches traditionally maintain, that the question for the elucidation of the nature of the proposed theoretical entities is not a relevant issue.

The advent of modern science started precisely just with one of these polemics. Indeed, in 1543, Osiander, making a referee revision for the publication of the book *De Revolutionibus Orbium Coelestium* (On the revolutions of the celestial spheres) written by Copernicus, made an assumption of the same nature. In the preface of the book he said that the theoretical model, proposed by Copernicus, in which the Sun was still and the Earth moving, was not necessarily true, or even probable, but was useful tool for computational purposes.

About 200 years later another controversy of this type occurred. This relatively long controversy was related with the ontic nature of the atoms. To some scholars,

J. R. Croca (✉)
Faculty of Sciences, Department of Physics, Centre for Philosophy of Sciences of the University of Lisbon (CFCUL), University of Lisbon, Lisbon, Portugal
e-mail: jncroca@ciencias.ulisboa.pt

P. Castro et al. (eds.), *Advances in Pilot Wave Theory*, Boston Studies in the Philosophy and History of Science 344,
https://doi.org/10.1007/978-3-031-49861-9_17

namely, Mach, Ostwald, Avenarius and many others, the so-called positivist and neopositivists, the atoms were only mere conceptual useful tools and nothing more.

Another group of thinkers, on the contrary, believed that the atoms were much more than mere conceptual tools and indeed do have physical reality. From these, we may point out, Boltzmann, Maxwell and Einstein. Still, at the time there were no experiments that could allow the decision on the ontic nature atoms. Things changed radically when in 1905 Einstein (1905) published a work in which he explained the observed Brownian motion assuming the physical reality of the atoms.

These ideas were experimentally confirmed by the very important experiments that, in 1909, Jean Perrin (1909) performed. Since then on, for the scientific community the atoms are assumed to have a real physical existence.

Now, everybody agrees that without the belief in the real nature of the Copernicus model or of the in the real existence of the atoms and molecules our modern science and technology would be impossible.

A similar controversy started about the first quarter of the last century. It is related with the concept of quantum waves introduced at the 1927 Solvay Conference.

Orthodox quantum mechanics claims that quantum waves, solution to the Schrödinger equation, are conceptual tools, mere probability waves devoid of any physical reality.

De Broglie and his nonlinear causal school, sustain that quantum waves, or as they may now be named, subquantum waves, guiding waves, pilot waves, empty waves, zero-point field waves, theta waves, stand for the description of something having real physical existence.

Now, just as before, to the great majority of physicists, having a more practical pragmatic attitude before science, this controversy on the ontic nature of the quantum waves, whether they are mere probability waves or, on the contrary, do have physical reality is devoid of any sense. They say that they don't care, because the issue is not relevant to do the actual quantum calculations.

We believe that the clarification of the nature of the quantum waves is a very important issue. If indeed quantum waves do have real physical existence, as some experiments seem to indicate (Zou et al. 1991; Menzel et al. 2012, 2013), then a whole new universe of theoretical and technological possibilities will be open.

17.2 De Broglie Model for the Quantum Particle

In his thesis of 1924, De Broglie (1924, 1969) dared to advance, for the first time explicitly, a nonlinear complex model for understanding in a causal intuitive way the riddle posed by the dualism wave corpuscle, characteristic of the quantum particles. In this approach to better understand Nature, a quantum particle is assumed to be a very complex inter-relational entity, compose of an extended part, a wave, with a relatively minute energy plus a complex and highly localized and energetic part, the corpuscle, $\phi(\theta, \xi)$, as the next picture seeks to represent, Fig. 17.1

Fig. 17.1 Representation of de Broglie quantum particle

The wave, the extended part of the particle, θ, is also called, quantum wave, empty wave, de Broglie wave, pilot wave, subquantum wave, zero-point field wave, or theta wave.

The very localized structure, ξ, also a wave, a very high energetic localized wave, the corpuscle, also named acron, from the Greek word acropolis, the high pick.

It convenient, at this point, to stress the great contribution of de Broglie for a better understanding of the quantum realm. Against all tradition of physics, he dared to propose, a nonlinear relational complex process in which a minor action may, in adequate conditions, give rise to an enormous reaction. To this nonlinear process he gave the name of, *principe de guidage,* guiding principle. This basic principle of Nature is also known as pilot wave principle or, as is now called principle of eurhythmy (Croca 2006, 2012). This principle tells us that the theta wave, the subquantum wave, guides, pilots, in average, thought a nonlinear process the acron, the corpuscle, to the regions where the density of the global wave, in which it is immerged, is relatively greater.

17.3 Proposals for Clarification of the Nature of the Quantum Waves

For many years there were no experimental possibility of distinguishing whether quantum waves had physical reality or were mere probability conceptual entities. That is, from practical operational point of view, whether assuming that quantum waves were mere probability waves or real physical waves was not important. The concrete calculations performed to make quantum computations were just the same.

Things started to change only about half a century after the question was formulated. Indeed, only in the seventies of the last century were the first conceptual experiments for the first time advanced.

In 1972, P. Neves (1972), a student, of Andrade e Silva, a direct collaborator of de Broglie, in a scholar work, indicated an eventual possibility for developing such an experiment.

Nearly 10 years later, in 1980, J. and Maria Andrade e Silva (1980) further advanced the idea, "*Une experience possibile concernant la nature du dualismo onde-corpuscule*". This work was further developed by 1983, Andrade e Silva in collaboration with F. Selleri and J.P. Vigier (1983), "*Some possible experiments on quantum waves*".

However, the first concrete proposal for an eventual realizable, yes-no, experiment was presented only in 1985, by J. R. Croca (1988), *Can the existence of de Broglie's empty waves be proven experimentally?*

The first feasible proposed experiment with the then available technology (Croca et al. 1990), *Quantum-Optical predictions for an experiment on de Broglie waves*, was presented in 1990 by J. R. Croca, A. Garuccio, V. L. Lepore and R. N. Moreira. A variant of this proposal was done in 1992, at the University of Rochester, by X.Y. Zou, T. Grayson and L. Mandel (1992).

17.4 Basics of the Experiments

The basic idea behind those yes-no type experiments is indicated in the experimental setup shown in Fig. 17.2, in which a quantum monoparticle source, emits, for instance, photos one by one. This quantum particle, described by the wave function, ϕ, impinges on a 50% beamsplitter in which is transmitted, ϕ_t, and reflected, ϕ_r, with equal probability.

Consider the situation in which the detector, placed in the reflected path, clicks and an observer may, if he wants, see the lamp on.

Now a question arises: What happens in the transmitted path?

For this question there are two opposite answers:

(a) Quantum waves are probability waves, ψ.

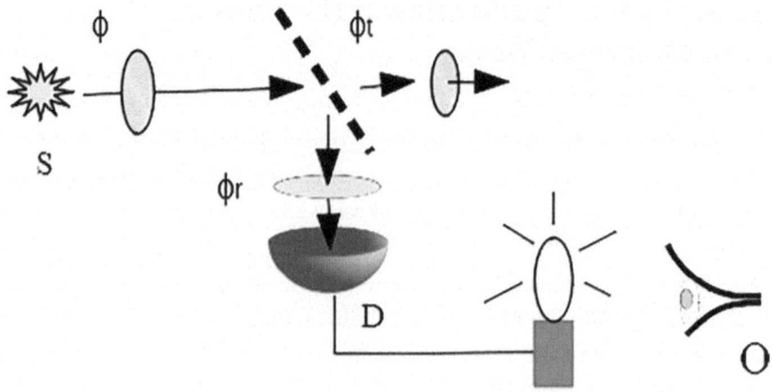

Fig. 17.2 A source emits quantum particles, one-by-one, towards a beamsplitter

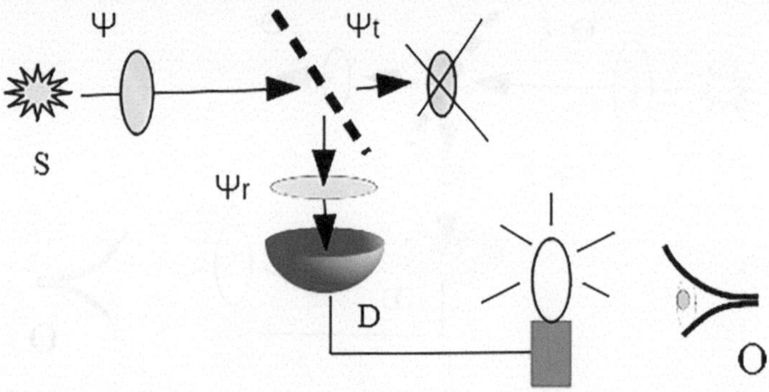

Fig. 17.3 Collapse of the quantum wave function

The quantum monoparticle source, photons in this case, emits one single particle at a time. According to orthodox quantum mechanics, this quantum entity is described by a probability wave ψ, which in its temporal evolution impinges, interacts with a beamsplitter giving thus origin to two probability waves, one transmitted ψ_t the other reflected ψ_r, so that

$$\psi = \psi_t + \psi_r. \tag{17.1}$$

If it happens that a detector, placed in the reflected path is activated, the Observer sees the lamp on. Then, there occurs the so-called collapse of the wave function and consequently

$$\psi \longrightarrow \psi_r,$$

that is,

$$\psi = \psi_r, \quad \psi_t = 0. \tag{17.2}$$

Next drawing, Fig. 17.3, tries to represent this situation.

Concluding: If quantum waves are mere probability waves, then, after measurement, that is, when we know for sure which-way the particle went, there occurs the collapse of the wave function and consequently nothing remains in the transmission path.

(b) Quantum waves are real waves

In this approach for better understanding Nature it is assumed that quantum waves, de Broglie waves, empty waves, pilot waves, theta wave or subquantum waves are real physical entities. Furthermore, it is assumed that a quantum particle is a very complex relational physical entity.

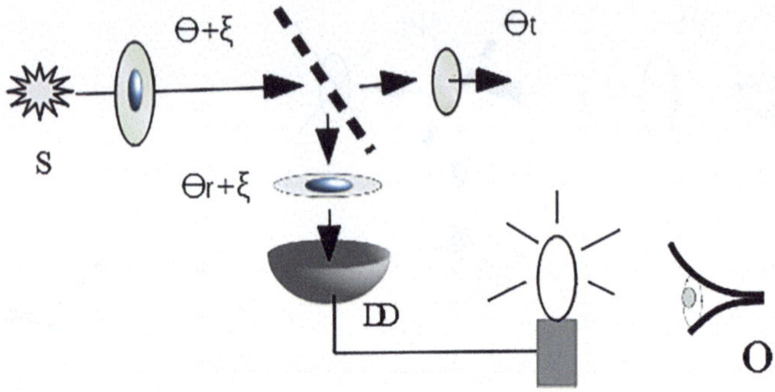

Fig. 17.4 In the transmission path follows a real quantum wave

A quantum particle is assumed to be composed of, a real physical extended wave, the theta wave, plus a very high energetic and relatively well localized structure, the acron.

The acron moves in the global theta wave field in a stochastic way so that it has an average propensity to move to the regions where the field is higher. That is, the feeble intensity theta wave guides, pilots, through a nonlinear process, the average motion of the corpuscle. So, the probability of localizing the acron is proportional to the intensity of the theta wave field

$$P \propto |\theta|^2 = I. \tag{17.3}$$

Let us now consider the same experiment situation in the light of the nonlinear quantum physics (Croca 2003) in which quantum waves are assumed to have a real nature.

The causal realistic approach is depicted next drawing, Fig. 17.4,

In this way of looking at Nature, it is assumed that the source emits a complex quantum particle composed of a wave with a kind of kernel inside it, the acron, $\phi = \theta + \xi$. Upon arriving at the beamsplitter the wave is divided in two real waves, one transmitted, θ_t, the other reflected, θ_r. The acron, ξ, being indivisible, may follow one path or the other. In this case is reflected and reaches the detector. In consequence of the interaction the detector is triggered and consequently the light is on. Then the Observer sees the light on and concludes that the acron has travelled along the reflected path. Consequently, by the transmitted path, follows the real quantum wave, the theta wave, θ_t undisturbed by the measurement.

Had the acron been transmitted and the single theta wave followed along the reflected path the detector would not be triggered. This is a consequence of the fact that the amount of energy carried by the subquantum wave is very low, therefore unable to trigger the detector.

This device may further be improved giving origin to what we may call a theta wave generator, TWG, as shown in next drawing (Fig. 17.5).

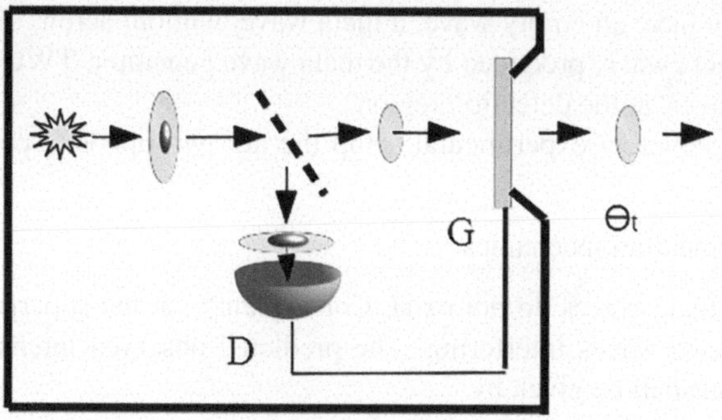

Fig. 17.5 TWG – Theta wave generator

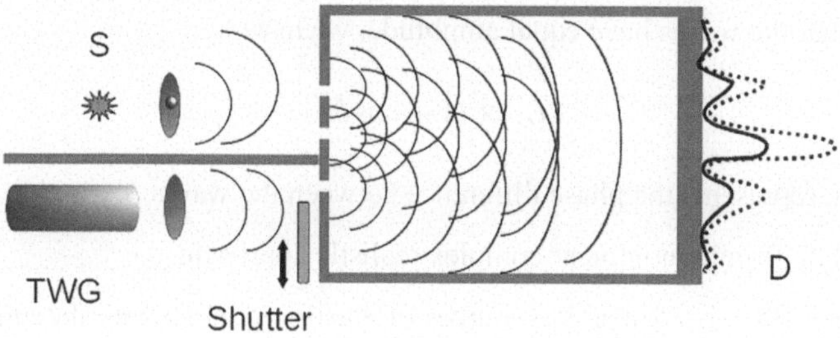

Fig. 17.6 The theta wave emitted by the TWG changes the interferometric pattern at the detection region

In this device the detector is connected to a gate. Each time the detector registers the arrival of a photon, it sends a signal that opens the gate so that the theta wave, following along the transmission path may leave the apparatus. In all other cases the gate remains closed. Consequently, from this apparatus only, theta waves, pilot waves may be emitted, thus the name of theta wave generator, TWG.

Now the most important question. How can we be sure that something real is emitted by the device?

To answer the question, we need to build a special device, a very special detector, able to detect waves with a very, very minute amount of energy. In plain words, we need to devise a subquantum wave detector.

There are a multiplicity of possibilities for developing such a device, as can be seen in the book, *Towards a Nonlinear Quantum* Physics (Croca 2003). Here we just consider one single possibility, outlined in next picture, Fig. 17.6.

In the picture a quantum monoparticle source emits complex particles one-by-one directed to the beamsplitter that upon interaction gives origin two waves, one reflected, ϕ_r and the other transmitted, ϕ_t, one of which carrying the acron. By means of a full mirror the two waves are led to superimpose. This superposition of the waves giving rise to an interferometric pattern. The theta wave, with or without corpuscle, the acron, is represented by the symbol ϕ. Only in the case when we are sure that we have an empty wave, a theta wave, without acron, the symbol θ is utilized. The theta wave, produced by the theta wave generator, TWG, superimpose to these two waves, at the detection region.

So, for this concrete experimental setup the are two opposite possible predictions:

(a) Orthodox quantum mechanics

Empty waves, theta waves do not exist. Consequently, at the superposition region we have only two waves interfering. The predicted observed intensity, I_O, at the detection region shall be given by,

$$I_O = |\psi|^2 = |\psi_r + \psi_t|^2 = |\psi_r|^2 + |\psi_t|^2 + \psi_r^* \psi_t + \psi_r \psi_t^*, \tag{17.4}$$

assuming that the waves have equal amplitude, we have

$$I_O \propto (1 + \cos\delta), \tag{17.5}$$

in which, δ, represents the phase difference between the waves, ψ_r and ψ_t.

(b) Causal de Broglie nonlinear complex realistic approach:

Empty waves do exist. Since two sources of waves are incoherent the empty wave coming from the theta wave generator introduce noise thus blurring the interference pattern.

In such conditions, at the detection region we have to consider the two waves, coming from the common source, one of which carrying the acron or the corpuscle, plus the empty, or theta wave coming from the theta wave generator, TWG.

The intensity, I_c, predicted by the de Broglie causal approach shall be given by

$$I_C = |\phi_r + \phi_t + \theta|^2. \tag{17.6}$$

Assuming, that the waves have equal intensity and that the two waves, coming from the monophotonic source, are coherent and incoherent relative to the empty wave, θ, coming from the TWG, we have after some relatively easy calculations, see, for instance, Croca (2003).

$$I_C \propto \left(1 + \frac{2}{3}\cos\delta_\phi\right), \tag{17.7}$$

in which, δ_ϕ, represents the phase difference between the waves, ϕ_r and ϕ_t.

Concluding: According to the Copenhagen interpretation, since the empty waves do not exist, a clear interference pattern with visibility one shall be observed. De Broglie causal approach predicts that the observed interference pattern shall be blurred, with a visibility factor of 2/3.

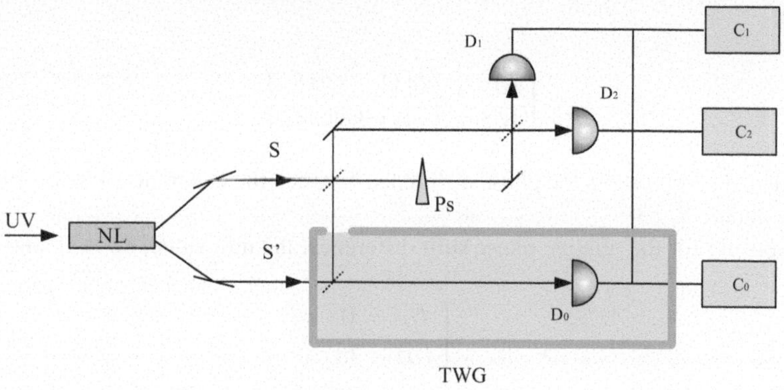

Fig. 17.7 Experimental setup for detection of theta waves

17.5 Concrete Experiment

Actual concrete experiment could be done with the present available technology according for example with to the experimental sketch shown in Fig. 17.7.

Form the sketch we see a UV source acting upon a nonlinear crystal, NL, producing by a parametric down conversion process two incoherent photons at the same time. Ps represents the phase shifting device. The signal photon coming from source S, that behaves like a single photon source, enters a Mach-Zehnder. The idler photon, from S', stands for the monophotonic source of the theta wave generator, TWG. Only when the detector D_0 is triggered by the arrival of the photonic acron, the corpuscle, the gate is opened allowing the empty wave, the theta wave, to enter the Mach-Zehnder interferometer. Naturally, as indicated in the drawing, all detectors are in coincidence so that only when detector D_0 is activated the counters are able to register the eventual arrival of photons.

The expected results of this experiment are:

(a) Copenhagen interpretation of quantum mechanics:

In this case, since the idler photon coming from S' was detected at D_0 there occurs the collapse of the wave function. In such conditions nothing, from source S' enters the interferometer. Then, at the interferometer we have only the photons coming from S. For this situation, the predicted intensity seen at detectors D_1 and D_2 and registered at counters C_1 and C_2 are respectively,

$$\begin{cases} I_{O1} = |\phi_{TR} + \phi_{RT}|^2 \\ I_{O2} = |\phi_{RR} + \phi_{TT}|^2 \end{cases} \tag{17.8}$$

giving,

$$\begin{cases} I_{O1} = \frac{1}{2} I_0 \left(1 - \cos \delta_\psi\right) \\ I_{O2} = \frac{1}{2} I_0 \left(1 + \cos \delta_\psi\right) \end{cases}, \tag{17.9}$$

in which, δ_ψ, represents the phase difference between the coherent waves originated by the source S.

Choosing for the relative phase shift difference, the null value, $\delta_\psi = 0$, one gets,

$$\begin{cases} I_{O1} = 0 \\ I_{O2} = I_0 \end{cases} \tag{17.10}$$

(b) Causal de Broglie realistic approach:

In this approach it is necessary to consider, the waves from source S, ϕ_{ij}, plus the theta waves θ_{ij}, coming from the theta wave generator, TWG.

$$\begin{cases} I_{C1} \propto |\phi_{TR} + \phi_{RT} + \theta_{TT} + \theta_{RR}|^2 \\ I_{C2} \propto |\phi_{RR} + \phi_{TT} + \theta_{RT} + \theta_{TR}|^2 \end{cases} \tag{17.11}$$

The sub-indices stand for the reflection and transmission at the beamsplitters. For instance, θ_{RT} indicates that the subquantum wave entering the interferometer is first reflected at the first beamsplitter and transmitted at the second.

Assuming, that we are in a situation that all mixing subquantum waves have equal intensity

$$|\theta_{RT}|^2 = |\theta_{TR}|^2 = |\theta_{TT}|^2 = |\theta_{RR}|^2 \tag{17.12}$$

After some easy calculations and recalling that the two sources have a random relative phase, we have for the predicted intensities at each output port

$$\begin{cases} I_{C1} = \frac{1}{2} I_0 \left(1 + \cos \delta_\phi - \cos \delta_\theta\right) \\ I_{C2} = \frac{1}{2} I_0 \left(1 - \cos \delta_\phi + \cos \delta_\theta\right) \end{cases} \tag{17.13}$$

For equal optical pathlengths, we may write

$$\delta_\phi = \delta_\theta = 0 \tag{17.14}$$

which by substitution in the previous expression we have

$$\begin{cases} I_{C1} = \frac{1}{2} I_0 \\ I_{C2} = \frac{1}{2} I_0 \end{cases} \tag{17.15}$$

These results imply, that for this particular choice of the experimental parameters the predictions of the two approaches are quite different:

(a) Orthodox approach:

Subquantum waves do not exist, therefore only detector D_2 counts that is:

$$I_{O1} = 0, \quad I_{O2} = I_0.$$

(b) Causal de Broglie approach.:

Subquantum waves do have physical reality. In this case both detector count at the same rate:

$$I_{O1} = I_{O2} = {}^1\!/_2\, I_0.$$

Concluding, for these experimental conditions, the real observable physical action of the empty waves is manifested by the change the overall probability distribution at the output ports of the interferometer.

$$\begin{matrix} \text{Orthodox} & \text{De Broglie} \\ \begin{cases} I_{O1} = 0 \\ I_{O2} = I_0 \end{cases} & \begin{cases} I_{C1} = \frac{1}{2} I_0 \\ I_{C2} = \frac{1}{2} I_0 \end{cases} \end{matrix}$$

17.5.1 *Performed Experiments*

The first experiment to test on the reality of de Broglie waves was performed by Mandel and collaborators at Rochester in 1995.

He did an experiment, in a way opposed to the one presented above. The full wave acts by blurring of the interference pattern due to the coherent superposition the two empty waves produced by the TWG. Even if theoretically is precisely the same situation, from the practical operational point of view is quite different. This happens, because in every concrete experiment the noise is ever-present, so we have to minimize its effects as much as we can.

In the proposed experiments the clear expected interference pattern, produced by the superposition of traditional waves, was to be blurred by the presence of the empty waves.

While, in Mandel's experiment, on the contrary, the interference pattern was produced by the subquantum waves and the noise introduced by the full wave with the corpuscle (Fig. 17.8).

For this experimental setup, the causal de Broglie approach predicts an interference pattern given by,

$$I_C \propto \left(1 + \frac{1}{3}\cos\delta_\theta\right), \tag{17.16}$$

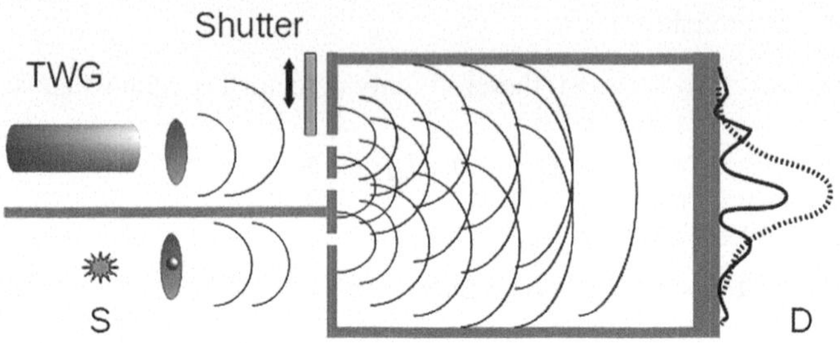

Fig. 17.8 Sketch of Mandel's experiment

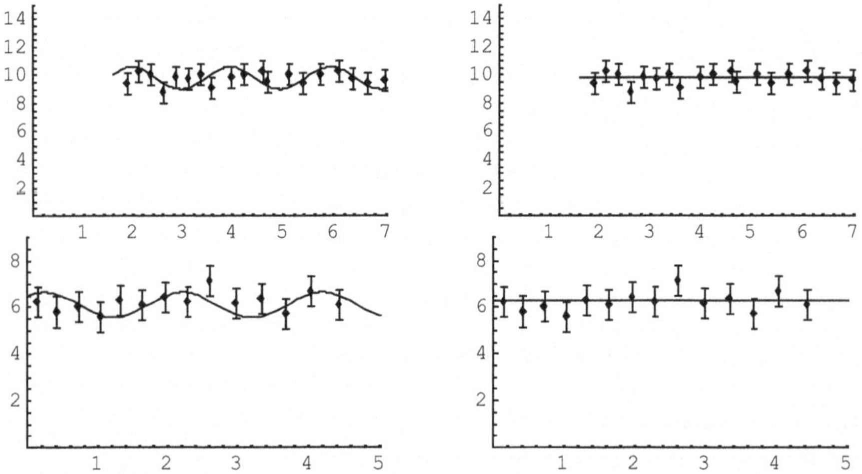

Fig. 17.9 Results of the experiment performed at Rochester for the detection of theta waves

while the orthodox approach predicts, no interferences,

$$I_O \propto const. \tag{17.17}$$

That is, orthodox approach predicts no interference whereas de Broglie expects interferences with a visibility factor of 1/3.

The results of the experiment are shown in next picture, Fig. 17.9.

The first line shows the published results, while the second line are the result privately communicated by Mandel.

The same data points are represented in two columns:

The first column first of data id fitted with an interference sinusoidal line, corresponding to a visibility factor of about 10%.

The second column, the data is fitted with a strait constant line drawn by Mandel and collaborators.

Mandel and collaborators concluded in their paper that the performed experiment disproved the real existence of de Broglie waves. We think that this conclusion is not very sound because as shown the experimental data could also be fitted with a sinusoidal line of visibility of 10%. Even if, in theoretically ideal conditions, the expected visibility would be of about 33% in real concrete experiments this value is never attained due to the ever-present noise. Any way, we think that the experiment needs to be redone, in better conditions, to clarify this important problem.

17.5.2 Another Type of Experiments

Recent advances in optics lead to the development of a very interesting imaging process called ghost imaging (Shih 2009). This process is based in the strong temporal and spatial correlation between to twin photons produced in a parametric down conversion.by a nonlinear crystal

The basic sketch of the experimental setup is shown in Fig. 17.10.

A UV pump laser beam impinges on a nonlinear crystal NL given origin to a pair of photons emitted at the same time and furthermore angular correlated. So, if the upper photon is seen at one instant in time at one point, S_1', we are sure that the twin photon is to be seen at the correlated point, S_1.

In the end, the basics of the process are: If one knows, at a certain instant in time, the position of one photon, then the position of the other correlated twin photon at the same moment is also known. That is, this ghost imaging technic allows us to know, according to Wheeler which-way (Wheeler 1978) the photon goes.

Inspired in these ideas, R. Menzel, A. Heuer and other researchers (Croca et al. 2015; Menzel et al. 2012, 2013) did a beautiful experiment confirming these which-way predictions. Their experiments correspond to a slight modification of the typical ghost imaging technique. Instead of making the traditional observation by placing the both detectors in the near field, they instead fixed one detector D' in the near field at position S_1', and placed the other D at the far field as shown in Fig. 17.11

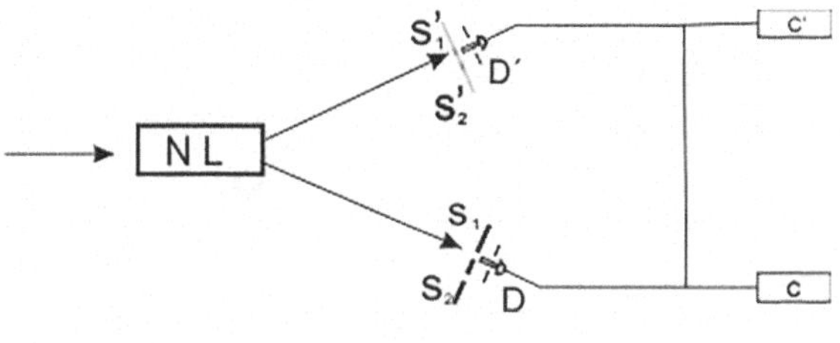

Fig. 17.10 Setup for ghost imaging

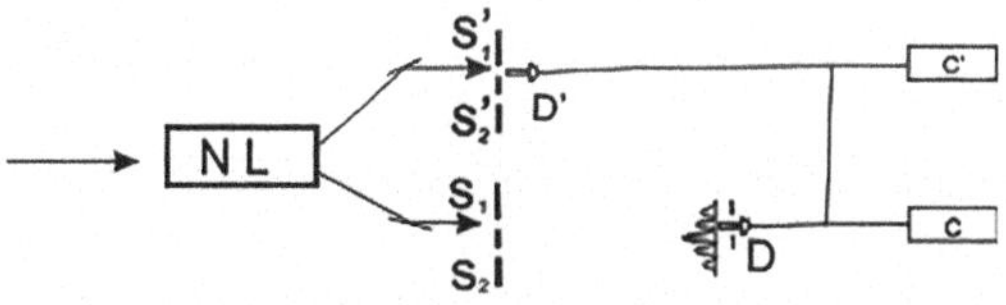

Fig. 17.11 Double-slit interference when the which-way is known

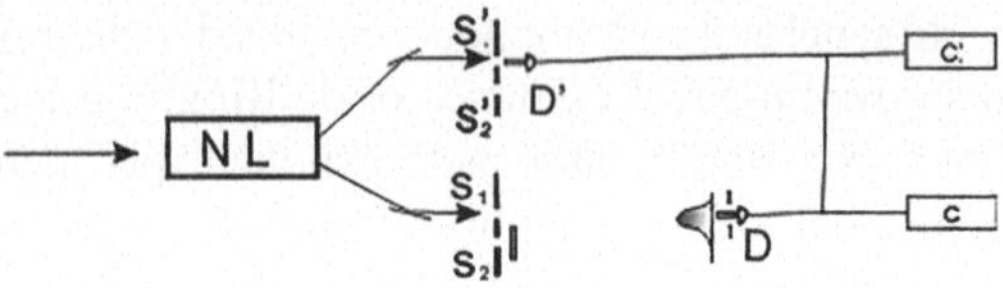

Fig. 17.12 If slit S_2 is closed, then the interference pattern disappears

In this experimental situation even if we know which-way, in this case through slit S_1, when the detector D, scans the far field plane a net interference pattern appears.

Naturally, when the slit S_2 is closed, see Fig. 17.12, the interference pattern disappears given, as expected, origin to non-interference, to a Gaussian distribution.

This experimental evidence confirms that even we know for sure that the photon highly energetic part went by slit S_1, yet something unable to trigger a common detector, has passed through slit S_2. This something is indeed an empty wave, a theta wave. Other implementations of these technics (Croca 1988; Croca et al. 1990) may eventually be developed.

17.6 Concrete Practical Technological Utilization of the Subquantum Waves

A first and immediate consequences of the clarification of the real physical nature of de subquantum waves is the possibility of allowing the development of concrete technological applications namely, subquantum wave telecommunications.

Next picture seeks to indicate a possible subquantum wave emitter, Fig. 17.13.

In the drawing, stimulated by the pump laser beam the nonlinear crystal in a parametric down process produces two photons at the same time. To make sure that we have a single photon impinging the beamsplitter BS at a time the two detectors D_1 and D_2 need to be activated at the same time. When the detector D_1 clicks it means that along that path, the reflection path, went the subquantum wave with

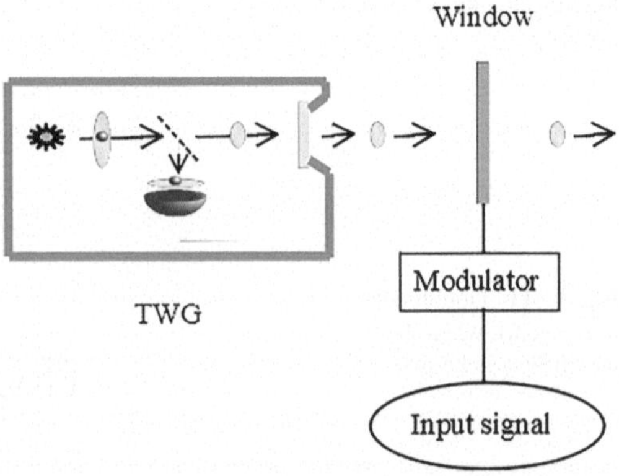

Fig. 17.13 Sketch for a subquantum wave emitter

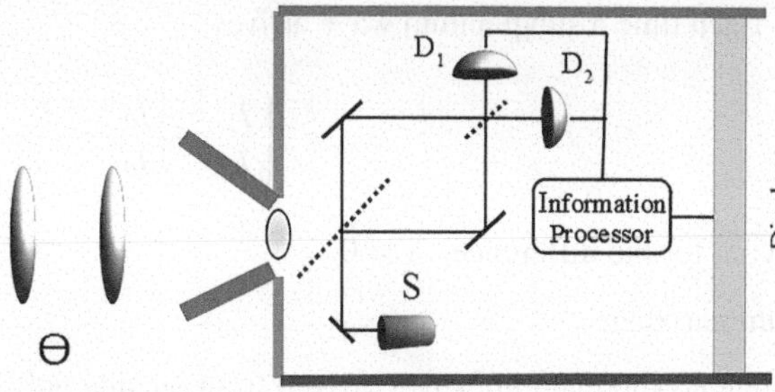

Fig. 17.14 Subquantum wave detector

the high energetic part of the photon. In this situation along the transmission path follows only the subquantum wave with a minute amount of energy unable to trigger common quadratic detectors.

A window connected electronically to the detector, D_1, opens at the same time allowing the passage of the subquantum wave.

In front of this window is placed a Shutter operated by Modulator. This Modulator commands the opening of the Shutter allowing the subquantum waves to pass according to the binary input signal for the information we wish to transmit.

Naturally, as expected, due to the very feeble energy the subquantum waves, the common detectors are unable to detect these waves. In order to be able to detect these subtle subquantum waves practically devoid of energy a very special kind of detector is needed. A possible device for doing such a job is shown in next picture, Fig. 17.14.

The device consists basically in a well-calibrated Mack-Zehnder interferometer, a very stable high intensity monophotonic source plus the detectors and the decoder. The interferometer is adjusted in such a way so that the waves are in phase at output port that leads to detector D_1 and in phase opposition in the other output port. For such arrangement only detector D_1 counts.

When the subquantum waves, coming from the emitter, enter the input port add to the other waves in the interferometer. Since there is no phase correlation with the other waves their contribution acts like a kind of noise. So, according to the strength and overlapping degree the influence of the subquantum waves is felt by the decreasing in the counts at detector D_1 and a concomitantly an increasing counting at detector D_2.

Calculations for the working of the subquantum wave detector are about the same as the those made in Sect. 17.5. Analytically this can be represented in the following way:

No subquantum waves arrive at the detector:

In this situation, we have

$$\begin{cases} I_1 = I_0 \\ I_2 = 0 \end{cases}$$

and the difference between the two output ports is, $\Delta = I_0$.

Each time a subquantum wave arrives

$$\begin{cases} I_1 = \frac{1}{2} I_0 \\ I_2 = \frac{1}{2} I_0 \end{cases}$$

giving for the difference, $\Delta = 0$.

Summarizing:

When no subquantum waves arrived from outside only detector D_1 counts.

When there are subquantum waves arriving from the outside emitter, and assuming the ideal case of perfect overlapping and same intensity, both detector count at the same rate. That is, the action of the entering subquantum wave is to change the overall probabilities scheme in such a way that now both detectors register the same number of photons.

These predictions, as stated, are valid only when the two sources, the emitter and the detector source, emit precisely at the same time and furthermore the waves have the same amplitude.

Still, this does not correspond to the real situation in which the two sources emit waves independently. Even if the detector source S is very steady, the rate waves arriving from the subquantum emitter depend on the input signal related with the information we wish to transmit.

In these conditions, sometimes the two different waves arrive precisely at the same time, which corresponds to a complete overlapping, while other times they don't even partially mix. If it happens that the two independent waves do not overlap at the mixing region then there is no output sign.

Between these two extreme situations there are, of course, all the intermediate cases of partial superposition as shown in next picture, Fig. 17.15.

The strait line stands for no subquantum waves arrival. Each time a subquantum wave arrives and superimposes partially with the waves from the stable high intensity photonic source S a dip appears. The strength of the dip depends on the overlapping degree. The next step is to transform these dip bit rate into meaningful information by the decoder.

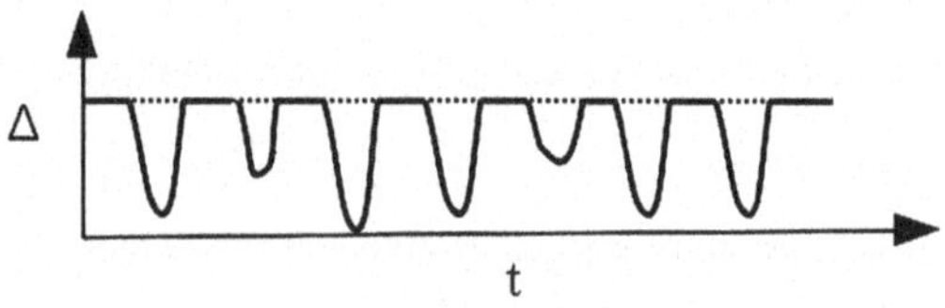

Fig. 17.15 Observed difference in the counting rate

17.7 Conclusion

From the above presented experiments, we may conclude that the quest for the detection of these real physical yet so elusive subquantum waves, after nearly a century, has finally reach its goal.

Still, as good scientific practice teaches us, these experiments need to be redone under the most diverse situations to confirm these results.

Yet even more important is the fact that these recent experimental achievements have opened a whole new realm for important conceptual and technological developments.

The confirmation of the reality of the subquantum waves, or theta waves or de Broglie waves, leads us to the conclusion the principle of complementarity and consequently that orthodox quantum mechanics, has indeed reached it limits of adequacy in the description and prediction of the subquantum realm.

Naturally, in any way do we wish to deny the great merits of orthodox quantum mechanics. Orthodox quantum mechanics, just like classical mechanics, is a very good theory at his scale of application. Still, as any human construct, it has its inherent limits of adequacy in the description and prediction of Reality.

Unsurprisingly, as happens with classical mechanics, by changing here and there the basic assumptions of the theory it is possible to fit a posteriori any experimental data. In a way very similar to the one applied in the old times when researchers tried to save at all costs the Ptolemaic system against the Copernicus by adding epicycles upon epicycles it is was possible to fit any observed data. Indeed, by suitable redefinition of the basic assumptions of the indeterministic orthodox quantum mechanics it is now also possible to derive a posterior, post factum, all new and unexpected experimental results. Naturally, what is difficult is to predict beforehand new experimental realms. By introducing completely ad hoc intricate mathematical and conceptual structures one upon the other, the final construct is something completely against the ideal of simplicity given by Occam's razor.

Acknowledgments I wish to thank the Department of Physics of the Faculty of Sciences of the University of Lisbon and the Centre for Philosophy of Sciences of the University of Lisbon (CFCUL).

This work has been funded by FCT (Portugal) through CFCUL (project UID/FIL/00678/2019).

References

Andrade e Silva, J., and Maria Andrade e Silva. 1980. *Une experience possibile concernant la nature du dualismo onde-corpuscule*, 290. Paris: Comptes rendus de l'Académie des Sciences.

Andrade e Silva, J., F. Selleri, and J.P. Vigier. 1983. Some possible experiments on quantum waves. *Lettere al Nuovo Cimento* 36 (15): 503.

Croca, J.R. 1988. Can the existence of de Broglie's empty waves be proven experimentally? In *Microphysical reality and quantum formalism*, ed. A. Van der Merwe et al., 285–287. London: Kluwer Academic Publishers. The idea was presented in 1985 at the International Conference Microphysical Reality and Quantum Formalism and only published in 1988.

———. 2003. *Towards a nonlinear quantum physics*. London: World Scientific.

———. 2006. The principle of eurhythmy a key to the unity of physics. In *Unity of science, nontraditional approaches*, Lisbon, October, 25–28, 2006.

———. 2012. The principle of eurhythmy a key to the unity of physics. In *Special sciences and the unity of sciences*, ed. O. Pombo, J.M. Torres, J. Symons, and S. Rahman. Dordrecht: Springer.

Croca, J.R., A. Garuccio, V.L. Lepore, and R.N. Moreira. 1990. Quantum-optical predictions for an experiment on de Broglie waves detection. *Foundations of Physics Letters* 3 (6): 557–564.

Croca, J.R., A. Garuccio, M. Gatta, Milena D'Angelo, R.N. Moreira, and A. Rica da Silva. 2015. Experimental evidence on the real physical existence of the subquantum waves. *Quantum Matter* 4.

de Broglie, L. 1924. *Recherches sur la théorie des quanta* (Researches on the quantum theory). Thesis, Paris, 1924, Annals de Physique 3(10): 22 (1925).

———. 1969. *The current interpretation of wave mechanics: A critical study*. Amsterdam: Elsevier.

Einstein, A. 1905. Über die von der molekularkinetischen Theorie der Wärme geforderte Bewegung von in ruhenden Flüssigkeiten suspendierten Teilchen. *Annalen der Physik, series* 4 (17): 549–560.

Menzel, R., A. Heuer, D. Puhlmann, and W.P. Schleich. 2012. Wave-particle dualism and complementarity unraveled by a different mode. *Proceedings of the National Academy of Sciences* 109 (24): 9314–9319.

Menzel, R., A. Heuer, D. Puhlmann, K. Dechoum, M. Hillery, M.J.A. Spähn, and W.P. Schleich. 2013. A two-photon double-slit experiment. *Journal of Modern Optics* 60 (1): 86–94.

Neves, P. 1972. *Incerteza e Indeterminação, Interpretação das Relações de Heisenberg*. Seminário do Departamento de Física da Faculdade de Ciências da Universidade de Lisboa.

Perrin, J. 1909. Mouvement brownien et realité moléculaire. *Annales de Chimie et de Physique, series* 8 (18): 1–114.

Shih, Yanhua. 2009, September 29. *The physical principles of ghost imaging*. arXiv:0805.1166v5 [quant-ph].

Wheeler, J.A. 1978. Delayed choice experiments. In *Proceedings of the New Orleans conference on the mathematical foundations of quantum theory*, ed. A.R. Marlow. New York: Academic.

Zou, X.Y., T. Grayson, and L. Mandel. 1991. *Physical Review Letters* 66: 111.

Zou, X.Y., T. Grayson, L.J. Wang, and L. Mandel. 1992. Can an 'empty' de Broglie pilot wave induce coherence? *Physical Review Letters* 68: 3667–3669.

Index

P. Castro et al. (eds.), *Advances in Pilot Wave Theory*, Boston Studies in the Philosophy and History of Science 344,
https://doi.org/10.1007/978-3-031-49861-9

The manufacturer's authorised representative in the EU is Springer Nature Customer Service Centre GmbH, Europaplatz 3, 69115 Heidelberg, Germany. If you have any concerns regarding our products, please contact ProductSafety@springernature.com

Printed and bound by CPI Group (UK) Ltd, Croydon, CR0 4YY
01/10/2026
02225936-0010